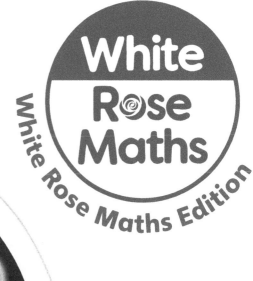

White Rose Maths

White Rose Maths Edition

Year 6A
A Guide to Teaching for Mastery

Series Editor: Tony Staneff
Lead author: Josh Lury

Contents

Introduction to the author team

Power Maths arises from the work of maths mastery experts who are committed to proving that, given the right mastery mindset and approach, **everyone can do maths**. Based on robust research and best practice from around the world, *Power Maths* was developed in partnership with a group of UK teachers to make sure that it not only meets our children's wide-ranging needs but also aligns with the National Curriculum in England.

Power Maths – White Rose Maths edition

This edition of *Power Maths* has been developed and updated by:

Tony Staneff, Series Editor and Author

Vice Principal at Trinity Academy, Halifax, Tony also leads a team of mastery experts who help schools across the UK to develop teaching for mastery via nationally recognised CPD courses, problem-solving and reasoning resources, schemes of work, assessment materials and other tools.

Josh Lury, Lead Author

Josh is a specialist maths teacher, author and maths consultant with a passion for innovative and effective maths education.

The first edition of *Power Maths* was developed by a team of experienced authors, including:

- **Tony Staneff and Josh Lury**

- **Trinity Academy Halifax** (Michael Gosling CEO, Emily Fox, Kate Henshall, Rebecca Holland, Stephanie Kirk, Stephen Monaghan and Rachel Webster)

- **David Board, Belle Cottingham, Jonathan East, Tim Handley, Derek Huby, Neil Jarrett, Stephen Monaghan, Beth Smith, Tim Weal, Paul Wrangles** – skilled maths teachers and mastery experts

- **Cherri Moseley** – a maths author, former teacher and professional development provider

- **Professors Liu Jian and Zhang Dan**, Series Consultants and authors, and their team of mastery expert authors: **Wei Huinv, Huang Lihua, Zhu Dejiang, Zhu Yuhong, Hou Huiying, Yin Lili, Zhang Jing, Zhou Da and Liu Qimeng**

  Used by over 20 million children, Professor Liu Jian's textbook programme is one of the most popular in China. He and his author team are highly experienced in intelligent practice and in embedding key maths concepts using a C-P-A approach.

- **A group of 15 teachers and maths co-ordinators**

  We consulted our teacher group throughout the development of *Power Maths* to ensure we are meeting their real needs in the classroom.

What is *Power Maths*?

Created especially for UK primary schools, and aligned with the new National Curriculum, *Power Maths* is a whole-class, textbook-based mastery resource that empowers every child to understand and succeed. *Power Maths* rejects the notion that some people simply 'can't do' maths. Instead, it develops growth mindsets and encourages hard work, practice and a willingness to see mistakes as learning tools.

Best practice consistently shows that mastery of small, cumulative steps builds a solid foundation of deep mathematical understanding. *Power Maths* combines interactive teaching tools, high-quality textbooks and continuing professional development (CPD) to help you equip children with a deep and long-lasting understanding. Based on extensive evidence, and developed in partnership with practising teachers, *Power Maths* ensures that it meets the needs of children in the UK.

Power Maths and Mastery

Power Maths makes mastery practical and achievable by providing the structures, pathways, content, tools and support you need to make it happen in your classroom.

To develop mastery in maths, children must be enabled to acquire a deep understanding of maths concepts, structures and procedures, step by step. Complex mathematical concepts are built on simpler conceptual components and when children understand every step in the learning sequence, maths becomes transparent and makes logical sense. Interactive lessons establish deep understanding in small steps, as well as effortless fluency in key facts such as tables and number bonds. The whole class works on the same content and no child is left behind.

Power Maths

- Builds every concept in small, progressive steps
- Is built with interactive, whole-class teaching in mind
- Provides the tools you need to develop growth mindsets
- Helps you check understanding and ensure that every child is keeping up
- Establishes core elements such as intelligent practice and reflection

The *Power Maths* approach

Everyone can!

Founded on the conviction that every child can achieve, *Power Maths* enables children to build number fluency, confidence and understanding, step by step.

Child-centred learning

Children master concepts one step at a time in lessons that embrace a concrete-pictorial-abstract (C-P-A) approach, avoid overload, build on prior learning and help them see patterns and connections. Same-day intervention ensures sustained progress.

Continuing professional development

Embedded teacher support and development offer every teacher the opportunity to continually improve their subject knowledge and manage whole-class teaching for mastery.

Whole-class teaching

An interactive, whole-class teaching model encourages thinking and precise mathematical language and allows children to deepen their understanding as far as they can.

What's different in the new edition?

If you have previously used the first editions of *Power Maths*, you might be interested to know how this edition is different. All of the improvements described below are based on feedback from *Power Maths* customers.

Changes to units and the progression

⚡ The order of units has been slightly adjusted, creating closer alignment between adjacent year groups, which will be useful for mixed age teaching.

⚡ The flow of lessons has been improved within units to optimise the pace of the progression and build in more recap where needed. For key topics, the sequence of lessons gives more opportunities to build up a solid base of understanding. Other units have fewer lessons than before, where appropriate, making it possible to fit in all the content.

⚡ Overall, the lessons put more focus on the most essential content for that year, with less time given to non-statutory content.

⚡ The progression of lessons matches the steps in the new White Rose Maths schemes of learning.

Lesson resources

⚡ There is a Quick recap for each lesson in the Teacher Guide, which offers an alternative lesson starter to the Power Up for cases where you feel it would be more beneficial to surface prerequisite learning than general number fluency.

⚡ In the **Discover** and **Share** sections there is now more of a progression from 1 a) to 1 b). Whereas before, 1 b) was mainly designed as a separate question, now 1 a) leads directly into 1 b). This means that there is an improved whole-class flow, and also an opportunity to focus on the logic and skills in more detail. As a teacher, you will be using 1 a) to lead the class into the thinking, then 1 b) to mould that thinking into the core new learning of the lesson.

⚡ In the **Share** section, for KS1 in particular, the number of different models and representations has been reduced, to support the clarity of thinking prompted by the flow from 1 a) into 1 b).

⚡ More fluency questions have been built into the guided and independent practice.

⚡ Pupil pages are as easy as possible for children to access independently. The pages are less full to provide greater focus on key ideas and instructions. Also, more freedom is offered around answer format, with fewer boxes scaffolding children's responses; squared paper backgrounds are used in the Practice Books where appropriate. Artwork has also been revisited to ensure the highest standards of accessibility.

New components

480 Individual Practice Games are available in *ActiveLearn* for practising key facts and skills in Years 1 to 6. These are designed in an arcade style, to feel like fun games that children would choose to play outside school. They can be accessed via the Pupil World for homework or additional practice in school – and children can earn rewards. There are Support, Core and Extend levels to allocate, with Activity Reporting available for the teacher. There is a Quick Guide on *ActiveLearn* and you can use the Help area for support in setting up child accounts.

There is also a new set of lesson video resources on the Professional Development tile, designed for in-school training in 10- to 20-minute bursts. For each part of the *Power Maths* lesson sequence, there is a slide deck with embedded video, which will facilitate discussions about how you can take your *Power Maths* teaching to the next level.

Your *Power Maths* resources

Pupil Textbooks

Discover, Share and Think together sections promote discussion and introduce mathematical ideas logically, so that children understand more easily.

Using a Concrete-Pictorial-Abstract approach, clear mathematical models help children to make connections and grasp concepts.

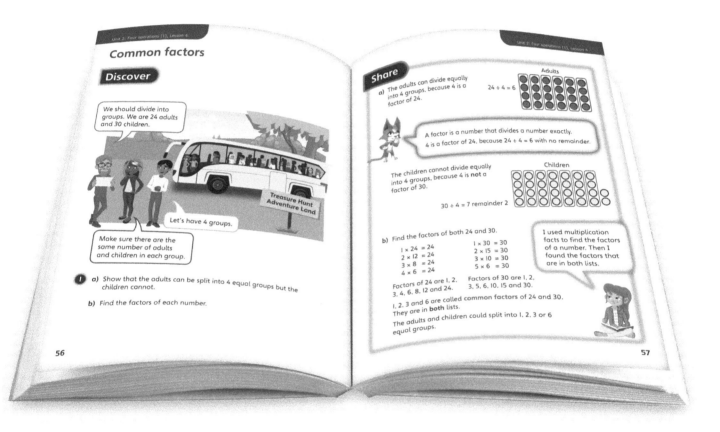

Appealing scenarios stimulate curiosity, helping children to identify the maths problem and discover patterns and relationships for themselves.

Friendly, supportive characters help children develop a growth mindset by prompting them to think, reason and reflect.

To help you teach for mastery, *Power Maths* comprises a variety of high-quality resources.

The coherent *Power Maths* lesson structure carries through into the vibrant, high-quality textbooks. Setting out the core learning objectives for each class, the lesson structure follows a carefully mapped journey through the curriculum and supports children on their journey to deeper understanding.

Pupil Practice Books

The Practice Books offer just the right amount of intelligent practice for children to complete independently in the final section of each lesson.

Practice questions are finely tuned to move children forward in their thinking and to reveal misconceptions.

The practice questions are for everyone – each question varies one small element to move children on in their thinking.

Calculations are connected so that children think about the underlying concept.

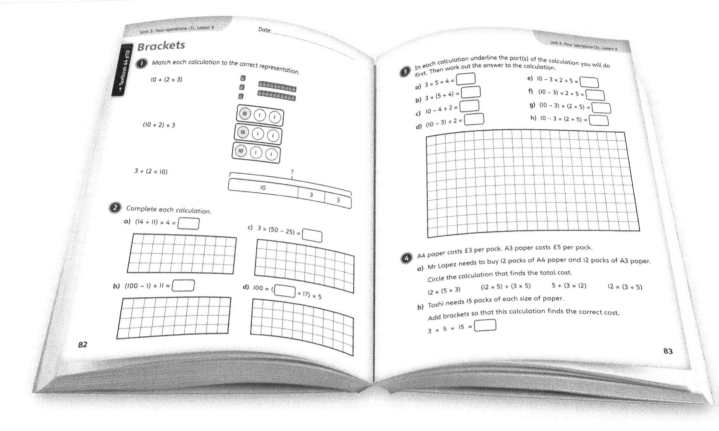

CHALLENGE

Challenge questions allow children to delve deeper into a concept.

Think differently questions encourage children to use reasoning as well as their mathematical knowledge to reach a solution.

Reflect questions reveal the depth of each child's understanding before they move on.

The *Power Maths* characters support and encourage children to think and work in different ways.

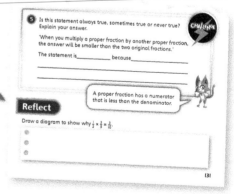

Online subscription

The online subscription will give you access to additional resources and answers from the Textbook and Practice Book.

eTextbooks

Digital versions of *Power Maths* Textbooks allow class groups to share and discuss questions, solutions and strategies. They allow you to project key structures and representations at the front of the class, to ensure all children are focusing on the same concept.

Teaching tools

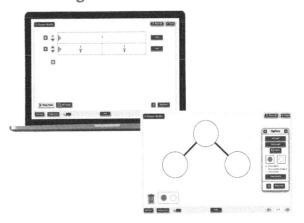

Here you will find interactive versions of key *Power Maths* structures and representations.

Power Ups

Use this series of daily activities to promote and check number fluency.

Online versions of Teacher Guide pages

PDF pages give support at both unit and lesson levels. You will also find help with key strategies and templates for tracking progress.

Unit videos

Watch the professional development videos at the start of each unit to help you teach with confidence. The videos explore common misconceptions in the unit, and include intervention suggestions as well as suggestions on what to look out for when assessing mastery in your students.

End of unit Strengthen and Deepen materials

The Strengthen activity at the end of every unit addresses a key misconception and can be used to support children who need it. The Deepen activities are designed to be low ceiling/high threshold and will challenge those children who can understand more deeply. These resources will help you ensure that every child understands and will help you keep the class moving forward together. These printable activities provide an optional resource bank for use after the assessment stage.

Individual Practice Games

These enjoyable games can be used at home or at school to embed key number skills (see page 6).

Professional Development videos and slides

These slides and videos of *Power Maths* lessons can be used for ongoing training in short bursts or to support new staff.

The *Power Maths* teaching model

At the heart of *Power Maths* is a clearly structured teaching and learning process that helps you make certain that every child masters each maths concept securely and deeply. For each year group, the curriculum is broken down into core concepts, taught in units. A unit divides into smaller learning steps – lessons. Step by step, strong foundations of cumulative knowledge and understanding are built.

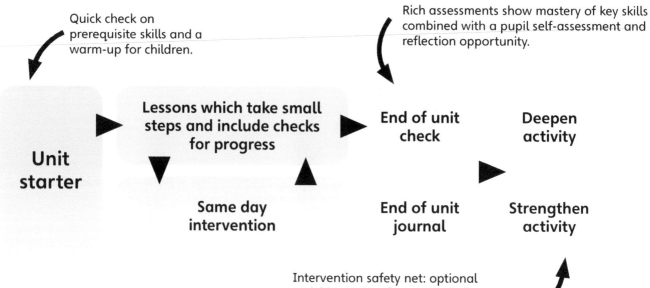

Quick check on prerequisite skills and a warm-up for children.

Rich assessments show mastery of key skills combined with a pupil self-assessment and reflection opportunity.

Unit starter → **Lessons which take small steps and include checks for progress** → **End of unit check** **Deepen activity**

Same day intervention **End of unit journal** **Strengthen activity**

Intervention safety net: optional activities to use if assessment shows some children still have misconceptions.

Unit starter

Each unit begins with a unit starter, which introduces the learning context along with key mathematical vocabulary and structures and representations.

- The Textbooks include a check on readiness and a warm-up task for children to complete.

- Your Teacher Guide gives support right from the start on important structures and representations, mathematical language, common misconceptions and intervention strategies.

- Unit-specific videos develop your subject knowledge and insights so you feel confident and fully equipped to teach each new unit. These are available via the online subscription.

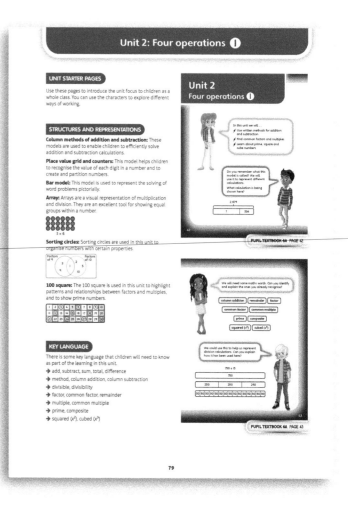

Lesson

Once a unit has been introduced, it is time to start teaching the series of lessons.

- Each lesson is scaffolded with Textbook and Practice Book activities and begins with a Power Up activity (available via online subscription) or the Quick recap activity in the Teacher Guide (see page 15).

- *Power Maths* identifies lesson by lesson what concepts are to be taught.

- Your Teacher Guide offers lots of support for you to get the most from every child in every lesson. As well as highlighting key points, tricky areas and how to handle them, you will also find question prompts to check on understanding and clarification on why particular activities and questions are used.

Same-day intervention

Same-day interventions are vital in order to keep the class progressing together. This can be during the lesson as well as afterwards (see page 28). Therefore, *Power Maths* provides plenty of support throughout the journey.

- Intervention is focused on keeping up now, not catching up later, so interventions should happen as soon as they are needed.

- Practice section questions are designed to bring misconceptions to the surface, allowing you to identify these easily as you circulate during independent practice time.

- Child-friendly assessment questions in the Teacher Guide help you identify easily which children need to strengthen their understanding.

End of unit check and journal

For each unit, the End of unit check in the Textbook lets you see which children have mastered the key concepts, which children have not and where their misconceptions lie. The Practice Books also include an End of unit journal in which children can reflect on what they have learned. Each unit also offers Strengthen and Deepen activities, available via the online subscription.

> The Teacher Guide offers different ways of managing the End of unit assessments as well as giving support with handling misconceptions.

> The End of unit check presents multiple-choice questions. Children think about their answer, decide on a solution and explain their choice.

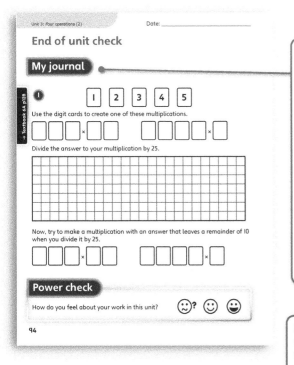

> The End of unit journal is an opportunity for children to test out their learning and reflect on how they feel about it. Tackling the 'journal' problem reveals whether a child understands the concept deeply enough to move on to the next unit.

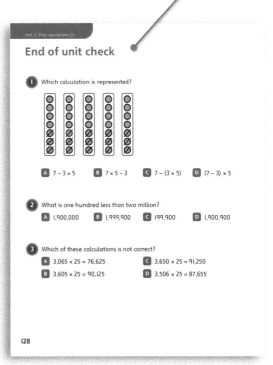

> In KS2, the End of unit assessment will also include at least one SATs-style question.

The *Power Maths* lesson sequence

At the heart of *Power Maths* is a unique lesson sequence designed to empower children to understand core concepts and grow in confidence. Embracing the National Centre for Excellence in the Teaching of Mathematics' (NCETM's) definition of mastery, the sequence guides and shapes every *Power Maths* lesson you teach.

Flexibility is built into the *Power Maths* programme so there is no one-to-one mapping of lessons and concepts and you can pace your teaching according to your class. While some children will need to spend longer on a particular concept (through interventions or additional lessons), others will reach deeper levels of understanding. However, it is important that the class moves forward together through the termly schedules.

Power Up ⏱ 5 minutes

Each lesson begins with a Power Up activity (available via the online subscription) which supports fluency in key number facts.

The whole-class approach depends on fluency, so the Power Up is a powerful and essential activity.

The Quick recap is an alternative starter, for when you think some or all children would benefit more from revisiting pre-requisite work (see page 15).

TOP TIP
If the class is struggling with the task, revisit it later and check understanding.

Power Ups reinforce the two key things that are essential for success: times-tables and number bonds.

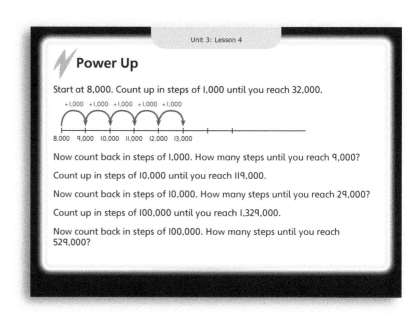

Unit 3: Lesson 4

⚡ Power Up

Start at 8,000. Count up in steps of 1,000 until you reach 32,000.

+1,000 +1,000 +1,000 +1,000 +1,000

8,000 9,000 10,000 11,000 12,000 13,000

Now count back in steps of 1,000. How many steps until you reach 9,000?

Count up in steps of 10,000 until you reach 119,000.

Now count back in steps of 10,000. How many steps until you reach 29,000?

Count up in steps of 100,000 until you reach 1,329,000.

Now count back in steps of 100,000. How many steps until you reach 529,000?

Discover ⏱ 10 minutes

A practical, real-life problem arouses curiosity. Children find the maths through story telling.

A real-life scenario is provided for the **Discover** section but feel free to build upon these with your own examples that are more relevant to your class, or get creative with the context.

TOP TIP
Discover works best when run at tables, in pairs with concrete objects.

Question **1** a) tackles the key concept and question **1** b) digs a little deeper. Children have time to explore, play and discuss possible strategies.

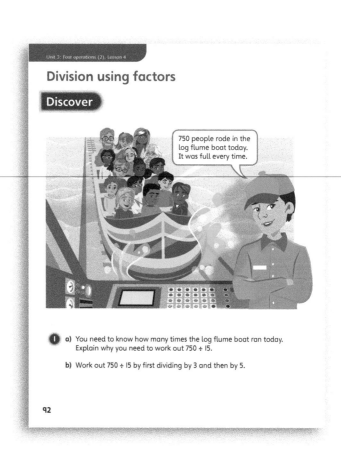

Unit 3: Four operations (2), Lesson 4

Division using factors

Discover

750 people rode in the log flume boat today. It was full every time.

1 a) You need to know how many times the log flume boat ran today. Explain why you need to work out 750 ÷ 15.

b) Work out 750 ÷ 15 by first dividing by 3 and then by 5.

92

Share ⏱ 10 minutes

Teacher-led, this interactive section follows the **Discover** activity and highlights the variety of methods that can be used to solve a single problem.

TOP TIP
Pairs sharing a textbook is a great format for **Share**!

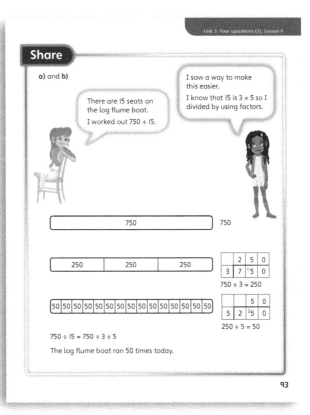

Your Teacher Guide gives target questions for children. The online toolkit provides interactive structures and representations to link concrete and pictorial to abstract concepts.

Bring children to the front to share and celebrate their solutions and strategies.

Think together

⏱ 10 minutes

Children work in groups on the carpet or at tables, using their textbooks or eBooks.

TOP TIP
Make sure children have mini whiteboards or pads to write on if they are not at their tables.

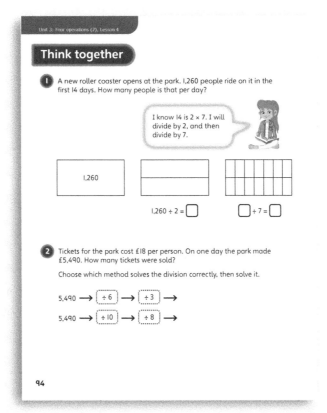

Using the Teacher Guide, model question ❶ for your class.

Question ❷ is less structured. Children will need to think together in their groups, then discuss their methods and solutions as a class.

In question ❸ children try working out the answer independently. The openness of the **Challenge** question helps to check depth of understanding.

Practice ⏱ 15 minutes

Using their Practice Books, children work independently while you circulate and check on progress.

Questions follow small steps of progression to deepen learning.

TOP TIP
Some children could work separately with a teacher or assistant.

Division using factors

1 **a)** Reena has shared 3,500 ml of juice equally between 14 glasses. How much juice is in each glass?

→ Textbook 6A p92

| 3,500 |

$3,500 \div 7 =$ ⬚   ⬚ $\div 2 =$ ⬚

| 7 | 3 | 5 | 0 | 0 |

| 2 | |

$3,500 \div 14 =$ ⬚

b) Aki has 360 g of clay. He makes small clay shells. Each shell weighs 24 g. How many shells can he make?

| 360 |

$360 \div 6 =$ ⬚   ⬚ $\div 4 =$ ⬚

67

Are some children struggling? If so, work with them as a group, using mathematical structures and representations to support understanding as necessary.

There are no set routines: for real understanding, children need to think about the problem in different ways.

Reflect ⏱ 5 minutes

'Spot the mistake' questions are great for checking misconceptions.

The **Reflect** section is your opportunity to check how deeply children understand the target concept.

4 **CHALLENGE**

Ambika: If I double the number that I am dividing by, the answer to the division will be halved.

Bella: I think that means that if I double both numbers in a division, the answer will be halved and then halved again.

Do you agree with both Ambika and Bella?

_____

Show examples to help you explain why.

Reflect

Show two ways to work out $6,440 \div 20$ using division by factors.

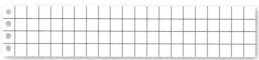

69

The Practice Books use various approaches to check that children have fully understood each concept.

Looking like they understand is not enough! It is essential that children can show they have grasped the concept.

14

Using the *Power Maths* Teacher Guide

Think of your Teacher Guides as *Power Maths* handbooks that will guide, support and inspire your day-to-day teaching. Clear and concise, and illustrated with helpful examples, your Teacher Guides will help you make the best possible use of every individual lesson. They also provide wrap-around professional development, enhancing your own subject knowledge and helping you to grow in confidence about moving your children forward together.

There is a Teacher Guide per year group for every term, with unit and lesson level guidance and support.

Never feel stuck! You will find ideas for introducing every unit and lesson and questions to encourage teacher reflection before and after each lesson.

Tips and advice on key elements such as C-P-A approaches, misconceptions, language, modelling growth mindsets and same day intervention.

Annotations for every Textbook and Practice Book page, providing prompts for key questions to ask to expose understanding and explanations as to why key questions have been chosen.

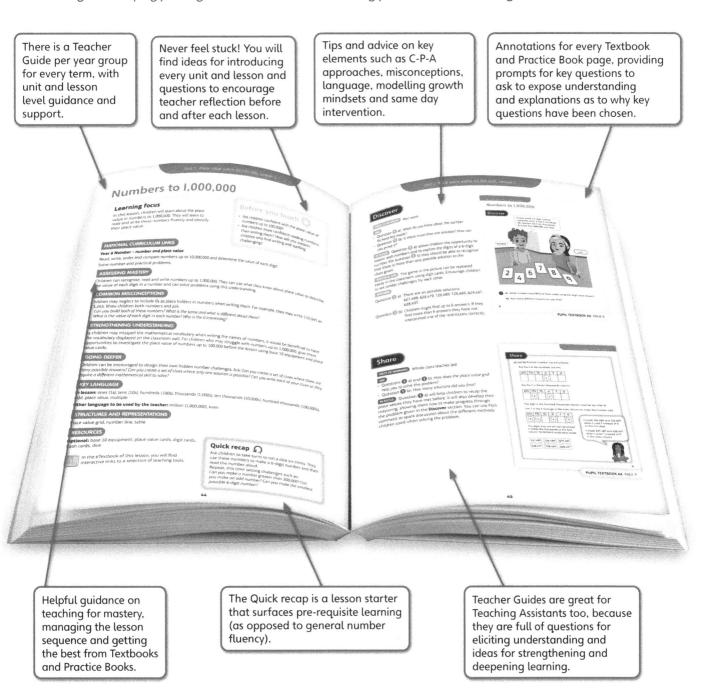

Helpful guidance on teaching for mastery, managing the lesson sequence and getting the best from Textbooks and Practice Books.

The Quick recap is a lesson starter that surfaces pre-requisite learning (as opposed to general number fluency).

Teacher Guides are great for Teaching Assistants too, because they are full of questions for eliciting understanding and ideas for strengthening and deepening learning.

At the end of each unit, your Teacher Guide helps you identify who has fully grasped the concept, who has not and how to move every child forward. This is covered later in the Assessment strategies section.

Power Maths Year 6, yearly overview

Textbook	Strand	Unit		Number of lessons
Textbook A / Practice Workbook A (Term 1)	Number – number and place value	1	Place value within 10,000,000	8
	Number – addition, subtraction, multiplication and division	2	Four operations (1)	8
	Number – addition, subtraction, multiplication and division	3	Four operations (2)	12
	Number - fractions	4	Fractions (1)	9
	Number - fractions	5	Fractions (2)	9
	Measurement	6	Measure – imperial and metric measures	5
Textbook B / Practice Workbook B (Term 2)	Ratio and proportion	7	Ratio and proportion	9
	Algebra	8	Algebra	11
	Number - fractions (including decimals and percentages)	9	Decimals	9
	Number - fractions (including decimals and percentages)	10	Percentages	8
	Measurement	11	Measure – perimeter, area and volume	11
Textbook C / Practice Workbook C (Term 3)	Statistics	12	Statistics	11
	Geometry – properties of shapes	13	Geometry – properties of shapes	12
	Geometry – position and direction	14	Geometry – position and direction	5
	Number – addition, subtraction, multiplication and division	15	Problem solving	14

Power Maths Year 6, Textbook 6A (Term I) overview

Strand	Unit		Lesson number	Lesson title	NC Objective 1	NC Objective 2
Number – number and place value	Unit 1	Place value within 10,000,000	1	Numbers to 1,000,000	Read, write, order and compare numbers up to 10,000,000 and determine the value of each digit	Solve number and practical problems
Number – number and place value	Unit 1	Place value within 1,000,000 (1)	2	Numbers to 10,000,000	Read, write, order and compare numbers up to 10,000,000 and determine the value of each digit	Solve number and practical problems
Number – number and place value	Unit 1	Place value within 1,000,000 (1)	3	Partition numbers to 10,000,000	Read, write, order and compare numbers up to 10,000,000 and determine the value of each digit	Solve number and practical problems
Number – number and place value	Unit 1	Place value within 1,000,000 (1)	4	Powers of 10	Read, write, order and compare numbers up to 10,000,000 and determine the value of each digit	Solve number and practical problems
Number – number and place value	Unit 1	Place value within 1,000,000 (1)	5	Number line to 10,000,000	Read, write, order and compare numbers up to 10,000,000 and determine the value of each digit	Solve number and practical problems
Number – number and place value	Unit 1	Place value within 1,000,000 (1)	6	Compare and order any number	Read, write, order and compare numbers up to 10,000,000 and determine the value of each digit	Solve number and practical problems
Number – number and place value	Unit 1	Place value within 1,000,000 (1)	7	Round any number	Round any whole number to a required degree of accuracy	
Number – number and place value	Unit 1	Place value within 1,000,000 (1)	8	Negative numbers	Use negative numbers in context, and calculate intervals across zero	
Number – addition, subtraction, multiplication and division	Unit 2	Four operations (1)	1	Add integers	Solve addition and subtraction multi-step problems in contexts, deciding which operations and methods to use and why	
Number – addition, subtraction, multiplication and division	Unit 2	Four operations (1)	2	Subtract integers	Solve addition and subtraction multi-step problems in contexts, deciding which operations and methods to use and why	
Number – addition, subtraction, multiplication and division	Unit 2	Four operations (1)	3	Problem solving – addition and subtraction	Solve addition and subtraction multi-step problems in contexts, deciding which operations and methods to use and why	
Number – addition, subtraction, multiplication and division	Unit 2	Four operations (1)	4	Common factors	Identify common factors, common multiples and prime numbers	
Number – addition, subtraction, multiplication and division	Unit 2	Four operations (1)	5	Common multiples	Identify common factors, common multiples and prime numbers	
Number – addition, subtraction, multiplication and division	Unit 2	Four operations (1)	6	Rules of divisibility	Identify common factors, common multiples and prime numbers	Use their knowledge of the order of operations to carry out calculations involving the four operations
Number – addition, subtraction, multiplication and division	Unit 2	Four operations (1)	7	Primes to 100	Identify common factors, common multiples and prime numbers	
Number – addition, subtraction, multiplication and division	Unit 2	Four operations (1)	8	Squares and cubes	Recognise and use square numbers and cube numbers, and the notation for squared (2) and cubed (3) (year 5)	

Strand	Unit		Lesson number	Lesson title	NC Objective 1	NC Objective 2
Number – addition and subtraction	Unit 3	Four operations (2)	1	Multiply by a 1-digit number	Multiply multi-digit numbers up to 4 digits by a two-digit whole number using the formal written method of long multiplication	
Number – addition and subtraction	Unit 3	Addition and subtraction	2	Multiply up to a 4-digit number by a 2-digit number	Multiply multi-digit numbers up to 4 digits by a two-digit whole number using the formal written method of long multiplication	
Number – addition and subtraction	Unit 3	Addition and subtraction	3	Short division	Divide numbers up to 4 digits by a two-digit number using the formal written method of short division where appropriate, interpreting remainders according to the context	
Number – addition and subtraction	Unit 3	Addition and subtraction	4	Division using factors	Identify common factors, common multiples and prime numbers	Divide numbers up to 4 digits by a two-digit number using the formal written method of short division where appropriate, interpreting remainders according to the context
Number – addition and subtraction	Unit 3	Addition and subtraction	5	Divide a 3-digit number by 2-digit (long division)	Divide numbers up to 4 digits by a two-digit number using the formal written method of short division where appropriate, interpreting remainders according to the context	
Number – addition and subtraction	Unit 3	Addition and subtraction	6	Divide a 4-digit number by 2-digit (long division)	Divide numbers up to 4 digits by a two-digit number using the formal written method of short division where appropriate, interpreting remainders according to the context	Divide numbers up to 4 digits by a two-digit whole number using the formal written method of long division, and interpret remainders as whole number remainders, fractions, or by rounding, as appropriate for the context
Number – addition and subtraction	Unit 3	Addition and subtraction	7	Long division with remainders	Divide numbers up to 4 digits by a two-digit number using the formal written method of short division where appropriate, interpreting remainders according to the context	Divide numbers up to 4 digits by a two-digit whole number using the formal written method of long division, and interpret remainders as whole number remainders, fractions, or by rounding, as appropriate for the context
Number – addition and subtraction	Unit 3	Addition and subtraction	8	Order of operations	Use their knowledge of the order of operations to carry out calculations involving the four operations	
Number – addition and subtraction	Unit 3	Addition and subtraction	9	Brackets	Use their knowledge of the order of operations to carry out calculations involving the four operations	
Number – addition and subtraction	Unit 3	Addition and subtraction	10	Mental calculations (1)	Perform mental calculations, including with mixed operations and large numbers	
Number – addition and subtraction	Unit 3	Addition and subtraction	11	Mental calculations (2)	Perform mental calculations, including with mixed operations and large numbers	
Number – addition and subtraction	Unit 3	Addition and subtraction	12	Reason from known facts	Use their knowledge of the order of operations to carry out calculations involving the four operations	Solve problems involving addition, subtraction, multiplication and division
Number – fraction	Unit 4	Fractions (1)	1	Equivalent fractions and simplifying	Use common factors to simplify fractions; use common multiples to express fractions in the same denomination	
Number – fraction	Unit 4	Fractions (1)	2	Equivalent fractions on a number line	Compare and order fractions, including fractions > 1	
Number – fraction	Unit 4	Fractions (1)	3	Compare and order fractions (	Compare and order fractions, including fractions > 1	

Strand	Unit		Lesson number	Lesson title	NC Objective 1	NC Objective 2
Number – fraction	Unit 4	Fractions (1)	4	Add and subtract simple fractions	Add and subtract fractions with different denominators and mixed numbers, using the concept of equivalent fractions	
Number – fraction	Unit 4	Fractions (1)	5	Add and subtract any two fractions	Add and subtract fractions with different denominators and mixed numbers, using the concept of equivalent fractions	
Number – fraction	Unit 4	Fractions (1)	6	Add mixed numbers	Add and subtract fractions with different denominators and mixed numbers, using the concept of equivalent fractions	
Number – fraction	Unit 4	Fractions (1)	7	Subtract mixed numbers	Add and subtract fractions with different denominators and mixed numbers, using the concept of equivalent fractions	
Number – fraction	Unit 4	Fractions (1)	8	Multi-step problems	Add and subtract fractions with different denominators and mixed numbers, using the concept of equivalent fractions	
Number – fraction	Unit 4	Fractions (1)	9	Problem solving - add and subtract fractions	Add and subtract fractions with different denominators and mixed numbers, using the concept of equivalent fractions	
Number – fractions (including decimals and percentages)	Unit 5	Fractions (2)	1	Multiply fractions by integers	Multiply proper fractions and mixed numbers by whole numbers, supported by materials and diagrams	
Number – fractions (including decimals and percentages)	Unit 5	Fractions (2)	2	Multiply fractions by fractions (1)	Multiply simple pairs of proper fractions, writing the answer in its simplest form [for example, $\frac{1}{4} \times \frac{1}{2} = \frac{1}{8}$]	
Number – fractions (including decimals and percentages)	Unit 5	Fractions (2)	3	Multiply fractions by fractions (2)	Multiply simple pairs of proper fractions, writing the answer in its simplest form [for example, $\frac{1}{4} \times \frac{1}{2} = \frac{1}{8}$]	
Number – fractions (including decimals and percentages)	Unit 5	Fractions (2)	4	Divide a fraction by an integer (1)	Divide proper fractions by whole numbers [for example, $\frac{1}{3} \div 2 = \frac{1}{6}$]	
Number – fractions (including decimals and percentages)	Unit 5	Fractions (2)	5	Divide a fraction by an integer (2)	Divide proper fractions by whole numbers [for example, $\frac{1}{3} \div 2 = \frac{1}{6}$]	
Number – fractions (including decimals and percentages)	Unit 5	Fractions (2)	6	Divide a fraction by an integer (2)	Divide proper fractions by whole numbers [for example, $\frac{1}{3} \div 2 = \frac{1}{6}$]	
Number – fractions (including decimals and percentages)	Unit 5	Fractions (2)	7	Mixed questions with fractions	Add and subtract fractions with different denominators and mixed numbers, using the concept of equivalent fractions	Multiply simple pairs of proper fractions, writing the answer in its simplest form [for example, $\frac{1}{4} \times \frac{1}{2} = \frac{1}{8}$]
Number – fractions (including decimals and percentages)	Unit 5	Fractions (2)	8	Fraction of an amount	Use written division methods in cases where the answer has up to two decimal places	

Strand	Unit		Lesson number	Lesson title	NC Objective 1	NC Objective 2
Number – fractions (including decimals and percentages)	Unit 5	Fractions (2)	9	Fraction of an amount – find the whole	Use written division methods in cases where the answer has up to two decimal places	
Measurement	Unit 6	Measure – imperial and metric measures	1	Metric measures	Use, read, write and convert between standard units, converting measurements of length, mass, volume and time from a smaller unit of measure to a larger unit, and vice versa, using decimal notation to up to three decimal places	
Number – fractions (including decimals and percentages)	Unit 6	Fractions (2)	2	Convert metric measures	Use, read, write and convert between standard units, converting measurements of length, mass, volume and time from a smaller unit of measure to a larger unit, and vice versa, using decimal notation to up to three decimal places	Solve problems involving the calculation and conversion of units of measure, using decimal notation up to three decimal places where appropriate
Number – fractions (including decimals and percentages)	Unit 6	Fractions (2)	3	Calculate with metric measures	Solve problems involving the calculation and conversion of units of measure, using decimal notation up to three decimal places where appropriate	
Number – fractions (including decimals and percentages)	Unit 6	Fractions (2)	4	Miles and kilometres	Convert between miles and kilometres	
Number – fractions (including decimals and percentages)	Unit 6	Fractions (2)	5	Imperial measures	Use, read, write and convert between standard units, converting measurements of length, mass, volume and time from a smaller unit of measure to a larger unit, and vice versa, using decimal notation to up to three decimal places	

Mindset: an introduction

Global research and best practice deliver the same message: learning is greatly affected by what learners perceive they can or cannot do. What is more, it is also shaped by what their parents, carers and teachers perceive they can do. Mindset – the thinking that determines our beliefs and behaviours – therefore has a fundamental impact on teaching and learning.

Everyone can!

Power Maths and mastery methods focus on the distinction between 'fixed' and 'growth' mindsets (Dweck, 2007).[1] Those with a fixed mindset believe that their basic qualities (for example, intelligence, talent and ability to learn) are pre-wired or fixed: 'If you have a talent for maths, you will succeed at it. If not, too bad!' By contrast, those with a growth mindset believe that hard work, effort and commitment drive success and that 'smart' is not something you are or are not, but something you become. In short, everyone can do maths!

Key mindset strategies

A growth mindset needs to be actively nurtured and developed. *Power Maths* offers some key strategies for fostering healthy growth mindsets in your classroom.

It is okay to get it wrong

Mistakes are valuable opportunities to re-think and understand more deeply. Learning is richer when children and teachers alike focus on spotting and sharing mistakes as well as solutions.

Praise hard work

Praise is a great motivator, and by focusing on praising effort and learning rather than success, children will be more willing to try harder, take risks and persist for longer.

Mind your language!

The language we use around learners has a profound effect on their mindsets. Make a habit of using growth phrases, such as, 'Everyone can!', 'Mistakes can help you learn' and 'Just try for a little longer'. The king of them all is one little word, 'yet'... I can't solve this...yet!' Encourage parents and carers to use the right language too.

Build in opportunities for success

The step-by-small-step approach enables children to enjoy the experience of success. In addition, avoid ability grouping and encourage every child to answer questions and explain or demonstrate their methods to others.

[1]Dweck, C (2007) *The New Psychology of Success*, Ballantine Books: New York

The *Power Maths* characters

The *Power Maths* characters model the traits of growth mindset learners and encourage resilience by prompting and questioning children as they work. Appearing frequently in the Textbooks and Practice Books, they are your allies in teaching and discussion, helping to model methods, alternatives and misconceptions, and to pose questions. They encourage and support your children, too: they are all hardworking, enthusiastic and unafraid of making and talking about mistakes.

Meet the team!

Creative Flo is open-minded and sometimes indecisive. She likes to think differently and come up with a variety of methods or ideas.

Determined Dexter is resolute, resilient and systematic. He concentrates hard, always tries his best and he'll never give up – even though he doesn't always choose the most efficient methods!

'Let's try again.'

'Mistakes are cool!'

'Have I found all of the solutions?'

'Let's try it this way...'

'Can we do it differently?'

'I've got another way of doing this!'

'I'm going to try this!'

'I know how to do that!'

'Want to share my ideas?'

Curious Ash is eager, interested and inquisitive, and he loves solving puzzles and problems. Ash asks lots of questions but sometimes gets distracted.

'What if we tried this...?'

'I wonder...'

'Is there a pattern here?'

Miaow!

Sparks the Cat

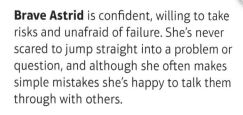

Brave Astrid is confident, willing to take risks and unafraid of failure. She's never scared to jump straight into a problem or question, and although she often makes simple mistakes she's happy to talk them through with others.

Mathematical language

Traditionally, we in the UK have tended to try simplifying mathematical language to make it easier for young children to understand. By contrast, evidence and experience show that by diluting the correct language, we actually mask concepts and meanings for children. We then wonder why they are confused by new and different terminology later down the line! *Power Maths* is not afraid of 'hard' words and avoids placing any barriers between children and their understanding of mathematical concepts. As a result, we need to be deliberate, precise and thorough in building every child's understanding of the language of maths. Throughout the Teacher Guides you will find support and guidance on how to deliver this, as well as individual explanations throughout the pupil Textbooks.

Use the following key strategies to build children's mathematical vocabulary, understanding and confidence.

Precise and consistent

Everyone in the classroom should use the correct mathematical terms in full, every time. For example, refer to 'equal parts', not 'parts'. Used consistently, precise maths language will be a familiar and non-threatening part of children's everyday experience.

Full sentences

Teachers and children alike need to use full sentences to explain or respond. When children use complete sentences, it both reveals their understanding and embeds their knowledge.

Stem sentences

These important sentences help children express mathematical concepts accurately, and are used throughout the *Power Maths* books. Encourage children to repeat them frequently, whether working independently or with others. Examples of stem sentences are:

'4 is a part, 5 is a part, 9 is the whole.'

'There are .... groups. There are .... in each group.'

Key vocabulary

The unit starters highlight essential vocabulary for every lesson. In the pupil books, characters flag new terminology and the Teacher Guide lists important mathematical language for every unit and lesson. New terms are never introduced without a clear explanation.

Mathematical signs

Mathematical signs are used early on so that children quickly become familiar with them and their meaning. Often, the *Power Maths* characters will highlight the connection between language and particular signs.

The role of talk and discussion

When children learn to talk purposefully together about maths, barriers of fear and anxiety are broken down and they grow in confidence, skills and understanding. Building a healthy culture of 'maths talk' empowers their learning from day one.

Explanation and discussion are integral to the *Power Maths* structure, so by simply following the books your lessons will stimulate structured talk. The following key 'maths talk' strategies will help you strengthen that culture and ensure that every child is included.

Sentences, not words

Encourage children to use full sentences when reasoning, explaining or discussing maths. This helps both speaker and listeners to clarify their own understanding. It also reveals whether or not the speaker truly understands, enabling you to address misconceptions as they arise.

Working together

Working with others in pairs, groups or as a whole class is a great way to support maths talk and discussion. Use different group structures to add variety and challenge. For example, children could take timed turns for talking, work independently alongside a 'discussion buddy', or perhaps play different *Power Maths* character roles within their group.

Think first – then talk

Provide clear opportunities within each lesson for children to think and reflect, so that their talk is purposeful, relevant and focused.

Give every child a voice

Where the 'hands up' model allows only the more confident child to shine, *Power Maths* involves everyone. Make sure that no child dominates and that even the shyest child is encouraged to contribute – and praised when they do.

Assessment strategies

Teaching for mastery demands that you are confident about what each child knows and where their misconceptions lie; therefore, practical and effective assessment is vitally important.

Formative assessment within lessons

The **Think together** section will often reveal any confusions or insecurities; try ironing these out by doing the first **Think together** question as a class. For children who continue to struggle, you or your Teaching Assistant should provide support and enable them to move on.

▶ Performance in practice can be very revealing: check Practice Books and listen out both during and after practice to identify misconceptions.

▶ The **Reflect** section is designed to check on the all-important depth of understanding. Be sure to review how the children performed in this final stage before you teach the next lesson.

End of unit check – Textbook

Each unit concludes with a summative check to help you assess quickly and clearly each child's understanding, fluency, reasoning and problem solving skills. Your Teacher Guide will suggest ideal ways of organising a given activity and offer advice and commentary on what children's responses mean. For example, 'What misconception does this reveal?'; 'How can you reinforce this particular concept?'

Assessment with young children should always be an enjoyable activity, so avoid one-to-one individual assessments, which they may find threatening or scary. If you prefer, the End of unit check can be carried out as a whole-class group using whiteboards and Practice Books.

End of unit check – Practice Book

The Practice Book contains further opportunities for assessment, and can be completed by children independently whilst you are carrying out diagnostic assessment with small groups. Your Teacher Guide will advise you on what to do if children struggle to articulate an explanation – or perhaps encourage you to write down something they have explained well. It will also offer insights into children's answers and their implications for next learning steps. It is split into three main sections, outlined below.

My journal is designed to allow children to show their depth of understanding of the unit. It can also serve as a way of checking that children have grasped key mathematical vocabulary. The question children should answer is first presented in the Textbook in the Think! section. This provides an opportunity for you to discuss the question first as a class to ensure children have understood their task. Children should have some time to think about how they want to answer the question, and you could ask them to talk to a partner about their ideas. Then children should write their answer in their Practice Book, using the word bank provided to help them with vocabulary.

The **Power check** allows pupils to self-assess their level of confidence on the topic by colouring in different smiley faces. You may want to introduce the faces as follows:

I am starting to understand. I need more practice

I don't yet know how to do this

I know how to do this, but I will keep practising

Each unit ends with either a Power play or a Power puzzle. This is an activity, puzzle or game that allows children to use their new knowledge in a fun, informal way.

Progress Tests

There are *Power Maths* Progress Tests for each half term and at the end of the year, including an Arithmetic test and Reasoning test in each case. You can enter results in the online markbook to track and analyse results and see the average for all schools' results. The tests use a 6-step scale to show results against age-related expectation.

How to ask diagnostic questions

The diagnostic questions provided in children's Practice Books are carefully structured to identify both understanding and misconceptions (if children answer in a particular way, you will know why). The simple procedure below may be helpful:

Ask the question, offering the selection of answers provided.

Children take time to think about their response.

Each child selects an answer and shares their reasoning with the group.

Give minimal and neutral feedback (for example, 'That's interesting', or 'Okay').

Ask, 'Why did you choose that answer?', then offer an opportunity to change their mind by providing one correct and one incorrect answer.

Note which children responded and reasoned correctly first time and everyone's final choices.

Reflect that together, we can get the right answer.

Keeping the class together

Traditionally, children who learn quickly have been accelerated through the curriculum. As a consequence, their learning may be superficial and will lack the many benefits of enabling children to learn with and from each other.

By contrast, *Power Maths'* mastery approach values real understanding and richer, deeper learning above speed. It sees all children learning the same concept in small, cumulative steps, each finding and mastering challenge at their own level. Remember that when you teach for mastery, EVERYONE can do maths! Those who grasp a concept easily have time to explore and understand that concept at a deeper level. The whole class therefore moves through the curriculum at broadly the same pace via individual learning journeys.

For some teachers, the idea that a whole class can move forward together is revolutionary and challenging. However, the evidence of global good practice clearly shows that this approach drives engagement, confidence, motivation and success for all learners, and not just the high flyers. The strategies below will help you keep your class together on their maths journey.

Mix it up

Do not stick to set groups at each table. Every child should be working on the same concept, and mixing up the groupings widens children's opportunities for exploring, discussing and sharing their understanding with others.

Recycling questions

Reuse the Textbook and Practice Book questions with concrete materials to allow children to explore concepts and relationships and deepen their understanding. This strategy is especially useful for reinforcing learning in same-day interventions.

Strengthen at every opportunity

The next lesson in a *Power Maths* sequence always revises and builds on the previous step to help embed learning. These activities provide golden opportunities for individual children to strengthen their learning with the support of Teaching Assistants.

Prepare to be surprised!

Children may grasp a concept quickly or more slowly. The 'fast graspers' won't always be the same individuals, nor does the speed at which a child understands a concept predict their success in maths. Are they struggling or just working more slowly?

Same-day intervention

Since maths competence depends on mastering concepts one by one in a logical progression, it is important that no gaps in understanding are ever left unfilled. Same-day interventions – either within or after a lesson – are a crucial safety net for any child who has not fully made the small step covered that day. In other words, intervention is always about keeping up, not catching up, so that every child has the skills and understanding they need to tackle the next lesson. That means presenting the same problems used in the lesson, with a variety of concrete materials to help children model their solutions.

We offer two intervention strategies below, but you should feel free to choose others if they work better for your class.

Within-lesson intervention

The **Think together** activity will reveal those who are struggling, so when it is time for practice, bring these children together to work with you on the first practice questions. Observe these children carefully, ask questions, encourage them to use concrete models and check that they reach and can demonstrate their understanding.

After-lesson intervention

You might like to use the **Think together** questions to recap the lesson with children who are working behind expectations during assembly time. Teaching Assistants could also work with these children at other convenient points in the school day. Some children may benefit from revisiting work from the same topic in the previous year group. Note also the suggestion for recycling questions from the Textbook and Practice Book with concrete materials on page 27.

The role of practice

Practice plays a pivotal role in the *Power Maths* approach. It takes place in class groups, smaller groups, pairs, and independently, so that children always have the opportunities for thinking as well as the models and support they need to practise meaningfully and with understanding.

Intelligent practice

In *Power Maths*, practice never equates to the simple repetition of a process. Instead we embrace the concept of intelligent practice, in which all children become fluent in maths through varied, frequent and thoughtful practice that deepens and embeds conceptual understanding in a logical, planned sequence. To see the difference, take a look at the following examples.

Traditional practice

- Repetition can be rote – no need for a child to think hard about what they are doing

- Praise may be misplaced

- Does this prove understanding?

Intelligent practice

- Varied methods – concrete, pictorial and abstract

- Equation expressed in different ways, requiring thought and understanding

- Constructive feedback

All practice questions are designed to move children on and reveal misconceptions.

Simple, logical steps build onto earlier learning.

C-P-A runs throughout – different ways of modelling and understanding the same concept.

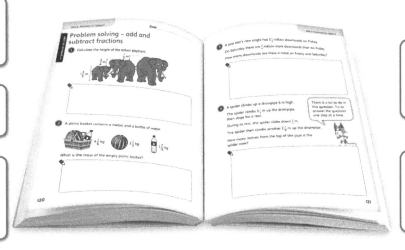

Conceptual variation – children work on different representations of the same maths concept.

Friendly characters offer support and encourage children to try different approaches.

A carefully designed progression

The Practice Books provide just the right amount of intelligent practice for children to complete independently in the final sections of each lesson. It is really important that all children are exposed to the practice questions, and that children are not directed to complete different sections. That is because each question is different and has been designed to challenge children to think about the maths they are doing. The questions become more challenging so children grasping concepts more quickly will start to slow down as they progress. Meanwhile, you have the chance to circulate and spot any misconceptions before they become barriers to further learning.

Homework and the role of parents and carers

While *Power Maths* does not prescribe any particular homework structure, we acknowledge the potential value of practice at home. For example, practising fluency in key facts, such as number bonds and times-tables, is an ideal homework task. You can share the Individual Practice Games for homework (see page 6), or parents and carers could work through uncompleted Practice Book questions with children at either primary stage.

However, it is important to recognise that many parents and carers may themselves lack confidence in maths, and few, if any, will be familiar with mastery methods. A Parents' and Carers' evening that helps them understand the basics of mindsets, mastery and mathematical language is a great way to ensure that children benefit from their homework. It could be a fun opportunity for children to teach their families that everyone can do maths!

Structures and representations

Unlike most other subjects, maths comprises a wide array of abstract concepts – and that is why children and adults so often find it difficult. By taking a concrete-pictorial-abstract (C-P-A) approach, *Power Maths* allows children to tackle concepts in a tangible and more comfortable way.

Non-linear stages

Concrete

Replacing the traditional approach of a teacher working through a problem in front of the class, the concrete stage introduces real objects that children can use to 'do' the maths – any familiar object that a child can manipulate and move to help bring the maths to life. It is important to appreciate, however, that children must always understand the link between models and the objects they represent. For example, children need to first understand that three cakes could be represented by three pretend cakes, and then by three counters or bricks. Frequent practice helps consolidate this essential insight. Although they can be used at any time, good concrete models are an essential first step in understanding.

Pictorial

This stage uses pictorial representations of objects to let children 'see' what particular maths problems look like. It helps them make connections between the concrete and pictorial representations and the abstract maths concept. Children can also create or view a pictorial representation together, enabling discussion and comparisons. The *Power Maths* teaching tools are fantastic for this learning stage, and bar modelling is invaluable for problem solving throughout the primary curriculum.

Abstract

Our ultimate goal is for children to understand abstract mathematical concepts, symbols and notation and of course, some children will reach this stage far more quickly than others. To work with abstract concepts, a child must be comfortable with the meaning of and relationships between concrete, pictorial and abstract models and representations. The C-P-A approach is not linear, and children may need different types of models at different times. However, when a child demonstrates with concrete models and pictorial representations that they have grasped a concept, we can be confident that they are ready to explore or model it with abstract symbols such as numbers and notation.

Use at any time and with any age to support understanding

Variation helps visualisation

Children find it much easier to visualise and grasp concepts if they see them presented in a number of ways, so be prepared to offer and encourage many different representations.

For example, the number six could be represented in various ways:

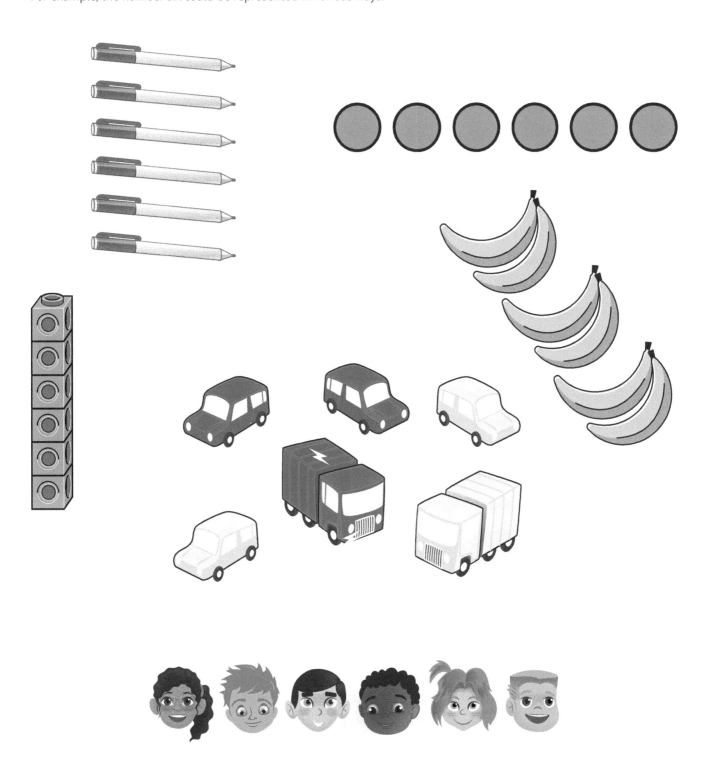

Practical aspects of *Power Maths*

One of the key underlying elements of *Power Maths* is its practical approach, allowing you to make maths real and relevant to your children, no matter their age.

Manipulatives are essential resources for both key stages and *Power Maths* encourages teachers to use these at every opportunity, and to continue the Concrete-Pictorial-Abstract approach right through to Year 6.

The Textbooks and Teacher Guides include lots of opportunities for teaching in a practical way to show children what maths means in real life.

Discover and Share

The **Discover** and **Share** sections of the Textbook give you scope to turn a real-life scenario into a practical and hands-on section of the lesson. Use these sections as inspiration to get active in the classroom. Where appropriate, use the **Discover** contexts as a springboard for your own examples that have particular resonance for your children – and allow them to get their hands dirty trying out the mathematics for themselves.

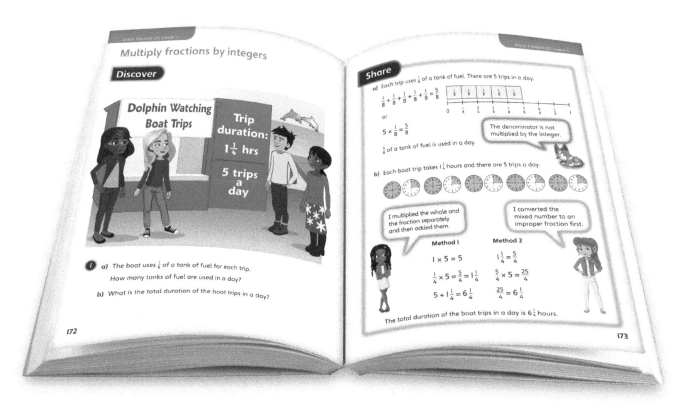

Unit videos

Every term has one unit video which incorporates real-life classroom sequences.

These videos show you how the reasoning behind mathematics can be carried out in a practical manner by showing real children using various concrete and pictorial methods to come to the solution. You can see how using these practical models, such as part-whole and bar models, helps them to find and articulate their answer.

Mastery tips

Mastery Experts give anecdotal advice on where they have used hands-on and real-life elements to inspire their children.

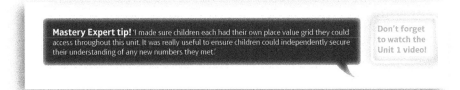

Mastery Expert tip! 'I made sure children each had their own place value grid they could access throughout this unit. It was really useful to ensure children could independently secure their understanding of any new numbers they met.'

Don't forget to watch the Unit 1 video!

Concrete-Pictorial-Abstract (C-P-A) approach

Each **Share** section uses various methods to explain an answer, helping children to access abstract concepts by using concrete tools, such as counters. Remember, this isn't a linear process, so even children who appear confident using the more abstract method can deepen their knowledge by exploring the concrete representations. Encourage children to use all three methods to really solidify their understanding of a concept.

Pictorial representation – drawing the problem in a logical way that helps children visualise the maths

Concrete representation – using manipulatives to represent the problem. Encourage children to physically use resources to explore the maths.

Abstract representation – using words and calculations to represent the problem.

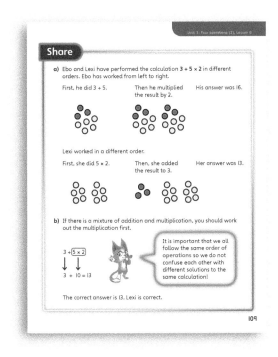

Practical tips

Every lesson suggests how to draw out the practical side of the **Discover** context.

You'll find these in the **Discover** section of the Teacher Guide for each lesson.

PRACTICAL TIPS The game in the picture can be repeated easily in the classroom using digit cards. Encourage children to set similar challenges for each other.

Resources

Every lesson lists the practical resources you will need or might want to use. There is also a summary of all of the resources used throughout the term on page 39 to help you be prepared.

RESOURCES

Mandatory: place value grids, counters
Optional: number lines

Working with children below age-related expectation

This section offers advice on using *Power Maths* with children who are significantly behind age-related expectation. Teacher judgement will be crucial in terms of where and why children are struggling, and in choosing the right approach. The suggestions can of course be adapted for children with special educational needs, depending on the specific details of those needs.

General approaches to support children who are struggling

Keeping the pace manageable

Remember, you have more teaching days than *Power Maths* lessons so you can cover a lesson over more than one day, and revisit key learning, to ensure all children are ready to move on. You can use the + and – buttons to adjust the time for each unit in the online planning. The NCETM's Ready-to-Progress criteria can be used to help determine what should be highest priority.

Same-day intervention

You could go over the Textbook pages or revisit the previous year's work if necessary (see Addressing gaps). Remember that same-day intervention can be within the lesson, as well as afterwards (see page 28). As children start their independent practice, you can work with those who found the first part of the lesson difficult, checking understanding using manipulatives.

Fluency sessions

Fit in as much practice as you can for number bonds and times-tables, etc., at other times of the day. If you can, plan a short 'maths meeting' for this in the afternoon. You might choose to use a Power Up you haven't used already.

Addressing gaps

Use material from the same topic in the previous year to consolidate or address gaps in learning, e.g. Textbook pages and Strengthen activities. The End of unit check will help gauge children's understanding.

Pre-teaching

Find a 5- to 10-minute slot before the lesson to work with the children you feel would benefit. The afternoon before the lesson can work well, because it gives children time to think in between. Recap previous work on the topic (addressing any gaps you're aware of) and do some fluency practice, targeting number facts etc. that will help children access the learning.

Focusing on the key concepts

If children are a long way behind, it can be helpful to take a step back and think about the key concepts for children to engage with, not just the fine detail of the objective for that year group (e.g. addition with a specific number of columns). Bearing that in mind, how could children advance their understanding of the topic?

Providing extra support within the lesson

Support in the Teacher Guide

First of all, use the Strengthen support in the Teacher Guide for guided and independent work in each lesson, and share this with Teaching Assistants, where relevant. As you read through the lesson content and corresponding Teacher Guide pages before the lesson, ask yourself what key idea or nugget of understanding is at the heart of the lesson. If children are struggling, this should help you decide what's essential for all children before they move on.

Annotating pages

You can annotate questions to provide extra scaffolding or hints if you need to, but aim to build up children's ability to access questions independently wherever you can. Children tend to get used to the style of the *Power Maths* questions over time.

Quick recap as lesson starter

The Quick recap for each lesson in the Teacher Guide is an alternative starter activity to the Power Up. You might choose to use this with some or all children if you feel they will need support accessing the main lesson.

Consolidation questions

If you think some children would benefit from additional questions at the same level before moving on, write one or two similar questions on the board. (This shouldn't be at the expense of reasoning and problem-solving opportunities: take longer over the lesson if you need to.)

Hard copy Textbooks

The Textbooks help children focus in more easily on the mathematical representations, read the text more comfortably, and revisit work from a previous lesson that you are building on, as well as giving children ownership of their learning journey. In main lessons, it can work well to use the e-Textbook for **Discover** and give out the books when discussing the methods in the **Share** section.

Reading support

It's important that all children are exposed to problem solving and reasoning questions, which often involve reading. For whole-class work you can read questions together. For independent practice you could consider annotating pages to help children see what the question is asking, and stem sentences to help structure their answer. A general focus on specific mathematical language and vocabulary will help children access the questions. You could consider pairing weaker readers with stronger readers, or read questions as a group if those who need support are on the same table.

Providing extra depth and challenge with *Power Maths*

Just as prescribed in the National Curriculum, the goal of *Power Maths* is never to accelerate through a topic but rather to gain a clear, deep and broad understanding. Here are some suggestions to help ensure all children are appropriately challenged as you work with the resources.

Overall approaches

First of all, remember that the materials are designed to help you keep the class together, allowing all children to master a concept while those who grasp it quickly have time to explore it in more depth. Use the Deepen support in the Teacher Guide (see below) to challenge children who work through the questions quickly. Here are some questions and ideas to encourage breadth and depth during specific parts of the lesson, or at any time (where no part of the lesson sequence is specified):

- **Discover**: 'Can you demonstrate your solution another way?'

- **Share**: Make sure every child is encouraged to give answers and engage with the discussion, not just the most confident.

- **Think together**: 'Can you model your answers using concrete materials? Can you explain your solution to a partner?'

- Practice: Allow all children to work through the full set of questions, so that they benefit from the logical sequence.

- **Reflect**: 'Is there another way of working out the answer? And another way?'
  'Have you found all the solutions?'
  'Is that always true?'
  'What's different between this question and that question? And what's the same?'

Note that the **Challenge** questions are designed so that all children can access and attempt them, if they have worked through the steps leading up to them. There may be some children in a given lesson who don't manage to do the **Challenge**, but it is not supposed to be a distinct task for a subset of the class. When you look through the lesson materials before teaching, think about what each question is specifically asking, and compare this with the key learning point for the lesson. This will help you decide which questions you feel it's essential for all children to answer, before moving on. You can at least aim for all children to try the **Challenge**!

Deepen activities and support

The Teacher Guide provides valuable support for each stage of the lesson. This includes Deepen tips for the guided and independent practice sections, which will help you provide extra stretch and challenge within your lesson, without having to organise additional tasks. If you have a Teaching Assistant, they can also make use of this advice. There are also suggestions for the lesson as a whole in the 'Going Deeper' section on the first page of the Teacher Guide section for that lesson. Every class is different, so you can always go a bit further in the direction indicated, if appropriate, and build on the suggestions given.

There is a Deepen activity for each unit. These are designed to follow on from the End of unit check, stretching children who have a firm understanding of the key learning from the unit. Children can work on them independently, which makes it easier for the teacher to facilitate the Strengthen activity for children who need extra support. Deepen activities could also be introduced earlier in the unit if the necessary work has been covered. The Deepen activities are on *ActiveLearn* on the Planning page for each unit, and also on the Resources page).

POWER MATHS Deepen Activities

1. Mrs Dean wants to fill a paddling pool using three different hoses. The first hose fills 75 litres per minute. The second hose fills 75 pints per minute. The third hose fills 750 ml per second.

   a) Which hose is the slowest?

   b) Mrs Dean fills the paddling pool using all three hoses. How much water would be in the pool after 1 hour?

   c) Mrs Dean decides not to use all the hoses and after exactly 1 minute, she removes one. After 5 minutes there is over 600 litres of water in the pool. Which hose did Mrs Dean remove?

 I will use the fact that 1 pint is about 560 ml to convert them all to the same unit of measure.

Using the questions flexibly to provide extra challenge

Sometimes you may want to write an extra question on the board or provide this on paper. You can usually do this by tweaking the lesson materials. The questions are designed to form a carefully structured sequence that builds understanding step by step, but, with careful thought about the purpose of each question, you can use the materials flexibly where you need to. Sometimes you might feel that children would benefit from another similar question for consolidation before moving on to the next one, or you might feel that they would benefit from a harder example in the same style. It should be quick and easy to generate 'more of the same' type questions where this is the case.

When you see a question like this one (from Unit 2, Lesson 1), it's easy to make extra examples to do afterwards if you need them, maximising the number of exchanges if you want to make it tricky. You can also blot out more digits and ask children if this makes it easier or harder.

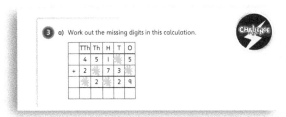

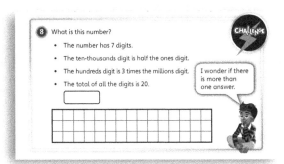

For this example (from Unit 1, Lesson 3), you could ask children to make up their own question(s) for a partner to solve, using 4 clues. (In fact, for any of these examples you could ask early finishers to create their own question for a partner.)

Here's an example (from Unit 4, Lesson 6) where the sum of two number cards is used in the question, but there are other combinations you could ask children to work out, for example, as an extra task at the end of the lesson.

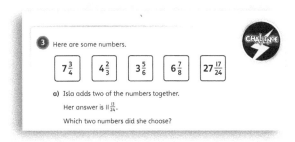

Besides creating additional questions, you should be able to find a question in the lesson that you can adapt into a game or open-ended investigation, if this helps to keep everyone engaged. It could simply be that, instead of answering 5 × 5 etc. on the page, they could build a robot with 5 lots of 5 cubes.

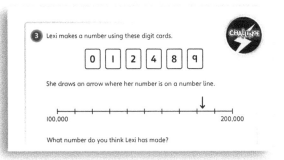

With a question like this (from Unit 1, Lesson 1), children could play a game where they have to guess their partner's mystery number, finding out each time if the guess is too high or too low.

See the bullets above for some general ideas that will help with 'opening out' questions in the books, e.g. 'Can you find all the solutions?' type questions.

Other suggestions

Another way of stretching children is through mixed ability pairs, or via other opportunities for children to explain their understanding in their own way. This is a good way of encouraging children to go deeper into the learning, rather than, for instance, tackling questions that are computationally more challenging but conceptually equivalent in level.

Using *Power Maths* with mixed age classes

Overall approaches

There are many variables between schools that would make it inadvisable to recommend a one-size-fits-all approach to mixed age teaching with *Power Maths*. These include how year groups are merged, availability of Teaching Assistants, experience and preference of teaching staff, range in pupil attainment across years, classroom space and layout, level of flexibility around timetables, and overall organisational structure (whether the school is part of a trust).

Some schools will find it best to timetable separate maths lessons for the different year groups. Others will aim to teach the class together as much as possible using the mixed age planning support on *ActiveLearn* (see the lesson exemplars for ways of organising lessons with strong/medium/weak correlation between year groups). There will also be ways of adapting these general approaches. For example, offset lessons where Year A start their lesson with the teacher, while Year B work independently on the practice from the previous lesson, and then start the next lesson with the teacher while Year A work independently; or teachers may choose to base their provision around the lesson from one year group and tweak the content up/down for the other group.

Key strategies for mixed age teaching

The mixed age teaching webinar on *ActiveLearn* provides advice on all aspects of mixed age teaching, including more detail on the ideas below.

Developing independence over time
Investing time in building up children's independence will pay off in the medium term.

Clear rationale
If someone asked, 'Why did you teach both Unit 3 and 4 in the same lesson/separate lessons?', what would your answer be?

Designing a lesson
1. Identify the core learning for each group
2. Identify any number skills necessary to access the core
3. Consider the flow of concepts and how one core leads to the other

Challenging all children
The questions are designed to build understanding step by step, but with careful thought about the purpose of each question you can tweak them to increase the challenge.

Multiple years combined
With more than two years together, teachers will inevitably need to use the resources flexibly if delivering a single lesson.

Enjoy the positives!

Comparison deepens understanding and there will be lots of opportunities for children, as well as misconceptions to explore. There is also in-built pre-teaching and the chance to build up a concept from its foundations. For teachers there is double the material to draw on! Mixed age teachers require a strong understanding of the progression of ideas across year groups, which is highly valuable for all teachers. Also, it is necessary to engage deeply with the lesson to see how to use the materials flexibly – this is recommended for all teachers and will help you bring your lesson to life!

List of practical resources

Year 6A Mandatory resources

Resource	Lesson
2D square	**Unit 2** Lesson 8
3D cube	**Unit 2** Lesson 8
counters	**Unit 1** Lesson 2 **Unit 2** Lesson 8
fraction strips	**Unit 5** Lesson 5
multilink cubes	**Unit 2** Lesson 8
paper plates	**Unit 5** Lesson 5
paper strips	**Unit 5** Lesson 5
place value equipment	**Unit 1** Lesson 4 **Unit 2** Lessons 1, 2 **Unit 3** Lesson 3
place value grids	**Unit 1** Lesson 2

Year 6A Optional resources

Resource	Lesson
100 square	**Unit 2** Lessons 7, 8
apples	**Unit 5** Lesson 8
bag of oats and other ingredients	**Unit 5** Lesson 2
base 10 equipment	**Unit 1** Lessons 1, 3 **Unit 2** Lesson 3 **Unit 3** Lessons 9, 10
baskets	**Unit 5** Lesson 8
bead strings	**Unit 1** Lesson 5 **Unit 3** Lessons 8, 9
bean bags	**Unit 2** Lesson 5
bingo game	**Unit 3** Lesson 11
bottle of liquid	**Unit 5** Lesson 4
card	**Unit 3** Lesson 9
concrete materials to represent lengths of bamboo (e.g. sticks, wool, ribbon, card, cubes)	**Unit 5** Lesson 6
containers (with different volumes)	**Unit 6** Lesson 3
counters (or other small objects, to make arrays)	**Unit 1** Lessons 6, 7 **Unit 2** Lessons 4, 7 **Unit 3** Lessons 4, 8, 9 **Unit 5** Lessons 3, 4, 6, 8, 9
cups	**Unit 5** Lesson 4
dice	**Unit 1** Lesson 1
digit cards	**Unit 1** Lessons 1, 6 **Unit 6** Lesson 2
flashcards	**Unit 1** Lesson 1
'Follow me' cards	**Unit 2** Lesson 4 **Unit 3** Lesson 11
fraction strips	**Unit 4** Lesson 4 **Unit 5** Lessons 1, 4, 6, 8
fraction walls	**Unit 5** Lesson 3
fraction wheels	**Unit 4** Lesson 4 **Unit 5** Lesson 1
graph paper	**Unit 6** Lesson 4
hoops	**Unit 2** Lesson 5
measuring jugs	**Unit 6** Lessons 1, 2, 3

Resource	Lesson
measuring jugs (millilitres)	**Unit 6** Lesson 5
measuring scales	**Unit 6** Lessons 1, 3
measuring scales (pounds and grams)	**Unit 6** Lesson 5
measuring scales (with a marked dial)	**Unit 6** Lesson 2
metre rulers	**Unit 6** Lesson 2
metre rulers (with two units of measurement (kg/g, l/ml, m/cm) Q	**Unit 6** Lesson 3
multilink cubes	**Unit 3** Lessons 8, 9 **Unit 5** Lesson 9
multiplication grids	**Unit 2** Lessons 4, 5, 8 **Unit 3** Lesson 12
number lines	**Unit 1** Lessons 2, 8 **Unit 3** Lesson 11
number lines (blank)	**Unit 3** Lessons 6, 7
number tracks	**Unit 2** Lesson 6
paper (large)	**Unit 3** Lesson 9
paper circles	**Unit 5** Lessons 4, 6
paper circles, squares and strips	**Unit 5** Lesson 5
paper squares	**Unit 5** Lesson 6
paper strips	**Unit 5** Lessons 4, 6, 8 **Unit 6** Lessons 4, 5
pictures of items from a building site	**Unit 6** Lesson 2
pictures of pandas	**Unit 5** Lesson 6
pint cartons	**Unit 6** Lesson 5
place value counters	**Unit 1** Lesson 3 **Unit 2** Lesson 3 **Unit 3** Lessons 1, 2, 5
place value equipment	**Unit 3** Lesson 2
place value grids	**Unit 1** Lessons 3, 6, 7 **Unit 3** Lesson 2
place value grids (printed)	**Unit 2** Lesson 3
plastic counters	**Unit 6** Lesson 4
play coins	**Unit 1** Lesson 3 **Unit 5** Lesson 9
real-life house sale advertisements	**Unit 3** Lesson 11
rulers (inches and centimetres)	**Unit 6** Lesson 5
sorting circles	**Unit 5** Lesson 6
square templates	**Unit 5** Lessons 2, 3
sticky notes (or small strips of paper)	**Unit 6** Lesson 3
string	**Unit 6** Lesson 5
ten frames	**Unit 3** Lesson 8
water	**Unit 6** Lesson 2
whiteboard pens	**Unit 5** Lessons 4, 6
whiteboards (mini)	**Unit 3** Lesson 12

Getting started with *Power Maths*

As you prepare to put *Power Maths* into action, you might find the tips and advice below helpful.

STEP 1: Train up!

A practical, up-front full day professional development course will give you and your team a brilliant head-start as you begin your *Power Maths* journey. You will learn more about the ethos, how it works and why.

STEP 2: Check out the progression

Take a look at the yearly and termly overviews. Next take a look at the unit overview for the unit you are about to teach in your Teacher Guide, remembering that you can match your lessons and pacing to match your class.

STEP 3: Explore the context

Take a little time to look at the context for this unit: what are the implications for the unit ahead? (Think about key language, common misunderstandings and intervention strategies, for example.) If you have the online subscription, don't forget to watch the corresponding unit video.

STEP 4: Prepare for your first lesson

Familiarise yourself with the objectives, essential questions to ask and the resources you will need. The Teacher Guide offers tips, ideas and guidance on individual lessons to help you anticipate children's misconceptions and challenge those who are ready to think more deeply.

STEP 5: Teach and reflect

Deliver your lesson — and enjoy!

Afterwards, reflect on how it went... Did you cover all five stages?
Does the lesson need more time? How could you improve it?
What percentage of your class do you think mastered the concept?
How can you help those that didn't?

Unit I
Place value within I0,000,000

Don't forget to watch the Unit 1 video!

WHY THIS UNIT IS IMPORTANT

This unit develops children's understanding of place value and properties of numbers up to 10,000,000. It is an important unit as the number sense they develop now will support their learning in future units.

Children will recap their understanding of place value and properties of numbers to 1,000,000 before investigating the same properties of numbers up to 10,000,000. They will investigate the partitioning of larger numbers and will use them in different contexts. Following this, children will develop their understanding and use of number lines up to 10,000,000 and will plot numbers on partially completed number lines.

Having deepened their understanding of the properties of numbers up to 10,000,000, children will use this to compare and order numbers and to round them to any degree of accuracy up to the nearest 1,000,000. Finally, children will investigate negative numbers, how they compare to positive numbers and their use in context.

WHERE THIS UNIT FITS

→ **Unit 1: Place value within 10,000,000**

→ Unit 2: Four operations (1)

In this unit, children extend their knowledge of numbers from within 1,000,000 to within 10,000,000, before they go on to work with the four operations in the next two units. This includes looking at place value, ordering and comparing numbers and rounding. They will also look at number lines and negative numbers. Before they start this unit, it is expected that children understand the place value of numbers within 1,000,000, can use number lines, including counting in 10s, 100s, 1,000s and 10,000s, and can round numbers within 1,000,000.

ASSESSING MASTERY

Children will demonstrate fluent understanding of place value within numbers up to 10,000,000. They will be able to read, write and partition numbers accurately and use this and their understanding of place value to compare and order numbers up to 10,000,000. They will demonstrate confident fluency using number lines and will be able to complete number lines using unlabelled intervals to calculate what each interval is worth, as well as calculating intervals across 0. Children will be able to use their understanding of place value to round to powers of 10, up to 1,000,000, and will be able to recognise and use negative numbers in real-life contexts.

COMMON MISCONCEPTIONS	STRENGTHENING UNDERSTANDING	GOING DEEPER
Children may assume that the only way to partition a number is into its place value headings (for example, only partitioning 136 as 100, 30 and 6).	Give children opportunities to create numbers using different representations, such as base 10 equipment, which they can then partition in different ways.	Children could be given, or create their own, missing number equations. For example: 1,456 = 500 + 500 + ___ + 156
Children may assume that negative numbers work in a similar way to positive numbers. For example, thinking that ⁻5 must be greater than ⁻1 because 5 is greater than 1.	Provide a number line from ⁻10 up to 10. Place the correct number of objects above the positive numbers. Ask: *What do you notice about the objects as you move down to 0? What will happen to the numbers below 0?*	Encourage children to count down from a given positive number. Ask: *What do you notice about the positive numbers as you count down? What will happen with the negative numbers?*

Unit I: Place value within 10,000,000

UNIT STARTER PAGES

Use these pages to introduce the focus of the unit to children. You can use the characters to explore different ways of working too.

STRUCTURES AND REPRESENTATIONS

Place value grid: Place value grids are used with both counters and numbers in this unit to help children read numbers and recognise the value of each digit in numbers up to 10,000,000.

M	HTh	TTh	Th	H	T	O

Part-whole model: Part-whole models are used to help children partition numbers.

Number line: Number lines are used to help children plot numbers from 0 to 10,000,000, work out differences and round numbers. Later in the unit, they are used to show negatives and work out intervals across 0.

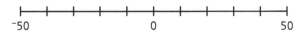

Bar chart: A bar chart is used in this unit to give context to children's learning about numbers.

Bar model: A bar model is used to help children understand how to work out the value of unlabelled intervals on a number line.

KEY LANGUAGE

There is some key language that children will need to know as a part of the learning in this unit.

→ ten thousands (10,000s), hundred thousands (100,000s), millions (1,000,000s), ten million (10,000,000)
→ place value
→ partition/partitioned/partitioning
→ interval
→ estimate
→ compare/comparison/comparing
→ order/ordering
→ less than (<), greater than (>), equal to (=)
→ rounding/rounded/round up/round down
→ negative, positive

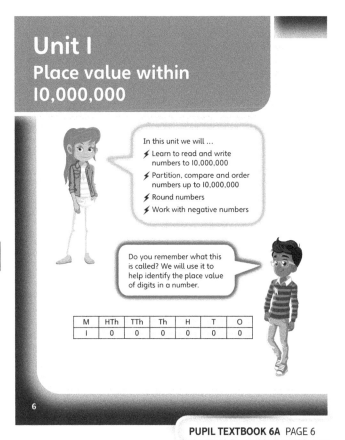

PUPIL TEXTBOOK 6A PAGE 6

PUPIL TEXTBOOK 6A PAGE 7

Numbers to 1,000,000

Learning focus

In this lesson, children will learn about the place value in numbers to 1,000,000. They will learn to read and write these numbers fluently and identify their place value.

Before you teach

- Are children confident with the place value of numbers up to 100,000?
- Are children more confident reading numbers than writing them? How will you support children who find writing and spelling challenging?

NATIONAL CURRICULUM LINKS

Year 6 Number – number and place value

Read, write, order and compare numbers up to 10,000,000 and determine the value of each digit.

Solve number and practical problems.

ASSESSING MASTERY

Children can recognise, read and write numbers up to 1,000,000. They can use what they know about place value to describe the value of each digit in a number and can solve problems using this understanding.

COMMON MISCONCEPTIONS

Children may neglect to include 0s as place holders in numbers when writing them. For example, they may write 132,045 as 13,245. Show children both numbers and ask:
- *Can you build both of these numbers? What is the same and what is different about them?*
- *What is the value of each digit in each number? Why is the 0 interesting?*

STRENGTHENING UNDERSTANDING

As children may misspell the mathematical vocabulary when writing the names of numbers, it would be beneficial to have the vocabulary displayed on the classroom wall. For children who may struggle with numbers up to 1,000,000, give them opportunities to investigate the place value of numbers up to 100,000 before the lesson using base 10 equipment and place value cards.

GOING DEEPER

Children can be encouraged to design their own hidden number challenges. Ask: *Can you create a set of clues where there are many possible answers? Can you create a set of clues where only one solution is possible? Can you write each of your clues so they require a different mathematical skill to solve?*

KEY LANGUAGE

In lesson: ones (1s), tens (10s), hundreds (100s), thousands (1,000s), ten thousands (10,000s), hundred thousands (100,000s), odd, place value, multiple

Other language to be used by the teacher: million (1,000,000), even

STRUCTURES AND REPRESENTATIONS

Place value grid, number line, table

RESOURCES

Optional: base 10 equipment, place value cards, digit cards, flash cards, dice

 In the eTextbook of this lesson, you will find interactive links to a selection of teaching tools.

Quick recap 🔁

Ask children to take turns to roll a dice six times. They use these numbers to make a 6-digit number and then read the number aloud.

Repeat, this time setting challenges such as:
Can you make a number greater than 300,000? Can you make an odd number? What is the smallest possible 6-digit number you can make?

Discover

Numbers to 1,000,000

Discover

WAYS OF WORKING Pair work

ASK

- Question ① a): *What do you know about the number Richard has made?*
- Question ① b): *Is there more than one solution? How can you prove it?*

IN FOCUS Question ① a) allows children the opportunity to reason with numbers and to explore the digits of a 6-digit number. For question ① b) they should be able to recognise that there is more than one possible solution to the clues given.

PRACTICAL TIPS The game in the picture can be repeated easily in the classroom using digit cards. Encourage children to set similar challenges for each other.

ANSWERS

Question ① a): There are six possible solutions:
627,489; 628,479; 726,489; 728,469; 629,487; 628,497.

Question ① b): Children might find up to 6 answers. If they find more than 6 answers they have not interpreted one of the restrictions correctly.

① a) What numbers could Richard have made using the digit cards shown?

b) How many different answers can you find?

8

PUPIL TEXTBOOK 6A PAGE 8

Share

WAYS OF WORKING Whole class teacher led

ASK

- Questions ① a) and ① b): *How does the place value grid help you to solve this problem?*
- Question ① b): *How many solutions did you find?*

IN FOCUS Question ① a) will help children to recap the place values they have met before. It will also develop their reasoning, showing them how to make progress through the problem given in the **Discover** section. You can use Flo's comment to spark discussion about the different methods children used when solving the problem.

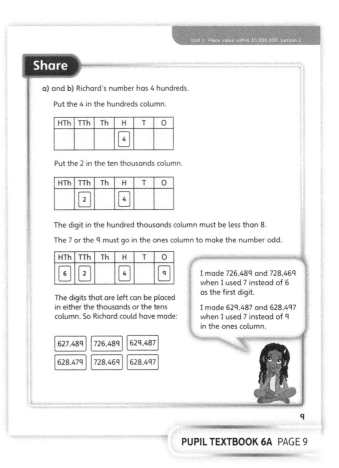

PUPIL TEXTBOOK 6A PAGE 9

Think together

Whole class teacher led (I do, We do, You do)

ASK

- Question ❶: *Can you say the number out loud?*
- Question ❶: *How can you tell what number to write?*
- Question ❷: *Do the place value columns help?*
- Question ❸: *How can you begin to work this out?*
- Question ❸: *Using the digit cards, what other numbers can you make that will appear on the number line? Prove it.*

IN FOCUS In question ❶ children practise writing a number that contains a place-holding 0 and then finding 1,000 more and 10 less. In question ❷, children practise their recognition of the place value of specific digits in given numbers. Question ❸ encourages children to begin ordering and comparing numbers up to 1,000,000. Observe carefully the numbers children create and which they choose as their solution, as this will give a good idea of their understanding of place value.

STRENGTHEN If children are struggling, it may help to provide flash cards showing pictorial representations of the different place values.

DEEPEN While solving question ❷, deepen children's reasoning about the place value of the digits. Ask: *How can you prove that your answers are correct? What representations could you use to prove your answers?*

ASSESSMENT CHECKPOINT Question ❶ assesses children's ability to read and write a number up to 1,000,000, while questions ❷ and ❸ assess their understanding of place value. Question ❸ also assesses whether children are able to compare and order numbers on a number line. In question ❸, look for children using reasoning to make sensible estimates. For example, children should be able to recognise that the number will be between 180,000 and 190,000. They may then identify that, as it is lower than half-way, the first three digits will be 182 or 184.

ANSWERS

Question ❶ a): 203,416 or two hundred and three thousand, four hundred and sixteen

Question ❶ b): 204,416

Question ❶ c): 203,406

Question ❷ a): 50,000 or 5 ten thousands

Question ❷ b): 5 or 5 ones

Question ❷ c): 500 or 5 hundreds

Question ❷ d): 500,000 or 5 hundred thousands

Question ❸: Lexi has made a 6-digit number starting with 182, for example 182,904, 182,094 or 182,409. Also accept a 6-digit number starting with 184.

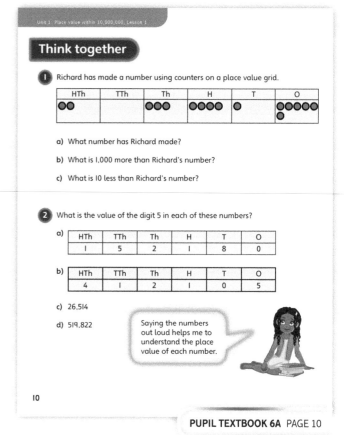

PUPIL TEXTBOOK 6A PAGE 10

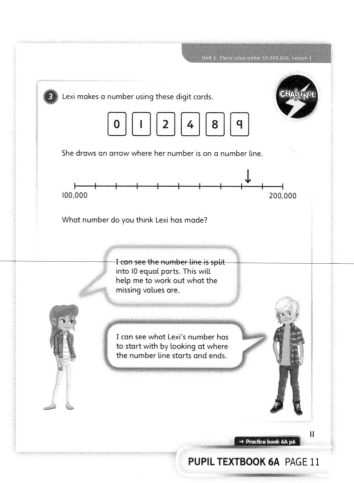

PUPIL TEXTBOOK 6A PAGE 11

Practice

WAYS OF WORKING Independent thinking

IN FOCUS Question **1** scaffolds children's independent understanding of, and ability to recognise, representations of the place value of different numbers up to 1,000,000. Question **2** is important, as it helps to build the link between two abstract ways of writing numbers. This allows you to check that children are not only able to read the numbers written as words, but also understand them in order to rewrite them in digits. In questions **4** and **5**, children then begin using their knowledge to reason and solve problems and puzzles about numbers.

STRENGTHEN In question **6**, if children are struggling to count on or back, ask: *What resources could you use to represent the numbers? How could you use them to find x more or x less than the given number?*

DEEPEN Question **7** deepens children's understanding of the properties of numbers by including clues that draw on other areas of mathematics. Children should be encouraged to give justifications for their solutions, potentially using resources or pictures as supporting evidence.

ASSESSMENT CHECKPOINT Children should understand the place value of different digits and be able to show this in multiple ways. In question **3**, children can be encouraged to justify how they know their answers are correct, in order to ensure their full understanding. Question **5** assesses children's ability to compare and order numbers up to 1,000,000. Look for their ability to recognise patterns in the numbers along a number line and use this to complete the unlabelled intervals. They should then be able to use their understanding of place value and number to sensibly estimate where the given number would fall between two intervals.

ANSWERS Answers for the **Practice** part of the lesson can be found in the *Power Maths* online subscription.

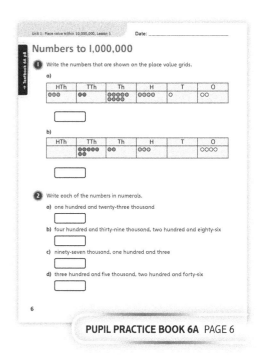

PUPIL PRACTICE BOOK 6A PAGE 6

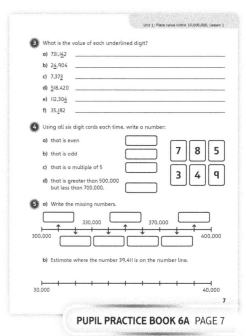

PUPIL PRACTICE BOOK 6A PAGE 7

Reflect

WAYS OF WORKING Independent thinking and pair work

IN FOCUS This question requires children to demonstrate their fluency with place value and properties of numbers up to 1,000,000. By comparing their facts about the number and discussing any differences in their ideas, children should strengthen their confidence and identify any remaining uncertainties.

ASSESSMENT CHECKPOINT Children should be able to identify the place value of each digit in the number. They may also be able to offer suggestions such as '10,000 more is …', '100 less is …' and so on.

ANSWERS Answers for the **Reflect** part of the lesson can be found in the *Power Maths* online subscription.

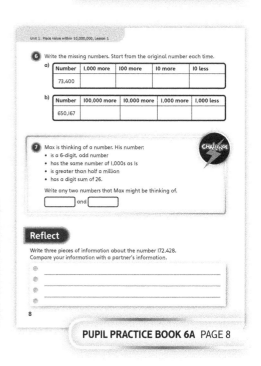

PUPIL PRACTICE BOOK 6A PAGE 8

After the lesson ⏸

- How confident and fluent were children at reading and writing numbers up to 1,000,000?
- Was this lesson as practical and hands on as it could have been? If not, how will you develop this next time you teach it?

Numbers to 10,000,000

Learning focus

In this lesson, children will learn about the place value of numbers to 10,000,000. They will learn to read and write these numbers fluently and identify their place value.

Before you teach

- Are children fluent at reading and writing numbers up to 1,000,000?

NATIONAL CURRICULUM LINKS

Year 6 Number – number and place value

Read, write, order and compare numbers up to 10,000,000 and determine the value of each digit.

Solve number and practical problems.

ASSESSING MASTERY

Children can recognise, describe, read and write numbers up to 10,000,000. They can use their knowledge of place value to describe the value of each digit in a number and can solve problems using this understanding.

COMMON MISCONCEPTIONS

As more digits appear in the numbers children are studying, it is increasingly likely that they may get muddled when they read the numbers, especially when there are place-holding 0s. Ask:
- *How might a place value grid help you to read this number accurately? Can you show me?*

STRENGTHENING UNDERSTANDING

Encourage children to count in steps of 100,000 up to 1,000,000 to help secure their fluency with the pattern of the numbers. Do this before they start looking at numbers up to 10,000,000 in this lesson. Also encourage them to write the name of each number in the count in words and numerals. This counting should start at different given numbers, not only 0. A wall display of vocabulary, as in the previous lesson, may also be useful.

GOING DEEPER

Children can use their knowledge of the numbers they have been studying to create 'guess my number' challenges with numbers up to 10,000,000, giving a partner a list of word clues, for example: *I have seven millions, thirty-eight thousands, three tens and no ones. In numerals, write what number I am.*

KEY LANGUAGE

In lesson: hundreds (100s), thousands (1,000s), hundred thousands (100,000s), millions (1,000,000s)

Other language to be used by the teacher: ones (1s), tens (10s), ten thousands (10,000s), ten million (10,000,000), multiple, place value, odd, even

STRUCTURES AND REPRESENTATIONS

Number line, place value grid

RESOURCES

Mandatory: place value grids, counters

Optional: number lines

 In the eTextbook of this lesson, you will find interactive links to a selection of teaching tools.

Quick recap 🔎

As a class, make a mind map of all the different names for numbers and kinds of numbers.

For example, thousands, tens, tenths, millions, halves, odds, evens, and so on.

Discover

WAYS OF WORKING Pair work

ASK

- Question ① a): *How can you prove how many 100,000s are in 1,000,000?*
- Question ① b): *How many 100,000s are in 4,590,124? How would you show this in a place value grid?*

IN FOCUS Question ① a) builds on the previous lesson by asking children to count up to 1,000,000 in 100,000s. They may find it helpful to use a number line. If children demonstrate they can do this confidently, encourage them to count beyond 1,000,000 by asking them to look at the price of the painting. Focus on the first two digits and ask them to continue counting in 100,000s until they reach £4,500,000. At this point, discuss what values the 4 and 5 have, before children move on to question ① b).

PRACTICAL TIPS A fun way to begin the lesson would be to host your own mock auction. Give children a set amount of money they can spend. To avoid them going for broke on the first item, you could make it a challenge that they must buy as many items as they can. Once they have finished bidding, discuss the sale prices and compare the numbers.

ANSWERS

Question ① a): There are ten 100,000s in one million.

Question ① b): The painting cost four million, five hundred and ninety thousand, one hundred and twenty-four pounds.

The clock cost two hundred and thirty-four thousand, five hundred pounds.

M	HTh	TTh	Th	H	T	O
4	5	9	0	1	2	4

M	HTh	TTh	Th	H	T	O
0	2	3	4	5	0	0

Share

WAYS OF WORKING Whole class teacher led

ASK

- Question ① a): *How did you prove how many 100,000s there are in 1,000,000?*
- Question ① b): *How do the place value grids help you to read the numbers? What do the grids tell you?*

IN FOCUS Question ① a) recaps children's understanding of place value and its link to multiples of 100,000. Continue this pattern of thought when children look at the place value grids in question ① b) by asking them which digit is a multiple of 10, 100, 1,000 and so on. It is worth spending time exploring the place value grids in this way, as it encourages children to not only begin reading and writing numbers beyond 1,000,000 up to 10,000,000, but to also think about what each digit represents.

Numbers to 10,000,000

Discover

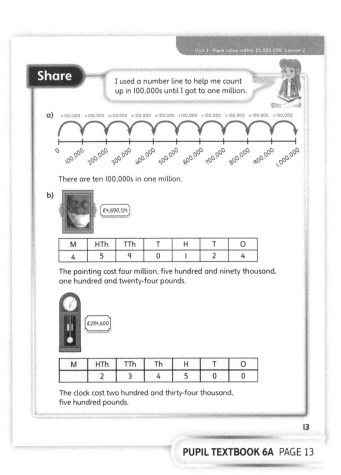

① a) The comic books sold for one million pounds.
    How many 100,000s are in one million?

   b) Write the sale prices of the painting and the clock on place value grids.
      Use the grids to help you to say the numbers out loud.

12

PUPIL TEXTBOOK 6A PAGE 12

PUPIL TEXTBOOK 6A PAGE 13

49

Think together

WAYS OF WORKING Whole class teacher led (I do, We do, You do)

ASK

- Question **1**: *What do you need to be careful with when writing a number in words?*
- Question **1**: *How many counters are in the 100,000s column? Why are there no counters in the 10,000s column?*
- Question **2**: *How would a place value grid help you?*
- Question **3**b): *Is there a way of knowing which numbers can be made without needing to test it?*

IN FOCUS Question **1** offers children the opportunity to practise writing numbers beyond 1,000,000 up to 10,000,000 in numerals and in words, while considering the meaning of place value. For question **3**, give children place value grids and place value counters so they can investigate the numbers using concrete resources. Use Flo's comment to prompt them to notice that the number of counters they need is the same as the sum of the digits.

STRENGTHEN While investigating the numbers in question **3** using place value grids and counters, ask: *How will you place the counters? What number would you make if you placed all your counters in the millions column? Explain. How do Ash and Flo's comments help you?*

DEEPEN Once children have solved question **3** b), deepen their reasoning by asking: *Has Lee listed all the possible numbers he can make or can you find any more? What patterns, similarities or differences can you find between the numbers?*

ASSESSMENT CHECKPOINT In question **1**, look for children demonstrating confidence when writing numbers in numerals and in words. Pay particular attention to whether they remember to include place-holding 0s or if these lead to errors. Question **2** assesses whether children can recognise place values in numbers up to 10,000,000, including finding 100,000 more than a given number. Question **3** assesses children's fluency with place value grids and their ability to reason with numbers up to 10,000,000. It will highlight whether they are equally comfortable working with small numbers and large numbers.

ANSWERS

Question **1** a): 462,305, four hundred and sixty-two thousand, three hundred and five

5,104,309, five million, one hundred and four thousand, three hundred and nine

Question **1** b): One million, four hundred and two thousand, three hundred and fifteen

Question **2** a): Three million, four hundred and five thousand, seven hundred and eighty-two

Question **2** b): 700 or 7 hundreds

Question **2** c): 5

Question **2** d): 3,505,782

Question **3** a): 2,411,301

Question **3** b): Lee can make 5,502 or 1,304,202 or 1,304,220 because the sum of their digits is 12.

PUPIL TEXTBOOK 6A PAGE 14

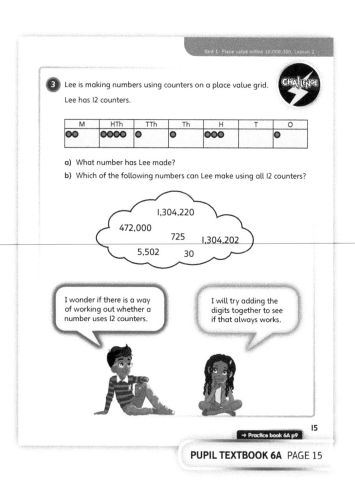

PUPIL TEXTBOOK 6A PAGE 15

Practice

WAYS OF WORKING Independent thinking

IN FOCUS Question **1** offers the opportunity for children to independently count in 100,000s to 1,000,000 and above. This will help to secure their understanding of the relationship these numbers share regarding place value. Question **2** allows children to develop their ability to write numbers in both numerals and words using their understanding of pictorial representations. This is reversed in question **3**, where children have to draw pictorial representations, using the scaffolding of a place value grid to help secure their understanding of the individual digits in numbers. This scaffolding is removed in question **4**, when children work only in the abstract, writing numbers in numerals using only the written names. Question **5** gives children facts about digits, which they have to arrange to create a number. Look for children remembering to include place-holding 0s for the 100,000s and 10s.

STRENGTHEN In question **3**, if children struggle to draw the counters on the place value grids, it may be beneficial to give them real counters to manipulate without the fear of making multiple mistakes in their drawings.

DEEPEN Question **6** gives children the opportunity to generalise using their knowledge of numbers and number patterns. Push their reasoning further by asking: *Can any digit be odd for it to be an odd number? Why must it be the 1s you look at to determine whether a number is odd or even? Can you prove your ideas with a picture?*

ASSESSMENT CHECKPOINT Question **1** assesses children's understanding of the values of multiples of 100,000 and how these link to 1,000,000s. It is important to observe children's understanding in question **2** to ensure they have secure knowledge of what each part of the place value grid is worth. At this point in the lesson, children should be confident and fluent when reading and writing numbers up to 10,000,000; this is further assessed in questions **3** and **4**.

ANSWERS Answers for the **Practice** part of the lesson can be found in the *Power Maths* online subscription.

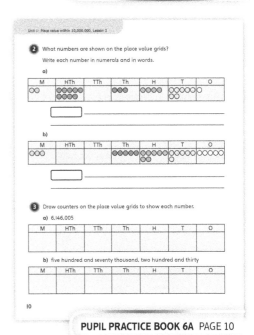

PUPIL PRACTICE BOOK 6A PAGE 9

PUPIL PRACTICE BOOK 6A PAGE 10

Reflect

WAYS OF WORKING Independent thinking

IN FOCUS The question requires children to demonstrate their knowledge and understanding of place value from ones to millions. Further draw out their understanding of the given number by asking them to draw it, say it, make it and write it.

ASSESSMENT CHECKPOINT Children should be able to correctly determine the place values of all digits and not be tripped up by any place-holding 0.

ANSWERS Answers for the **Reflect** part of the lesson can be found in the *Power Maths* online subscription.

After the lesson ⏸

- How confident were children with numbers in the millions?
- If some children still lack confidence, what other manipulatives could you use to help cement their conceptual understanding of numbers in the millions?

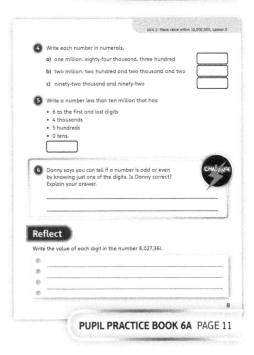

PUPIL PRACTICE BOOK 6A PAGE 11

Partition numbers to 10,000,000

Learning focus

In this lesson, children will use their understanding of place value and numbers up to 10,000,000 to partition numbers and solve problems in real-life contexts.

Before you teach

- Were children able to reason confidently about place value within numbers up to 10,000,000 when working with them out of context?

NATIONAL CURRICULUM LINKS

Year 6 Number – number and place value

Read, write, order and compare numbers up to 10,000,000 and determine the value of each digit.

Solve number and practical problems.

ASSESSING MASTERY

Children can use their fluent understanding of the place value of digits in numbers up to 10,000,000 to read, understand and solve problems with real-life contexts. They are able to explain the reasoning behind their solutions using the vocabulary and representations they have learnt about in the previous two lessons.

COMMON MISCONCEPTIONS

Children may assume that the only way to partition a number is into its place value headings (for example, 136 partitioned as 100, 30 and 6), which could lead to confusion in the **Challenge** questions in this lesson. Base 10 equipment or place value counters may help children to become more flexible in partitioning. Ask:
- *Is that the only way you can break this number into parts?*
- *Can you show me another way using a resource?*

STRENGTHENING UNDERSTANDING

Before starting the lesson, children can strengthen their fluency with the new numbers they have been learning about, and their place value, by counting up and down in different steps from a given number. Ask: *Can you count up in 10,000s from 360? Can you count down in 1,000,000s from 9,780,002?*

GOING DEEPER

Children can be encouraged to create their own missing number equations to challenge their partner with. For example, $300,000 + \boxed{} + 5,000 + 34 = 376,034$. Ask: *Can you write a question like this for your partner? How can you make your question easier or more challenging?*

KEY LANGUAGE

In lesson: ones (1s), tens (10s), hundreds (100s), thousands (1,000s), ten thousands (10,000s), hundred thousands (100,000s), millions (1,000,000s), partition/partitioned, place value

Other language to be used by the teacher: ten million (10,000,000)

STRUCTURES AND REPRESENTATIONS

Place value grid, part-whole model

RESOURCES

Optional: base 10 equipment, place value grids, place value counters, place value cards, play coins

 In the eTextbook of this lesson, you will find interactive links to a selection of teaching tools.

Quick recap 🔎

Write a 4-digit number on the board and ask children to partition it into its place value parts. Then write a 5-digit number on the board and ask children to partition this into its place value parts. Repeat for different numbers with up to five digits.

Discover

Pair work

ASK

- Questions ❶ a) and ❶ b): *What mathematical representation may be helpful when writing how much money each player has?*
- Questions ❶ a) and ❶ b): *Can you show the amounts with counters in a place value grid?*
- Questions ❶ a) and ❶ b): *Are there other ways to partition the numbers?*
- Question ❶ b): *What do you notice about the number of 1,000s?*

IN FOCUS Both questions give children the opportunity to recognise the place value of numbers up to 10,000,000 when used in a real-life context and how these numbers can be partitioned. It is important to discuss how the mathematics changes and stays the same when numbers are used in context. Question ❶ b) requires children to recognise where the denominations that Aki has will bridge into the next place value column. In this example, the eleven £1,000 notes bridge into the 10,000s.

PRACTICAL TIPS Children could be given toy money to investigate the numbers with. How many different ways can they make Jamie and Aki's amounts using different notes?

ANSWERS

Question ❶ a): Jamie has £4,520,123.

Question ❶ b): Aki has £2,071,000.

Partition numbers to 10,000,000

Discover

❶ a) In the game, how much money does Jamie have?

b) How much money does Aki have?

16

PUPIL TEXTBOOK 6A PAGE 16

Share

Whole class teacher led

ASK

- Questions ❶ a) and ❶ b): *Did you choose to use place value grids? Explain why or why not.*
- Questions ❶ a) and ❶ b): *How do the place value grids help to make writing the numbers easier?*
- Questions ❶ a) and ❶ b): *What is similar or different about the two numbers in the place value grids?*

IN FOCUS Use this opportunity to make sure children are ready to apply their mathematical knowledge and understanding to contextual scenarios. Discuss what mathematical information has been used from the **Discover** picture and how it has been manipulated as non-contextual numbers before being turned back into pounds. In question ❶ b), spend time exploring the bridging of 10 thousands into the ten thousands column, to ensure all children understand this before moving on.

DEEPEN If children are confident with the place value grids, challenge them to write the amounts of money using part-whole models. Ask: *Is the part-whole model clearer than the place value grid? Explain your ideas.*

Share

a) I used a place value grid to help me organise the numbers.

M	HTh	TTh	Th	H	T	O

4,000,000 + 500,000 + 20,000 + 100 + 20 + 3 = 4,520,123
Jamie has £4,520,123.

b) I know that 10 thousands are equal to 1 ten thousand.

M	HTh	TTh	Th	H	T	O
2	0	7	1	0	0	0

Aki has £2,071,000.

17

PUPIL TEXTBOOK 6A PAGE 17

Think together

ASK

- Question **1**: *How can you use the representations to work out the missing number?*
- Question **2**: *How can you prove your part-whole models are completed accurately?*
- Question **3**: *Is it always necessary to use a place value grid? Explain your ideas.*

IN FOCUS Questions **1** and **2** give children the opportunity to practise partitioning numbers up to 10,000,000, both in a contextual scenario and using different representations, with the introduction of part-whole models in question **2**. Question **3** ensures children's fluency with partitioned numbers by removing pictorial representations, although children may find it helpful to draw their own. It also recaps their understanding of numbers when written as words.

STRENGTHEN In question **3**, if children are struggling to write the numbers that have been partitioned, it may help to provide place value cards to enable them to build each number. Ask: *Can you find the right cards to help build this number? What does the number look like once you have built it? Can you read it to me?*

DEEPEN When solving question **3**, children could be asked to demonstrate their deep understanding of all the concepts covered in this lesson and the previous one by asking: *Choose one of the numbers in the list. How many different ways can you represent that number and the place values of its digits? How many different equipment, diagrams or number sentences to represent the number?*

ASSESSMENT CHECKPOINT Questions **1** and **2** assess children's understanding of partitioning and the place values of digits in numbers up to 10,000,000. In question **3**, assess children's fluency with place value and different abstract ways of representing numbers up to 10,000,000 by seeing if they are confident working entirely in the abstract, if they use pictorial representations or if they require concrete resources for additional support.

ANSWERS

Question **1**: 6,000

Question **2** a): Part-whole model showing
3,150,260 = 3,000,000 + 100,000 + 50,000
+ 200 + 60

Question **2** b): Part-whole model showing
3,500,000 = 3,000,000 + 500,000

Question **3**: 7,691,712; 570,209; 348,509; 4,038,200; 759,421; 4,300,916; 399,710

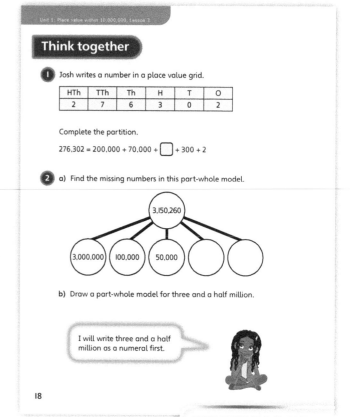

PUPIL TEXTBOOK 6A PAGE 18

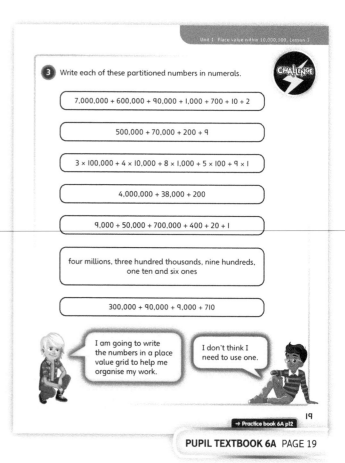

PUPIL TEXTBOOK 6A PAGE 19

Practice

WAYS OF WORKING Independent thinking, pair work

IN FOCUS Question ❶ gives children a further opportunity to develop their understanding of place value in numbers up 10,000,000. As they work through the question, scaffolding for partitioning is gradually reduced. It may be valuable for children to work through question ❺ in pairs, as pictorial representations are removed entirely and they are required to both partition numbers and find numbers using given partitions. Question ❺ g) in particular will test their fluency with the numbers they have been using. As the numbers are not in order, children will need to ensure they carefully check when finding the solution.

STRENGTHEN In question ❸, if children are struggling to recognise the value of the underlined digits, ask: *What mathematical representations have you used to help you recognise the place value of digits before? How could a place value grid help you now?*

DEEPEN Question ❻ requires children to demonstrate their understanding of how digits in a number change when they are adding or subtracting. Deepen their understanding by asking them to generalise based on what they observe. Ask: *What do you notice changes and stays the same? Does this happen every time? Why? Can you predict how a number will change before you add to, or subtract from, it?*

ASSESSMENT CHECKPOINT Questions ❶ and ❷ assess children's recognition of numbers when partitioned and whether they are able to use the partitions to create totals. By the time they reach question ❸, look for children being able to identify the value of each digit in a number without partitioning. Question ❽ allows you to assess children's fluency with different ways of partitioning numbers, not only into their separate place value headings.

ANSWERS Answers for the **Practice** part of the lesson can be found in the *Power Maths* online subscription.

Reflect

WAYS OF WORKING Independent thinking and pair work

IN FOCUS This question requires children to show flexibility with large numbers. They should now recognise that there are many ways to partition numbers, not only into their place value headings.

ASSESSMENT CHECKPOINT Look for children confidently discussing how the number can be made in different ways and making accurate references to place value.

ANSWERS Answers for the **Reflect** part of the lesson can be found in the *Power Maths* online subscription.

After the lesson ⏸

- How fluent are children with partitioning numbers up to 10,000,000?
- What learning aids could you display in your classroom to help secure children's use of the written vocabulary they have learnt?

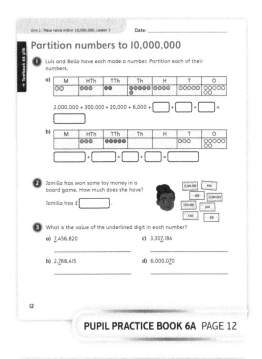

PUPIL PRACTICE BOOK 6A PAGE 12

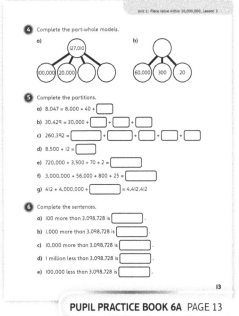

PUPIL PRACTICE BOOK 6A PAGE 13

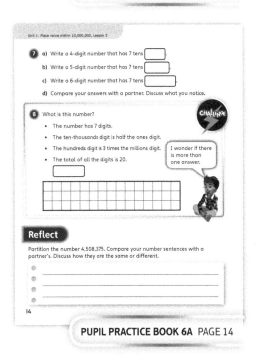

PUPIL PRACTICE BOOK 6A PAGE 14

Powers of 10

Learning focus

In this lesson, children deepen their understanding of the key place value units: 10s, 100s, and 1,000s.

NATIONAL CURRICULUM LINKS

Year 6 Number – number and place value

Read, write, order and compare numbers up to 10,000,000 and determine the value of each digit.

Solve number and practical problems.

ASSESSING MASTERY

Children can express a number in different place value units. For example, understanding that 20,000 is 200 hundreds, 2,000 tens or 20 thousands.

COMMON MISCONCEPTIONS

Children may find the many 'zero' digits in large powers of ten difficult to read confidently. Ask:
- *How could we break this number up in order to read it more easily?*

STRENGTHENING UNDERSTANDING

Model how to bunch numbers into groups of three digits from the right, using commas to separate the parts.

GOING DEEPER

Challenge children to explore number names for large powers of ten, such as millions, billions, trillions, quadrillions.

KEY LANGUAGE

In lesson: exchange, power of ten, ones, tens, hundred, thousands

STRUCTURES AND REPRESENTATIONS

Place value equipment, place value chart

RESOURCES

Mandatory: place value equipment

 In the eTextbook of this lesson, you will find interactive links to a selection of teaching tools.

Quick recap

Write five different 2-, 3- or 4-digit numbers on the board.

Challenge children to multiply each of these numbers by 10.

Discover

Unit 1: Place value within 10,000,000, Lesson 4

Powers of 10

WAYS OF WORKING Pair work

ASK

- Question ①: *What do you notice about the numbers in each row?*
- Question ①: *What do you notice about the numbers in each column?*
- Question ①: *Can you describe how you could use this chart to represent different numbers?*

IN FOCUS Understanding how the Gattegno chart is a helpful representation when thinking about the relationship between different powers of 10 in the place value system.

PRACTICAL TIPS Guide children to point at the chart and count along a row, moving their fingers to point to each number as they go.

Now ask children to count up a column and then back down again, pointing to each number as they go.

ANSWERS

Question ① a): 78,500

Question ① b): 785,000   7,850,000

Discover

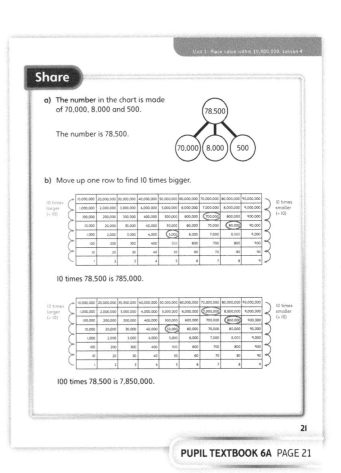

Gattegno chart

10,000,000	20,000,000	30,000,000	40,000,000	50,000,000	60,000,000	70,000,000	80,000,000	90,000,000
1,000,000	2,000,000	3,000,000	4,000,000	5,000,000	6,000,000	7,000,000	8,000,000	9,000,000
100,000	200,000	300,000	400,000	500,000	600,000	700,000	800,000	900,000
10,000	20,000	30,000	40,000	50,000	60,000	70,000	80,000	90,000
1,000	2,000	3,000	4,000	5,000	6,000	7,000	8,000	9,000
100	200	300	400	500	600	700	800	900
10	20	30	40	50	60	70	80	90
1	2	3	4	5	6	7	8	

① a) What number is shown on the chart?

   b) What is 10 times the number?

   What is 100 times the number?

   A Gattegno chart is made up of rows that show numbers 10, 100, 1,000 times bigger or smaller.

20

PUPIL TEXTBOOK 6A PAGE 20

Share

WAYS OF WORKING Whole class teacher led

ASK

- Question ① a): *How does the part-whole model relate to the numbers in the Gattegno chart?*
- Question ① a): *Why do you think there is a maximum of one number chosen per row?*
- Question ① b): *Describe the relationship between the numbers as you move up or down a column.*
- Question ① b): *How could this relationship between numbers in a column help you multiply by 10 or 100?*

IN FOCUS This exercise focuses on how the place value system allows for partitioning and also for multiplying and dividing by 10, 100 and 1,000.

DEEPEN Ask children to use three counters to represent their own number on a Gattegno chart. They can then move the counters by 1 or 2 rows, to multiply by 10 or 100, and challenge a partner to say the original number.

Unit 1: Place value within 10,000,000, Lesson 4

Share

a) The number in the chart is made of 70,000, 8,000 and 500.

The number is 78,500.

78,500 → 70,000, 8,000, 500

b) Move up one row to find 10 times bigger.

10 times 78,500 is 785,000.

100 times 78,500 is 7,850,000.

21

PUPIL TEXTBOOK 6A PAGE 21

Think together

Whole class teacher led (I do, We do, You do)

ASK

- Question **1**: *How does the value of each digit change as you multiply by 10? Or 100? How does the value of each digit change as you divide by 10? Or 100?*
- Question **2**: *What do you visualise as you think about each of these calculations?*
- Question **3**: *Do you notice any patterns as you move from left to right? Can you explain this pattern?*

IN FOCUS

Questions **1** and **2** allow children to practise using the Gattegno chart to multiply and divide by 10, 100, 1,000 or 10,000. Question **3** requires children to understand each of the place value units and their relationship to 1,000,000.

STRENGTHEN In question **3**, use place value equipment to enact exchanges of 10 tens for 1 hundred, 10 thousands for 1 ten thousand, 10 hundred thousands for 1 million.

DEEPEN Debate the two possible answers to this question: 'How many 10s are there in 2,500?'

Answer 1: 'There are zero tens because there is a 0 in the tens place.'

Answer 2: 'This number has 250 tens.'

ASSESSMENT CHECKPOINT Use question **2** to probe explanations for children's answers. They should be able to justify their process by explaining how it relates to the calculation. For example, a child may simply say 'When I multiply by 10, I just write another zero on the end'. This process is efficient when multiplying whole numbers by 10, but the child should be able to explain why it works when prompted.

ANSWERS

Question **1** a): Counters on: 3,000; 700; 50

Question **1** b): 37,500
375
375,000
3,750,000

Question **2** a): 2,300,000

Question **2** b): 9,300,000

Question **2** c): 6,240,000

Question **2** d): 23,000

Question **2** e): 93

Question **3** a): 100 × 10,000
1,000 × 1,000
10,000 × 100
100,000 × 10
1,000,000 × 1

Question **3** b): 6,000

Think together

1 a) Represent 3,750 on the Gattegno chart using counters.

10,000,000	20,000,000	30,000,000	40,000,000	50,000,000	60,000,000	70,000,000	80,000,000	90,000,000
1,000,000	2,000,000	3,000,000	4,000,000	5,000,000	6,000,000	7,000,000	8,000,000	9,000,000
100,000	200,000	300,000	400,000	500,000	600,000	700,000	800,000	900,000
10,000	20,000	30,000	40,000	50,000	60,000	70,000	80,000	90,000
1,000	2,000	3,000	4,000	5,000	6,000	7,000	8,000	9,000
100	200	300	400	500	600	700	800	900
10	20	30	40	50	60	70	80	90
1	2	3	4	5	6	7	8	9

b) Use the Gattegno chart to help you work these out.

3,750 × 10
3,750 ÷ 10
3,750 × 100
3,750 × 1,000

I know that multiplying by 1,000 is the same as multiplying by 10 three times.

2 Work out

a) 230,000 × 10
b) 93,000 × 100
c) 624 × 10,000
d) 230,000 ÷ 10
e) 93,000 ÷ 1,000

22

PUPIL TEXTBOOK 6A PAGE 22

3 a) Make 1,000,000 using different place value counters.
How many of each counter do you need?

1,000,000	100,000	10,000	1,000	100	10	1
1	10					

I don't have enough counters to work this out.

I think there is a pattern. I think the answers are on the Gattegno chart too.

b) How many 1,000s are there in 6 million?

23

PUPIL TEXTBOOK 6A PAGE 23

Practice

WAYS OF WORKING Independent thinking

IN FOCUS Question **1** is about partitioning numbers into place value parts on a Gattegno chart. Children use this to support them when multiplying by 10 and 100. Question **2** develops this to include multiplying and dividing by 10, 100 and 1,000. Question **3** prompts calculation fluency without the use of a chart. Question **4** challenges deep understanding of multiplicative reasoning based on place value.

STRENGTHEN Use place value equipment throughout.

DEEPEN Question **5** deepens understanding of the concept of exchange.

ASSESSMENT CHECKPOINT Question **3** will asses whether children can fluently multiply or divide a number by a given power of 10. This should help you to gauge their understanding of place value.

ANSWERS Answers for the **Practice** part of the lesson can be found in the *Power Maths* online subscription.

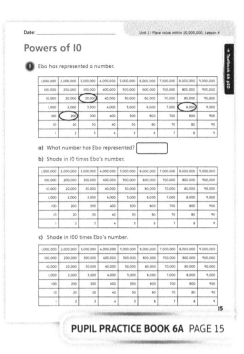

PUPIL PRACTICE BOOK 6A PAGE 15

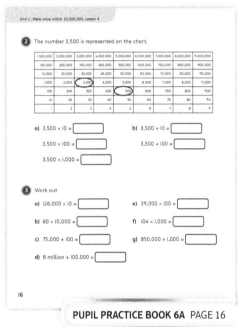

PUPIL PRACTICE BOOK 6A PAGE 16

Reflect

WAYS OF WORKING Pair work

IN FOCUS The **Reflect** part of the lesson gives children the opportunity to express their understanding of the pattern of exchange between powers of 10.

ASSESSMENT CHECKPOINT Assess whether children can justify the patterns they describe through exchange and can use different representations and examples to show this.

ANSWERS Answers for the **Reflect** part of the lesson can be found in the *Power Maths* online subscription.

After the lesson ⏸

- Can children write 1,250,000 as different multiples of 10, 100, 1000 and so on?
- What other opportunities can you find for children to experience and work with very large numbers like these?

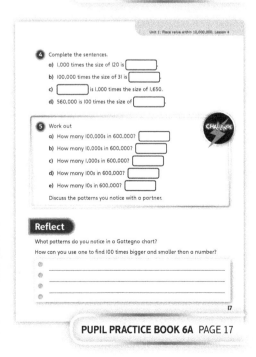

PUPIL PRACTICE BOOK 6A PAGE 17

Number line to 10,000,000

Learning focus

In this lesson, children will accurately identify and estimate where numbers up to 10,000,000 lie on a number line. They will use their understanding of place value to help them achieve this.

Before you teach

- How will you ensure children retain their confidence with number lines when working with very large numbers?
- How will you support abstract understanding when numbers are conceptually harder to envisage?
- How will you build awareness of what approximations are sensible when reading a number line marked in millions?

NATIONAL CURRICULUM LINKS

Year 6 Number – number and place value

Read, write, order and compare numbers up to 10,000,000 and determine the value of each digit.

Solve number and practical problems.

ASSESSING MASTERY

Children can use their understanding of place value to help them accurately identify, or estimate, where a number up to 10,000,000 lies on a number line.

COMMON MISCONCEPTIONS

Instead of using their knowledge of place value to help them estimate as accurately as possible, children may put numbers into three groups – big, small and medium – placing all low numbers very close to 0, all large numbers very close to 10,000,000 and any others near the centre. Ask:
- *Is that number really close to 0?*
- *Can you first estimate where 10 is? 100? 1,000?*

Children may miscalculate, and so misinterpret, the unlabelled intervals on a number line. Ask:
- *If you count in those steps, does your number line work? Show me.*

STRENGTHENING UNDERSTANDING

Children can be given cards with different numbers up to 10,000,000. Encourage them to place the cards along a piece of string in order of size, smallest first. Ask: *How do you know which number comes first?*

Discuss whether the gaps between numbers should be the same. If not, why not? To show this practically, use four or five numbers up to 50 along a bead string and discuss why the gaps between them are different.

GOING DEEPER

Children can be given more number lines where the intervals do not represent a power of 10. Find out, for example, if they can place numbers along a number line from 0 to 10,000 that has been divided into 8 equal intervals.

KEY LANGUAGE

In lesson: ten million (10,000,000), accurate/accurately, interval, half-way, approximately, exactly, scale, estimate, divide/dividing

Other language to be used by the teacher: accuracy, division, increase

STRUCTURES AND REPRESENTATIONS

Number line, bar chart

RESOURCES

Optional: place value cards, bead strings

 In the eTextbook of this lesson, you will find interactive links to a selection of teaching tools.

Quick recap

Draw a 0 to 1,000,000 number line on the board. Ask children to copy the number line and then to draw arrows to indicate the approximate or exact position of these numbers: 999,999; 500; 250,000; 862,194.

Discover

Unit 1: Place value within 10,000,000, Lesson 5

Number line to 10,000,000

Discover

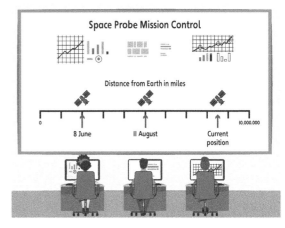

WAYS OF WORKING Pair work

ASK

- Question ① a): *What would be an unrealistic estimate for this point on the number line? What makes it an unrealistic estimate?*
- Question ① a): *What clues can you use to know how far the probe has travelled?*
- Question ① b): *How will you work out how far the probe has travelled?*
- Question ① b): *What do you need to know about the number line to answer this question?*

IN FOCUS Question ① a) encourages children to estimate. It is important to discuss what makes an 'accurate' estimate and what makes an unrealistic estimate. Discuss with children what clues they can use to help them make a more accurate estimate (for example, the end values and the number of intervals). Question ① b) gives children the opportunity to begin calculating using a number line.

PRACTICAL TIPS On the playground, draw a long number line with a child standing at either end, holding number cards 0 and 10,000,000. Ask the other children to stand at points on the line where given values (for example, 4,000,000) would be.

ANSWERS

Question ① a): The probe is approximately 8,500,000 miles from Earth. Because the scale is quite small, the answer can only be an estimate.

Question ① b): The probe travelled 3,000,000 miles between 8 June and 11 August.

① a) How far away from Earth is the space probe currently? How accurate is your answer?

b) How far did the space probe travel between 8 June and 11 August?

24

PUPIL TEXTBOOK 6A PAGE 24

Share

WAYS OF WORKING Whole class teacher led

ASK

- Question ① a): *Why is it important to know what each interval on the number line represents?*
- Question ① a): *How close were you with your estimate? Explain why you were close or far off.*
- Question ① b): *How did you use the number line to solve this?*
- Question ① b): *How did knowing the value of the intervals help?*

IN FOCUS When working through these questions, it is important that children recognise why it is essential to know the value of each interval on a number line. To illustrate this, count along the number line in different steps, such as 100,000s or 2,000,000s to show that you will not reach 10,000,000 at the end.

DEEPEN To help children's later fluency and reasoning, it may be beneficial to show the same number line but divided into different numbers of intervals, for example, 0 to 10,000,000 divided into four equal intervals.

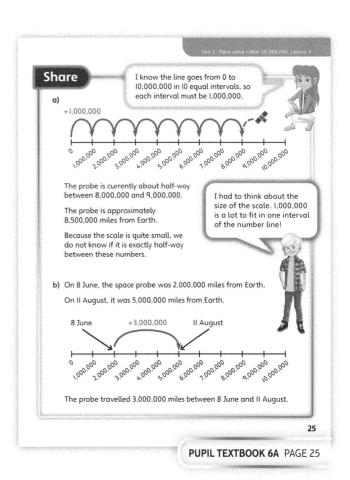

PUPIL TEXTBOOK 6A PAGE 25

Think together

WAYS OF WORKING Whole class teacher led (I do, We do, You do)

ASK

- Questions ❶ and ❷: *What do you need to know first?*
- Questions ❶ and ❷: *Will the answers be exact? Why?*
- Question ❸ a): *What is Kate's mistake?*

IN FOCUS Questions ❶ and ❷ allow children to begin interpreting different number lines and finding unlabelled numbers along them. Discuss the fact that the intervals are not the same. Question ❸ a) helps to tackle the potential problem of children misinterpreting the unlabelled intervals on a number line.

STRENGTHEN In question ❷ b), if children are struggling to interpret the value of the intervals, ask: *How many intervals are there? How can these facts help you to work out what each interval represents?*

DEEPEN In question ❸ b), as only the start and end numbers are labelled, children need to show a deeper understanding of how number lines work as they decide how best to go about solving the problems. Use Dexter and Ash's comments to help scaffold their reasoning. Ask: *How many intervals would be useful? What will each interval represent?*

ASSESSMENT CHECKPOINT Questions ❶, ❷ and ❸ b) assess children's ability to accurately place or estimate numbers on a number line. Look for children understanding why the large numbers on the scale mean readings can only be estimates of the precise numbers. Use question ❸ a) to draw out whether children are accurately calculating the value of intervals.

ANSWERS

Question ❶ a): A = 1,200,000; B = 1,650,000 approximately

Question ❶ b): A = 12,500 approximately; B = 18,000;
C = 19,100 approximately

Question ❶ c): A = 704,000 approximately; B = 740,000;
C = 792,000 approximately

Question ❷ a): 100,000

Question ❷ b): 200,000

Question ❸ a): Kate has made a mistake with place value. She thinks each interval on the line represents 10,000 instead of 1,000. The arrow is pointing at 260,500.

Question ❸ b): Look for children making a reasonable estimate for A = 250,000; B = 270,000; C = 280,000.

Question ❸ c): They should estimate the position of 210,573 as the same distance from 200,000 as point B is from point C.

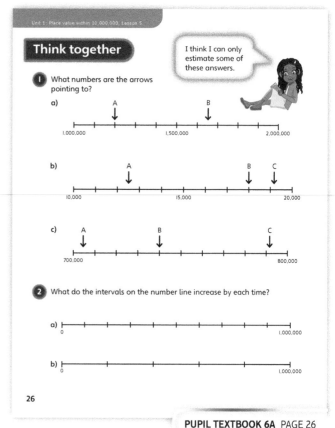

PUPIL TEXTBOOK 6A PAGE 26

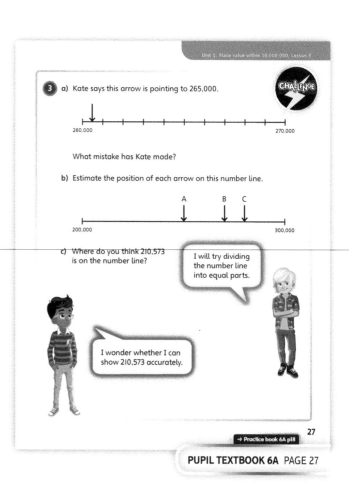

PUPIL TEXTBOOK 6A PAGE 27

Practice

WAYS OF WORKING Independent thinking

IN FOCUS In question **1**, ensure children pay attention to the start and end values, as well as the number of intervals, so they can work out what value each interval represents rather than making an assumption that it is the same for every number line. This will prepare them for question **2** where the number line again counts in different amounts.

STRENGTHEN In question **5**, ask: *What do you need to do if the number does not fall on a division? How did you estimate as accurately as possible previously?*

DEEPEN Question **6** deepens children's reasoning by presenting what they have learnt in a real-life context. They should recognise that the scale on the bar chart is a number line, but with a real-life interpretation. Ask: *What other facts can you tell me about the populations of the countries listed? How can you prove the facts you have noted?*

THINK DIFFERENTLY In question **3**, number lines are omitted. Children should continue with the same thought process as for question **2**, counting from number to number and recognising that the differences are consistent in each step. Question **3** c) may be challenging for some children, as the numbers decrease, but they should realise that the process of reasoning is the same as for increasing number lines.

ASSESSMENT CHECKPOINT Throughout, children should demonstrate independent ability to complete number lines. Look for children counting along the lines and arriving at the correct end numbers to demonstrate that they have correctly worked out what each interval represents. Children may assume that number lines always increase in a power of 10, so pay attention to question **5** b) as this does not. Questions **4** and **5** assess children's ability to place numbers on number lines and make accurate estimates for numbers that fall between divisions.

ANSWERS Answers for the **Practice** part of the lesson can be found in the *Power Maths* online subscription.

Reflect

WAYS OF WORKING Independent thinking

IN FOCUS This question requires children to apply what they have learnt to a number line with no divisions. They should work out an appropriate scale, before making a reasoned estimate.

ASSESSMENT CHECKPOINT Methods should show a clear attempt to divide the number line and make an accurate estimate. For example:
- *I would find half-way first.*
- *I would split the number line into 10 equal parts of 10,000.*
- *After that, I would count up to 240,000.*

ANSWERS Answers for the **Reflect** part of the lesson can be found in the *Power Maths* online subscription.

After the lesson ⏸

- What percentage of the class was able to use number lines that were not fully labelled?
- How will you support children who still rely on the intervals being labelled for them?
- What other real-life contexts could you bring into the lesson next time you teach this topic?

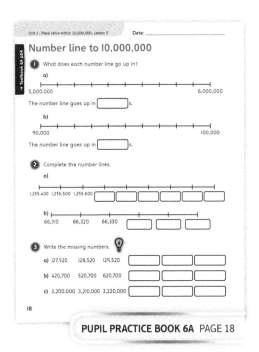

PUPIL PRACTICE BOOK 6A PAGE 18

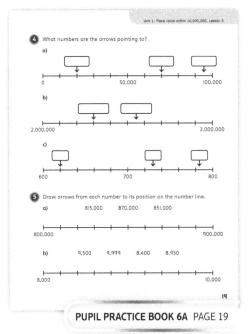

PUPIL PRACTICE BOOK 6A PAGE 19

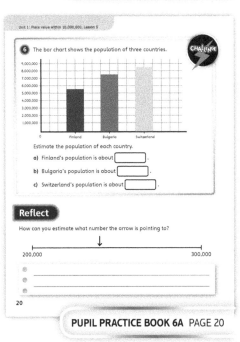

PUPIL PRACTICE BOOK 6A PAGE 20

Compare and order any number

Learning focus

In this lesson, children will use their understanding of place value and numbers up to 10,000,000 to compare and order numbers.

Before you teach

- How fluent are children with place value?

NATIONAL CURRICULUM LINKS

Year 6 Number – number and place value

Read, write, order and compare numbers up to 10,000,000 and the value of each digit.

Solve number and practical problems.

ASSESSING MASTERY

Children can confidently use their understanding of place value and numbers up to 10,000,000 to compare the value of numbers. They can explain how they know a number is greater or less than another number and can use this to make accurate comparisons and order numbers correctly.

COMMON MISCONCEPTIONS

Children may compare digits that do not have the same place value; for example, saying that 46,000 is greater than 360,000 because 4 is greater than 3. Children may order numbers in the wrong direction, such as in ascending order instead of descending. Ask:

- *How will you prove that you have put the numbers in the order the question has asked for?*

STRENGTHENING UNDERSTANDING

Give children opportunities to compare and order smaller numbers up to 1,000,000 (met in Year 5) to remind themselves of the processes involved, before moving on to bigger numbers with more place value headings.

GOING DEEPER

Set children a game to play in groups of two or three. Provide enough digit cards for each child to make a number in the millions up to 10,000,000. In turn, each child picks a digit card and places it in any position on a place value grid, with the aim of making the largest (or smallest) number compared with the rest of their group.

KEY LANGUAGE

In lesson: sort/sorted, most, least, compare/comparison/comparing, thousands (1,000s), ten thousands (10,000s), hundred thousands (100,000s), millions (1,000,000s), order/ordering, less than (<), greater than (>)

Other language to be used by the teacher: ones (1s), tens (10s), hundreds (100s), greatest, biggest, largest, smallest, lightest, equal to (=), decrease, place value, partition/partitioning, ten million, (10,000,000), increase, reduce, mass, heaviest, ascending, descending, rating

STRUCTURES AND REPRESENTATIONS

Place value grid, table

RESOURCES

Optional: digit cards, place value grids, counters

 In the eTextbook of this lesson, you will find interactive links to a selection of teaching tools.

Quick recap

Ask children to order this set of numbers from least to greatest:

345; 4,050; 12; 999; 5,000; 4;048.

Discuss the strategies they used to complete the ordering.

Discover

WAYS OF WORKING   Pair work

ASK

- Question ❶ a): *How will you know which is the most expensive?*
- Question ❶ b): *What mathematical structure can you use to compare the numbers? How will it help you?*

IN FOCUS   Question ❶ a) introduces the main concept of this lesson: comparing and ordering numbers up to 10,000,000. For question ❶ b), it is important to focus on the usefulness of place value grids and how they can help to show the comparative values of different numbers.

PRACTICAL TIPS   Children could research real sales listings for items with similar high prices (for example, house advertisements). Children could then sort the values from most to least expensive.

ANSWERS

Question ❶ a): B

Question ❶ b): B, A, D, E and C

Compare and order any number

Discover

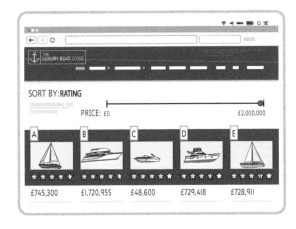

❶ a) The boats are currently sorted by star rating. Which boat is most expensive?

b) Sort the boats by price, starting with the most expensive.

28

PUPIL TEXTBOOK 6A PAGE 28

Share

WAYS OF WORKING   Whole class teacher led

ASK

- Question ❶ a): *What did you use to compare the numbers? Explain why. How did this help you to find the most expensive?*
- Question ❶ b): *Why is a place value grid a useful tool to help compare the numbers? Did your order match the correct answer? Explain.*

IN FOCUS   In question ❶ a), talk through the place value grid to ensure children understand how it has been completed. For question ❶ b), discuss how a systematic process – working from left to right to compare the digits in each place value column – has been used to work out the correct order.

Share

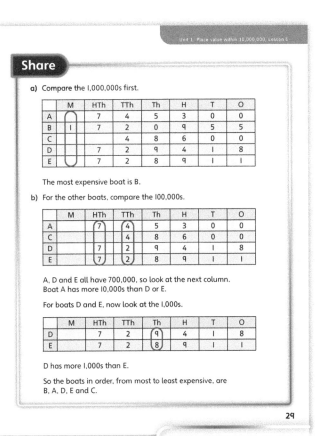

a) Compare the 1,000,000s first.

	M	HTh	TTh	Th	H	T	O
A		7	4	5	3	0	0
B	1	7	2	0	9	5	5
C			4	8	6	0	0
D		7	2	9	4	1	8
E		7	2	8	9	1	1

The most expensive boat is B.

b) For the other boats, compare the 100,000s.

	M	HTh	TTh	Th	H	T	O
A		7	4	5	3	0	0
C			4	8	6	0	0
D		7	2	9	4	1	8
E		7	2	8	9	1	1

A, D and E all have 700,000, so look at the next column. Boat A has more 10,000s than D or E.

For boats D and E, now look at the 1,000s.

	M	HTh	TTh	Th	H	T	O
D		7	2	9	4	1	8
E		7	2	8	9	1	1

D has more 1,000s than E.

So the boats in order, from most to least expensive, are B, A, D, E and C.

29

PUPIL TEXTBOOK 6A PAGE 29

Think together

WAYS OF WORKING Whole class teacher led (I do, We do, You do)

ASK

- Question ❶: *What do the signs < and > mean?*
- Question ❶: *How can you prove you are correct?*
- Question ❷: *What order do you have to put the numbers in? How will this change the way you solve the problem?*
- Question ❸: *Why is it difficult to compare the amounts in this question? What can you do to make this easier?*

IN FOCUS Question ❶ gives the opportunity to link the learning in this lesson to children's understanding of the mathematical notations for comparison. It also challenges children's understanding of place value: children may assume that 5,999,999 is bigger because it has more 9s than 7,000,000. Question ❷ will help develop children's flexibility with ordering numbers, ensuring they are able to order in ascending order, as well as descending. As the numbers include different combinations of the same five or six digits, it also challenges their understanding of place value.

STRENGTHEN When working on question ❶ b), if children find one solution for each digit and do not look for more, ask: *Is that the only solution that works? Can you prove to me that there are no other possible solutions?*

DEEPEN Question ❸ especially challenges children's understanding of place value and the values of numbers, as the highest value amount is written in ones and tens, while the lowest amount is written using hundreds of thousands. Extend this by asking children to write sets of three numbers, where one number is written in words, one in digits and one in ones and tens, to give to a partner to order.

ASSESSMENT CHECKPOINT The questions all assess children's ability to compare and order numbers up to 10,000,000. Look for children demonstrating their ability to use the mathematical notations of comparison accurately (< and >). Question ❸ particularly draws out children's fluency and reasoning with place value. It also assesses children's ability to understand and interpret the value of numbers written in different ways.

ANSWERS

Question ❶ a): 429,118 < 518,128; 392,271 > 392,098; 41,510 > 4,151; 7,000,000 kg > 5,999,999 kg; 900,000 kg < 2,000,000 kg

Question ❶ b): The first digit could be 8 or 9; the second digit could be 0, 1, 2, 3, 4, 5, 6, 7, 8 or 9.

Question ❷: £32,000; £302,400; £320,400; £412,500

Question ❸ a): Player 4

Question ❸ b): Player 3

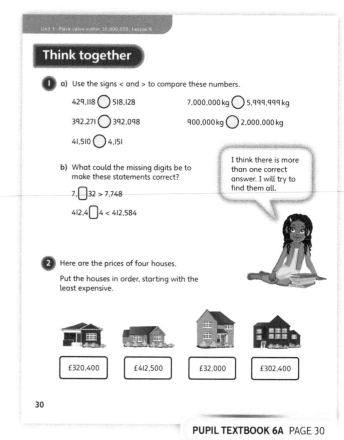

PUPIL TEXTBOOK 6A PAGE 30

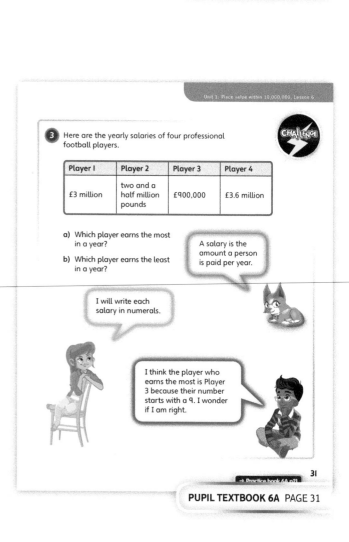

PUPIL TEXTBOOK 6A PAGE 31

Practice

WAYS OF WORKING Independent thinking

IN FOCUS In Question ❶, look at like values in turn on a place value grid in order to determine which number is greater. Question ❷ requires children to demonstrate their understanding of the mathematical notations for comparison (<, >). In questions ❸ and ❺, children may find it helpful to draw on the previous lesson by imagining they are ordering the numbers on number lines. However, in question ❻ children need to demonstrate their fluency and reasoning when comparing numbers up to 10,000,000. Note that question ❹ uses the term 'masses'; some children may be more comfortable with the term 'weight' when considering lightest and heaviest, but this should not detract from the task of comparing and ordering.

STRENGTHEN If children are struggling to compare numbers in the abstract, give them place value grids and counters to represent the numbers, as in question ❶. In question ❹, for children struggling to identify the third elephant to be fed, ask: *Why is this question trickier than asking for first or fourth place? How can you write down your thinking and check that you have compared accurately?*

DEEPEN When solving question ❼, deepen children's reasoning by asking: *How did you start this problem? Is there a specific order you have to solve this in? Explain.*

ASSESSMENT CHECKPOINT Questions ❶ to ❺ assess children's ability to compare and order numbers up to 10,000,000, based on their understanding of place value. Look for confident use of structures and representations to justify children's ideas, as well as the accurate use of the < and > signs. In questions ❻ and ❼, look for whether children can reason confidently when deciding what the missing numbers should be and whether there is more than one possible solution.

ANSWERS Answers for the **Practice** part of the lesson can be found in the *Power Maths* online subscription.

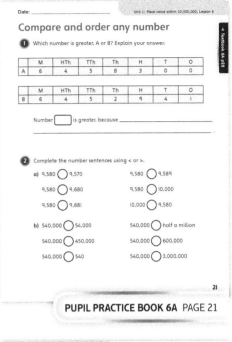

PUPIL PRACTICE BOOK 6A PAGE 21

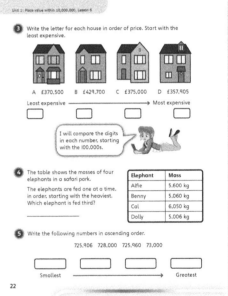

PUPIL PRACTICE BOOK 6A PAGE 22

Reflect

WAYS OF WORKING Independent thinking

IN FOCUS This question requires children to demonstrate their ability to reason by evaluating the place value of digits within numbers in order to accurately order them.

ASSESSMENT CHECKPOINT Look for children's references to place value or use of a place value grid to prove their ideas and provide a clear explanation of their comparisons. Ensure children understand the meaning of 'descending'.

ANSWERS Answers for the **Reflect** part of the lesson can be found in the *Power Maths* online subscription.

After the lesson ⏸

- How could you apply children's learning to other areas of the curriculum?
- What were the barriers to learning for those children who struggled to make progress?

PUPIL PRACTICE BOOK 6A PAGE 23

Round any number

Learning focus

In this lesson, children will use their understanding of place value to help them round numbers up to 10,000,000. They will discuss when rounding is appropriate and which power of 10 to round to in a given context.

Before you teach ⏸

- How fluent are children with the rules for rounding from their work in previous years?

NATIONAL CURRICULUM LINKS

Year 6 Number – number and place value

Round any whole number to a required degree of accuracy.

ASSESSING MASTERY

Children can recognise and explain how to round numbers fluently. They can apply this understanding to larger numbers, reliably rounding up and down to the nearest 10, 100, 1,000, 10,000, 100,000 and 1,000,000.

COMMON MISCONCEPTIONS

Children may assume that rounding only affects the place value column referenced and those below it (for example, assuming that rounding to the nearest 100 only affects the 100s, 10s and 1s digits). Ask:
- If you round 1,992 to the nearest 100, what happens?
- Can you show me using a picture or resources how the number will change?
- What happens when you count on one more 100 from 900?

STRENGTHENING UNDERSTANDING

Before the lesson, children who may need more help can be encouraged to recap rounding smaller numbers to the nearest 10,000 (or smaller powers of 10). Ask: *What does it mean to round a number? Can you show me how to round x to the nearest y?*

GOING DEEPER

Deepen children's understanding by inviting them to play a 'guess the number' game in pairs. Children could set questions for their partner such as: *I have a 5-digit number that when rounded to the nearest 1,000 is 567,000. What could my number be?*

KEY LANGUAGE

In lesson: rounding/rounded/round up/round down/rounds, maximum, minimum, nearest, tens (10s), hundreds (100s), thousands (1,000s), ten thousands (10,000s), hundred thousands (100,000s)

Other language to be used by the teacher: millions (1,000,000s), greater than (>), less than (<), equal to (=)

STRUCTURES AND REPRESENTATIONS

Number line, place value grid, table

RESOURCES

Optional: place value grids, counters

 In the eTextbook of this lesson, you will find interactive links to a selection of teaching tools.

Quick recap

Play a 'Guess my number' game together. Say: *I am thinking of a number that rounds to 600.* Ask children to suggest possible correct answers until they find your number (595). Repeat for different mystery numbers that have been rounded to the nearest 10 or 100 or 1,000.

Discover

Unit 1: Place value within 10,000,000, Lesson 7

Round any number

Discover

There are 500,000 termites in this mound.

Wow! There are only 76,392 people in my town.

Ambika   Zac

WAYS OF WORKING Pair work

ASK

- Question ① a): *How will you round this number to the nearest 10,000 and 1,000?*
- Question ① b): *Which digit do you need to look at?*

IN FOCUS In question ① a), children have to think about which digit to look at when rounding to different powers of 10 and then decide whether to round the given number up or down. Check whether children recall when to round up and when to round down. This knowledge is then required in question ① b) when children have to think about what the boundary figures for rounding up and down are.

PRACTICAL TIPS Discuss with children real-life examples of where it may be appropriate to give a rounded estimate, rather than an exact figure. For example, country populations, distances between planets, number of fish in the sea. Children could research and find the rounded estimates for things that interest them and present these to the class. The numbers could then be compared and ordered, linking this lesson to the previous one.

ANSWERS

Question ① a): The population of Zac's town rounded to the nearest 10,000 is 80,000.
The population of Zac's town rounded to the nearest 1,000 is 76,000.

Question ① b): Minimum 450,000; maximum 549,999

① a) Round the population of Zac's town to the nearest 10,000 and 1,000.

b) The number of termites has been rounded to the nearest 100,000.

What are the minimum and maximum numbers of termites there could actually be?

32

PUPIL TEXTBOOK 6A PAGE 32

Share

WAYS OF WORKING Whole class teacher led

ASK

- Question ① a): *Did you remember the rules for rounding? What are they?*
- Question ① a): *How do you know whether to round up or down?*
- Question ① b): *What is the first number that you would round up to 600,000?*
- Question ① b): *Why do you think the number has been rounded?*

IN FOCUS Question ① a) offers the opportunity to recap the rules for rounding. Ensure children recall that they round down if a digit is less than 5 and round up if a digit is greater than or equal to 5. It is important to link these rules to previous learning from this unit regarding place value. In question ① b), ensure children understand that they need to consider the ten thousands digit and encourage them to think what numbers would round to 400,000, 500,000 and 600,000. Use Ash's comment to prompt children to consider contexts in which rounding is useful (for example, when it is too difficult to count an amount exactly or when a rounded number is easier to understand).

DEEPEN Deepen children's understanding in question ① a) by encouraging them to consider how context can determine what degree of accuracy to round to. Ask: *Which rounded total is more appropriate to use and why?*

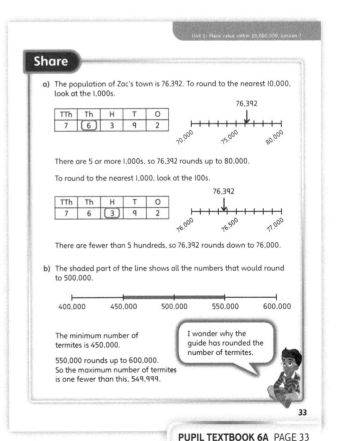

Share

a) The population of Zac's town is 76,392. To round to the nearest 10,000, look at the 1,000s.

TTh	Th	H	T	O
7	6	3	9	2

There are 5 or more 1,000s, so 76,392 rounds up to 80,000.

To round to the nearest 1,000, look at the 100s.

TTh	Th	H	T	O
7	6	3	9	2

There are fewer than 5 hundreds, so 76,392 rounds down to 76,000.

b) The shaded part of the line shows all the numbers that would round to 500,000.

The minimum number of termites is 450,000.

550,000 rounds up to 600,000. So the maximum number of termites is one fewer than this, 549,999.

I wonder why the guide has rounded the number of termites.

33

PUPIL TEXTBOOK 6A PAGE 33

Think together

Whole class teacher led (I do, We do, You do)

ASK

- Question **1** a): *Are you rounding up or rounding down?*
- Question **1** b): *What number is in the middle of your number line? How does this help you?*
- Question **2**: *Is there anything unexpected about some of the numbers that you have rounded to?*
- Question **3**: *Is there a way to work systematically?*
- Question **3**: *Is there anything you can say about your target number before you make it?*

IN FOCUS Question **1** uses the number line to scaffold children's understanding of how to round and to aid their verbal reasoning when sharing their explanations. Spend some time discussing question **2** to tackle the misconception that rounding a number to the nearest (for example) thousand will only change the digits in the 1,000s column and below. Also draw attention to the fact that rounding to two different powers of 10 can sometimes result in the same number.

STRENGTHEN In question **3** a), if children struggle to find numbers that can be made with the digit cards, ask: *What can you say about the digits in the thousands and ten thousands columns? Does it matter what digits are used for the smaller place values? Explain.*

DEEPEN When children have solved question **2** and noticed that some of the solutions are the same, deepen their reasoning and generalisations by asking: *Will this happen with every 6-digit number you try? Why?*

ASSESSMENT CHECKPOINT Questions **1** and **2** assess children's ability to round numbers up to 10,000,000 to the nearest 10, 100, 1,000, 10,000 and 100,000. Look for children making valid arguments about place value. Question **3** assesses children's fluency, reasoning and problem-solving with number and rounding; this should be particularly demonstrated in their explanations for part c).

ANSWERS

Question **1** a): 20,559 lies between 20,000 and 30,000.

Question **1** b): 20,559

Question **1** c): 20,559 rounded to the nearest 10,000 is 20,000.

Question **2**:

Rounded to the nearest...				
100,000	10,000	1,000	100	10
200,000	180,000	180,000	179,900	179,900

Question **3** a): The range is 45,000 – 54,999.
E.g. 49,572   54,972

Question **3** b): The range is 49,500 – 50,499.
E.g. 49,752   49,725

Question **3** c): No. The range is 49,950 to 50,049; either two 0s or two 9s are needed.

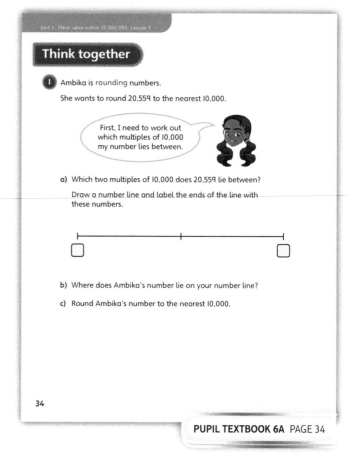

PUPIL TEXTBOOK 6A PAGE 34

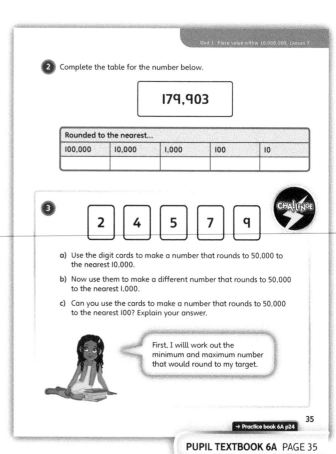

PUPIL TEXTBOOK 6A PAGE 35

70

Practice

WAYS OF WORKING Independent thinking

IN FOCUS Question ❶ a) reinforces the rules for rounding by encouraging children to diagnose where another child has made a mistake in their rounding. Question ❶ b) supports the idea of rounding visually by showing why numbers are rounded up or down depending on their proximity to the nearest specified power of 10. Question ❸ offers children the opportunity to spot patterns in rounding and, through this, generalise facts about rounding based on their understanding of numbers. Questions ❻ and ❼ give children an independent opportunity to develop their reasoning and problem-solving with rounding.

STRENGTHEN If children struggle to recognise which digits to round, they may find it helpful to represent numbers using place value grids and counters, as in question ❶. If they find question ❺ difficult, ask: *To round to the nearest 1,000, which column of a place value grid would you look at? How does this fact help you to solve the question?*

DEEPEN Question ❼ challenges children to pull together all of their learning from this lesson. Ask: *What does each sentence tell you? What facts do you know and what is missing? Is there only one answer to part b)?*

Deepen children's understanding by asking them to write similar sentences of their own to challenge a partner.

THINK DIFFERENTLY Question ❷ requires children to identify a number that is represented on a number line before explaining how to round that number to the nearest 1,000,000. Encourage children to describe the number line, and how it helps them when rounding.

ASSESSMENT CHECKPOINT Questions ❶ to ❺ assess children's ability to apply the rules for rounding to numbers up to 10,000,000. Look for children making mistakes with place value, particularly in question ❺. In questions ❻ and ❼, children should demonstrate their ability to reason, using their understanding of place value and rounding.

ANSWERS Answers for the **Practice** part of the lesson can be found in the *Power Maths* online subscription.

Reflect

WAYS OF WORKING Independent thinking

IN FOCUS This question not only assesses children's understanding and fluency when applying the rules for rounding to numbers but also their ability to represent their thinking in more than one way.

ASSESSMENT CHECKPOINT Children should be able to use concrete, pictorial or abstract representations to demonstrate their conceptual understanding of rounding in more than one way. For example, they may find the difference between the number and 16,000 and the number and 15,000, show it on a number line or explain that 7 hundreds will round the number up.

ANSWERS Answers for the **Reflect** part of the lesson can be found in the *Power Maths* online subscription.

After the lesson ⏸

- How will you continue to encourage children to see the usefulness of rounding in real-life contexts?
- How did children's reasoning skills improve the mathematics in this lesson? Could you offer more or improved opportunities for reasoning next time you teach this?

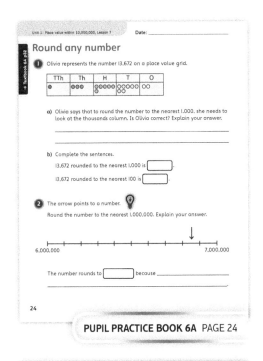

PUPIL PRACTICE BOOK 6A PAGE 24

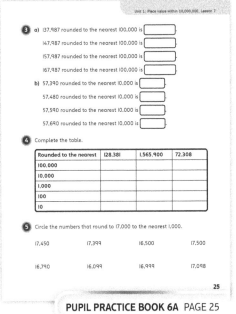

PUPIL PRACTICE BOOK 6A PAGE 25

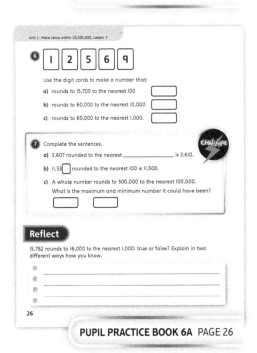

PUPIL PRACTICE BOOK 6A PAGE 26

Negative numbers

Learning focus

In this lesson, children will learn about negative numbers and their relationship with positive numbers. They will use negative numbers in context and use a number line to identify negative numbers and begin calculating with them.

Before you teach

- In which real-life contexts may children have experienced negative numbers before?

NATIONAL CURRICULUM LINKS

Year 6 Number – number and place value

Use negative numbers in context, and calculate intervals across zero.

ASSESSING MASTERY

Children can identify and explain what a negative number is and how negative numbers are similar to and different from positive numbers. They can use negative numbers in context and can reliably find negative numbers on a number line. They can also use a number line to begin calculating with negatives.

COMMON MISCONCEPTIONS

Children may assume that negative numbers work in a similar way to positive numbers. For example, assuming that ⁻5 must be greater than ⁻1 because 5 is greater than 1. Show children a number line and ask:

- *Can you plot the numbers from ⁻10 to 10 on this number line?*
- *Which end of the number line shows the greatest number? Which end shows the smallest?*
- *What can you say about how ⁻5 is different from ⁻1? Explain how you know.*

STRENGTHENING UNDERSTANDING

For children who struggled with finding unmarked intervals on a partially completed number line, it is essential to develop this skill before this lesson begins. Number lines can be linked to bar models. Identify the two known numbers and discuss and identify with children the difference between them. Use this difference as the total bar in a bar model and underneath show bars equal to the number of unlabelled intervals. Ask: *What does this bar model show? How will you use it to find how much each unlabelled interval is worth? What calculation do you need to use? Can you write it?*

GOING DEEPER

Discuss with children how negative numbers are used in real-life contexts (see the **Discover** practical tips section below) and encourage them to use these ideas to create word problems involving negative numbers.

KEY LANGUAGE

In lesson: trial and error, interval, divide/divided, halved/half/half-way, difference

Other language to be used by the teacher: negative, minus, positive

STRUCTURES AND REPRESENTATIONS

Number line, bar model, table

RESOURCES

Optional: number lines

 In the eTextbook of this lesson, you will find interactive links to a selection of teaching tools.

Quick recap

Count back together from 20 to 0. Then tell children to keep going with the count beyond zero. Show them how to draw a number line to show counting down from 5 to ⁻5.

Discover

Negative numbers

Discover

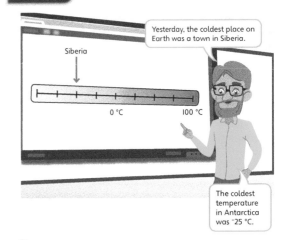

WAYS OF WORKING Pair work

ASK

- Question ❶ a): *Can you see how the scale on the thermometer is like a number line?*
- Question ❶ a): *How will you find out what each interval on the thermometer is worth?*
- Question ❶ a): *What is similar and different about where the arrow is pointing to on the thermometer compared with number lines you have seen before?*
- Question ❶ b): *Is ⁻25 °C colder or warmer than ⁻50 °C?*
- Question ❶ b): *Which end of the thermometer shows colder temperatures?*

IN FOCUS Question ❶ a) links the learning in this lesson to children's understanding of number lines from earlier in the unit. They should recognise the scale on the thermometer as a number line. It is important for children to be reminded of how they found the value of each interval on a number line. Question ❶ b) gives children their first opportunity to place a negative number on a number line. Be sure to discuss the number's position in relation to 0.

PRACTICAL TIPS Children could be encouraged to share where they have seen negative numbers before. They could research the different uses for negative numbers (for example, distance below sea level, debits on bank statements, floors below ground in lifts or department stores) and share these ideas with the class.

❶ a) What was the temperature in the town in Siberia?

   b) Where should the temperature in Antarctica be labelled on the thermometer?

36

PUPIL TEXTBOOK 6A PAGE 36

ANSWERS

Question ❶ a): The temperature in the town in Siberia was ⁻50 °C.

Question ❶ b): ⁻25 °C lies half-way between ⁻50 °C and 0 °C.

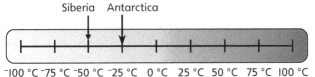

Share

WAYS OF WORKING Whole class teacher led

ASK

- Question ❶ a): *Can you explain Ash's method? Could you use this method to find other temperatures easily along the thermometer?*
- Question ❶ b): *Having found ⁻25 °C, can you estimate where other negative numbers are on the thermometer, such as ⁻10 °C?*

IN FOCUS In question ❶ a), talk through Ash's comment and link it to the bar model shown above the thermometer to help explain how to find the value of each interval. Question ❶ b) is important as it will prompt children to begin to realise that negative numbers work inversely to positives as they see that ⁻50 is smaller than ⁻25.

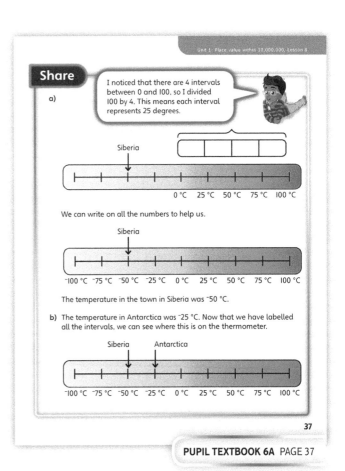

37

PUPIL TEXTBOOK 6A PAGE 37

Think together

WAYS OF WORKING Whole class teacher led (I do, We do, You do)

ASK

- Question **1** b): *What will you do to find the new temperature?*
- Question **1** d): *Will a subtraction help you here?*
- Questions **2** and **3**: *How will you find out the value of each interval?*
- Question **3**: *What is tricky about this question?*
- Question **3**: *Can you predict whether C is positive or negative? Explain how you know.*

IN FOCUS Question **1** provides children with further opportunity to work with negative numbers in context. It is also the first time they will consider the difference between a negative number and another given number. Children may count the intervals between two numbers on the number line provided. Look out for any children who get stuck as they count across 0. Question **1** b) gives children an opportunity to begin adding on to negative numbers. Again, they may use the number line provided, but it is important to discuss how the numbers change when adding on to a negative number, especially when the addition bridges 0. Question **2** helps children to secure their ability to find the value of intervals on a number line and find numbers greater and less than 0.

STRENGTHEN For children struggling with question **2**, ask: *How many intervals are there between 0 and 10? How will this help you to work out what each interval is worth? How will what you did for question **2** a) help you with question **2** b)?*

DEEPEN Question **3** deepens children's problem-solving, reasoning and fluency with number lines. Children should recognise that they need to count the intervals between points A and B to work out the value of each interval. To develop reasoning, ask: *How did you find out what each interval was worth? How was this question different from those you have tackled before? Was it more difficult? Why?*

ASSESSMENT CHECKPOINT Question **1** assesses children's ability to calculate with negative numbers, using a number line as scaffolding. Children should be able to place positive and negative numbers on the number line and work out difference and addition. Questions **2** and **3** assess children's fluency and reasoning with partially completed number lines and negative numbers. They should be able to place numbers even when intervals are unlabelled. Question **3** also assesses children's problem solving with negative numbers and partially completed number lines.

ANSWERS

Question **1** a): Helsinki and Moscow have temperatures below freezing.

Question **1** b): The temperature in Helsinki is now 3 °C.

Question **1** c): The temperature in Paris is now ⁻5 °C.

Question **1** d): The difference in temperature is 14 °C.

Question **2** a): A = ⁻6, B = 9.

Question **2** b):

Question **3**: C = ⁻24 (each interval is 6)

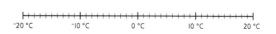

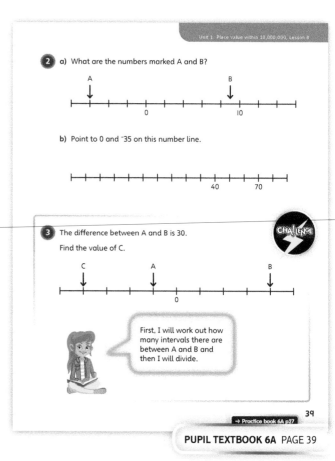

PUPIL TEXTBOOK 6A PAGE 38

PUPIL TEXTBOOK 6A PAGE 39

Practice

WAYS OF WORKING Independent thinking

IN FOCUS Questions ❶, ❷ and ❸ offer children the opportunity to develop their ability to calculate with negative numbers. Questions ❹ and ❺ develop children's independent fluency and reasoning with number lines that include negative numbers. In question ❺, decimals and fractions are introduced. Look out for children incorrectly assuming that negatives work the same way as positives, for example, thinking that ⁻11·1 is greater than ⁻11. In question ❻ children are calculating with negative numbers in context. It is important for them to see that they will meet negatives in contexts other than temperature.

STRENGTHEN For questions ❶, ❷ and ❸, it may help to provide children with a printed number line for them to use to represent their thinking. Ask: *Can you complete the number line so it has numbers on it that will help you to solve this question? How can you use the number line now to help solve this question?*

DEEPEN Question ❼ requires children to demonstrate their deep conceptual understanding of what they have learnt about number lines and negative numbers across this unit in order to solve this successfully. Ask: *How will you approach this question? What do you know? What do you not know? What do you need to find out?*

ASSESSMENT CHECKPOINT Questions ❶, ❷ and ❸ will demonstrate children's ability to apply their understanding of negative numbers. They should be able to count forwards and backwards and work out additions and differences involving negative numbers. Questions ❹, ❺ and ❻ assess children's ability to fluently apply their understanding of number lines and negative numbers when using partially complete number lines. They should be able to use reasoning to work out the values of intervals on number lines and place both positive and negative numbers accurately.

ANSWERS Answers for the **Practice** part of the lesson can be found in the *Power Maths* online subscription.

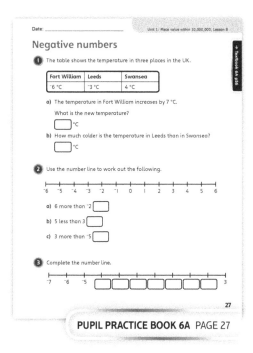

PUPIL PRACTICE BOOK 6A PAGE 27

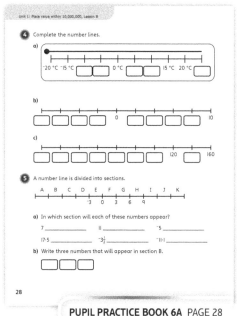

PUPIL PRACTICE BOOK 6A PAGE 28

Reflect

WAYS OF WORKING Independent thinking

IN FOCUS This question requires children to demonstrate their fluency with partially completed number lines and negative numbers. They first need to explain how they work out the value of intervals and then need to show they can count forwards and backwards across 0 on the number line.

ASSESSMENT CHECKPOINT Look for explanations of how children would calculate the value of each interval before working out the value of each letter.

ANSWERS Answers for the **Reflect** part of the lesson can be found in the *Power Maths* online subscription.

After the lesson

- Could children confidently and fluently explain how to find the value of intervals along a number line?
- How fluent were children's explanations of negative numbers? Could they understand and explain the differences between negative and positive numbers?

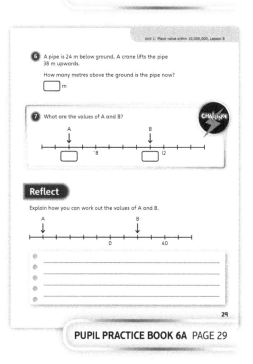

PUPIL PRACTICE BOOK 6A PAGE 29

End of unit check

Don't forget the unit assessment grid in your *Power Maths* online subscription.

Group work adult led

IN FOCUS

- Question ① assesses children's recognition of the place value of digits in a large number.
- Question ② assesses children's recognition of the written vocabulary of number names.
- Question ③ assesses children's understanding of place value and their ability to partition numbers.
- Question ④ assesses children's ability to round numbers.
- Question ⑤ assesses children's ability to use a number line fluently to identify where numbers are in relation to each other, including calculating the value of intervals and identifying numbers that fall between intervals. Number lines including negative numbers are assessed in question ⑧, which is a SATs-style question.
- Question ⑥ assesses children's understanding of place value and their ability to compare and order numbers. This is also assessed in question ⑦, which is a SATs-style question.

ANSWERS AND COMMENTARY

Children who have mastered the concepts in this unit will demonstrate fluent understanding of place value within numbers up to 10,000,000. They will be able to read, write and partition numbers accurately and compare and order numbers up to 10,000,000. They will demonstrate fluency when using a number line and will be able to complete partially completed number lines, confidently working out the value of unlabelled intervals. Children will be able to round up and down numbers up to 10,000,000 to different powers of 10. They will be able to recognise and use negative numbers on number lines and in real-life contexts.

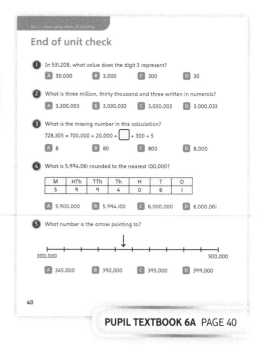

PUPIL TEXTBOOK 6A PAGE 40

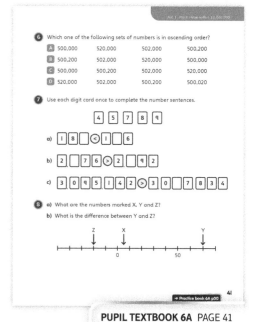

PUPIL TEXTBOOK 6A PAGE 41

Q	A	WRONG ANSWERS AND MISCONCEPTIONS	STRENGTHENING UNDERSTANDING
	A	Incorrect answers suggest children are struggling with place value or have misread the numbers.	Encourage children to use a place value grid. Ask: *Can you write the number in the place value grid? How can this help you to tell me what each digit is worth? Can you use this to partition the number? If you write all the numbers you are comparing into the place value grid, how can it help you to compare and order them?*
	C		
	D	A, B or C suggest problems with place value or partitioning.	
	C	A, B or D suggest problems with rounding or place value.	
5	B	A, C or D suggest difficulties finding the value of unlabelled intervals on a number line and using them to estimate numbers.	
6	C	Incorrect answers suggest children are struggling with place value or comparing and ordering numbers.	To help children with rounding, offer them a number line. Ask: *Where is the number on the number line? Is it closer to X or Y?*
7		Various combinations are possible. Here are two such: combination 1: 188 < 196; 2,**5**76 > 2,**4**92; 3,095,142 > 3,0**7**7,834 Combination 2: 18**4** < 18**6**; 2,**9**76 > 2,**7**92; 3,095,142 > 3,0**5**7,834 Errors suggest incomplete mastery of topic.	For labelling the intervals, ask: *What is the difference between the two labelled values? How can this help you to find what each interval is worth?*
8		**a)** X = 5, Y = 70, Z = ⁻20, **b)** 90. Errors suggest problems with unlabelled intervals or (errors for Z and b) with negative numbers.	

My journal

WAYS OF WORKING Independent thinking

ANSWERS AND COMMENTARY

Children should be able to demonstrate deep conceptual understanding of numbers through their reasoning. Children may share ideas such as:
- *I know 130,689 has 3 more hundreds than ten thousands as 6 is 3 more than 3.*
- *I know 6,985,310 is the greatest number less than 7,000,000 I can make, as I used 6 millions and then made sure that the rest of the digits go in descending order so the bigger digits are in the higher place value positions.*
- *I know 60,389 is 10,000 more than 50,389 as the ten thousands digit has increased by 1.*

If children are struggling to begin unpicking the reasoning in the question, ask: *What mathematical models might help you in this task? How would a place value grid help?*

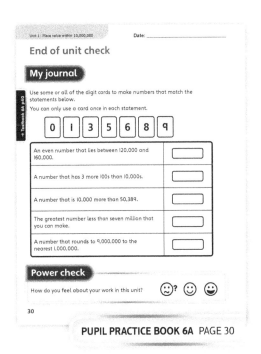

PUPIL PRACTICE BOOK 6A PAGE 30

Power check

WAYS OF WORKING Independent thinking

ASK
- *If I gave you a number in the millions, how many different ways could you write it? Could you place it on a number line? How about if the number is negative?*
- *How confident are you with the place value of numbers up to 10,000,000?*
- *How confident are you that you can compare and order numbers up to 10,000,000?*
- *Given a number up to 10,000,000, could you explain to someone how to round it to the nearest x?*

Power puzzle

WAYS OF WORKING Pair work

IN FOCUS Children need to use their reasoning to solve problems involving the numbers they have been studying in this unit. They should demonstrate their fluency with number by trying different solutions and adhering to the given clues. Once they have solved the puzzle, the suggested activity from Sparks will allow children to demonstrably deepen their understanding of the vocabulary and mathematical concepts by writing their own puzzles.

ANSWERS AND COMMENTARY Children should be able to confidently read and interpret the given clues, using their understanding of number and place value. If children are struggling, encourage them to use a place value grid to help them structure their ideas, writing all the possibilities for each place value column and then narrowing them down systematically until they reach the correct solution (5,293,187).

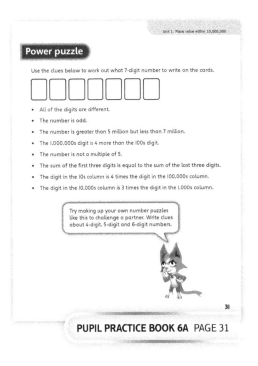

PUPIL PRACTICE BOOK 6A PAGE 31

After the unit ⏸

- To what extent did the learning in this unit deepen children's understanding of the place values of digits in a number?
- What concrete representations did you make most effective use of in this unit? What representation would you develop the use of next time you teach this? How and why?

Strengthen and **Deepen** activities for this unit can be found in the *Power Maths* online subscription.

Unit 2
Four operations ❶

Mastery Expert tip! 'Always ask children to explain the links between the written methods they are learning and using. This can really secure conceptual understanding and helps children to work with greater confidence!'

Don't forget to watch the Unit 2 video!

WHY THIS UNIT IS IMPORTANT

This unit allows children to develop fluency with efficient columnar written methods for addition and subtraction, without and with exchanges. Children will make links to methods they have met before and apply new learning to contextual word problems. They will learn to recognise and find common factors and multiples, before looking at prime numbers as a special example of numbers with specific factors. Next, children investigate the effects of squaring and cubing, linking this to what they know about the dimensions of the namesake shapes.

WHERE THIS UNIT FITS

→ Unit 1: Place value within 10,000,000
→ **Unit 2: Four operations (1)**
→ Unit 3: Four operations (2)

This unit builds on children's knowledge of using formal columnar written methods for addition and subtraction, including with real-life contexts. Children use their knowledge of the four operations to consider specific properties of numbers.

Before they start this unit, it is expected that children:
• recognise and understand the symbols for the four operations
• recognise and use written methods for addition and subtraction
• are fluent in their multiplication tables
• understand the terms, and are able to find factors and multiples
• understand and can use the four operations
• have had experience solving word problems.

ASSESSING MASTERY

Children will demonstrate mastery in this unit by efficiently and fluently solving addition and subtraction calculations using a compact columnar method, with exchanges where necessary, and explaining how this is linked to place value. Children will demonstrate mastery by fluently finding common factors and multiples of two or more numbers. They will be able to explain how prime numbers differ from other numbers and confidently square and cube numbers. They will also be able to use their understanding of the multiplicative properties of numbers to solve problems and share their reasoning.

COMMON MISCONCEPTIONS	STRENGTHENING UNDERSTANDING	GOING DEEPER
Children may confuse the definitions for 'factor' and 'multiple'.	New vocabulary and its meaning should be displayed prominently in the classroom.	Children could investigate the prime factors of different numbers. What patterns can they discover?
Children may misinterpret the information in a word problem and so choose the wrong operation to use when solving the calculation.	Ask children to look for key words that might indicate what operation is needed, for example total or difference, and to use visual representations such as bar models to show what information is given in the problem and what information they are being asked to find.	Children could write their own addition and subtraction story problems. Can they write a multi-step problem that requires both addition and subtraction?

UNIT STARTER PAGES

Use these pages to introduce the unit focus to children as a whole class. You can use the characters to explore different ways of working.

STRUCTURES AND REPRESENTATIONS

Column methods of addition and subtraction: These models are used to enable children to efficiently solve addition and subtraction calculations.

Place value grid and counters: This model helps children to recognise the value of each digit in a number and to create and partition numbers.

Bar model: This model is used to represent the solving of word problems pictorially.

Array: Arrays are a visual representation of multiplication and division. They are an excellent tool for showing equal groups within a number.

3×6

Sorting circles: Sorting circles are used in this unit to organise numbers with certain properties.

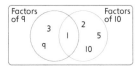

100 square: The 100 square is used in this unit to highlight patterns and relationships between factors and multiples, and to show prime numbers.

1	2	③	4	5	⑥	7	8	⑨	10
11	⑫	13	14	⑮	16	17	⑱	19	20
㉑	22	23	㉔	25	26	㉗	28	29	㉚

KEY LANGUAGE

There is some key language that children will need to know as part of the learning in this unit.

→ add, subtract, sum, total, difference
→ method, column addition, column subtraction
→ divisible, divisibility
→ factor, common factor, remainder
→ multiple, common multiple
→ prime, composite
→ squared (x^2), cubed (x^3)

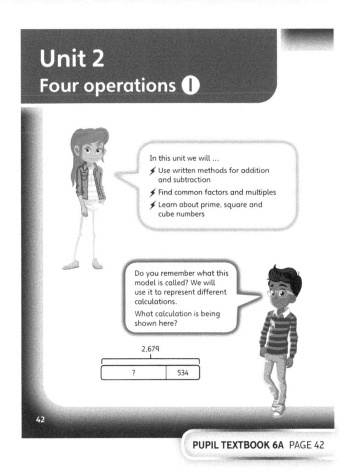

PUPIL TEXTBOOK 6A PAGE 42

PUPIL TEXTBOOK 6A PAGE 43

Add integers

Learning focus

In this lesson, children rehearse and refine efficient written methods for the addition of whole numbers.

NATIONAL CURRICULUM LINKS

Year 6 Number – addition, subtraction, multiplication and division

Solve addition and subtraction multi-step problems in contexts, deciding which operations and methods to use and why.

ASSESSING MASTERY

Children can use the columnar method of addition to add together whole numbers with up to five digits.

COMMON MISCONCEPTIONS

Children may make mistakes when a column adds up to more than 9 and an exchange is required. Ask:
- *What do the digits add to? If the total of the digits is 10 or more, how will you write that?*

STRENGTHENING UNDERSTANDING

Provide place value equipment for children to use alongside the calculations to support and check their arithmetic and answers.

GOING DEEPER

Challenge children to explore ways of checking additions, for example, adding the numbers in a different order.
Ask: *How many different ways can you suggest? Which do you prefer? Why?*

KEY LANGUAGE

In lesson: addition, exchange

Other language to be used by the teacher: integer, ten thousands (10,000s), thousands (1,000s), hundreds (100s), tens (10s), place value

STRUCTURES AND REPRESENTATIONS

Column methods

RESOURCES

Mandatory: place value equipment

 In the eTextbook of this lesson, you will find interactive links to a selection of teaching tools.

Quick recap

Ask children to lay out and complete column additions of these 3-digit numbers:

234 + 432

516 + 342

408 + 713

Discover

Add integers

WAYS OF WORKING Pair work

ASK

- Question ① a): *What do you notice about Kate's method?*
- *Has Kate lined up the digits correctly?*
- Question ① b): *Are there any other methods you could use to add these numbers?*
- Question ① b): *How do you check your calculations?*

IN FOCUS This question introduces children to the correct use of the columnar method of written addition. They first explore a common error made when using this method. They then consider how to use the method correctly to add two numbers, one with five digits and one with four digits, where two exchanges are required.

PRACTICAL TIPS Encourage children to use place value equipment to represent and check their calculations.

ANSWERS

Question ① a): Kate has set out the calculation incorrectly and has not aligned the digits in the correct place value columns.

Question ① b): 38,219 + 3,128 = 41,347

Discover

① a) What mistake has Kate made?

b) Use column addition to work out 38,219 + 3,128.
Explain each of your steps to a partner.

44

PUPIL TEXTBOOK 6A PAGE 44

Share

WAYS OF WORKING Whole class teacher led

ASK

- Question ① a): *How would you correct Kate's mistake?*
- Question ① a): *Have you ever made this mistake?*
- Question ① b): *What advice would you give to someone who is just learning this column method of addition for the first time?*

IN FOCUS Question ① a) will prompt discussion about using careful and correct layout for column additions. It explores the common place value error of misaligning the digits in the two numbers. Children should then take steps to correct this error in question ① b) and apply the correct layout for column addition. They should then notice that the digits in the 1s column and in the 1,000s column both add up to more than 9. Support them in representing and carrying out these exchanges.

Share

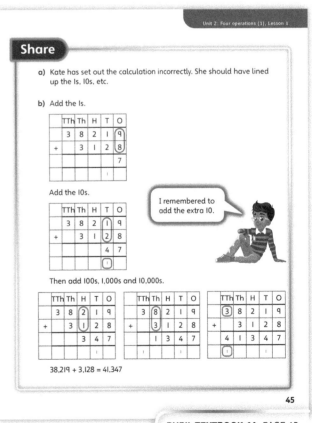

Think together

WAYS OF WORKING Whole class teacher led (I do, We do, You do)

ASK

- Question **1**: *Which column is the correct place to start?*
- Question **2**: *How will you avoid making common errors?*
- Question **3**: *For the ones column, are you finding*
  5 + 9 or 5 + ☐ *= 9?*

IN FOCUS In question **1**, children are demonstrating use of the column addition method for two numbers that both have five digits. For question **2**, they are using the column addition method, including for numbers that have different numbers of digits. Question **3** requires children to apply flexible thinking when using the column method to find missing numbers. They then progress to add three numbers.

STRENGTHEN Ensure children are using place value equipment alongside calculations to check their working and their answers.

DEEPEN Challenge children to make their own version of the missing digits puzzle in question **3** a) for a partner to solve.

ASSESSMENT CHECKPOINT Question **2** assesses whether children can correctly apply the procedure for the column method of addition. Question **3** assesses the extent to which children understand the mathematical structure that underlies the process of column addition.

ANSWERS

Question **1** a): Yes, he has.

Question **1** b): 40,371 + 28,753 = 69,124

Question **2** a): 135,704 + 417,915 = 553,619

Question **2** b): 205,918 + 25,126 = 231,044

Question **2** c): £1,300,000 + £2,570,000 = £3,870,000

Question **3** a): 45,1**9**5 + 2**7**,73**4** = **7**2,**9**29

Question **3** b): 312,057 + 48,903 + 2,510 = 363,470

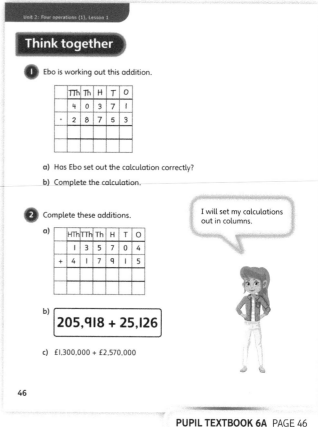

PUPIL TEXTBOOK 6A PAGE 46

PUPIL TEXTBOOK 6A PAGE 47

Practice

WAYS OF WORKING Independent thinking

IN FOCUS Question ① gives an image of place value equipment to support children's working when doing a column addition. In question ②, children solve a series of additions that are displayed on squared paper to demonstrate that they can correctly work through the column method of addition. Question ③ gives additions that children will need to lay out themselves before completing the column additions.

In question ④, children are using the column addition method to solve word problems. In question ⑤, children use what they have learned to calculate missing digits in written column additions. For question ⑥, children will need to demonstrate that they can understand the vocabulary of place value in the context of numbers that will be used in column additions.

STRENGTHEN Provide place value equipment and ensure children know how to use it to check their workings and their answers.

DEEPEN Give children fact-file figures for the populations of different counties to explore. Challenge them to add together the populations of two or three counties.

ASSESSMENT CHECKPOINT Question ③ assesses whether children can correctly lay out and complete additions using the column method of addition.

ANSWERS Answers for the **Practice** part of the lesson can be found in the *Power Maths* online subscription.

Reflect

WAYS OF WORKING Pair work

IN FOCUS The **Reflect** part of the lesson prompts a discussion about the steps that children will take to work through using the column method to add two numbers.

ASSESSMENT CHECKPOINT Assess whether children can explain the importance of laying out column additions carefully, in the context of place value.

ANSWERS Answers for the **Reflect** part of the lesson can be found in the *Power Maths* online subscription.

After the lesson ⏸

- Can children use column addition to solve calculations involving even bigger whole numbers, such as 308,945 + 2,559,678?
- Are there any errors or misconceptions that you will need to address before moving on to use the column method for subtraction?

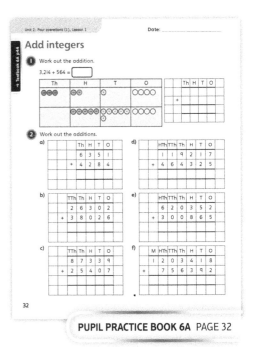

PUPIL PRACTICE BOOK 6A PAGE 32

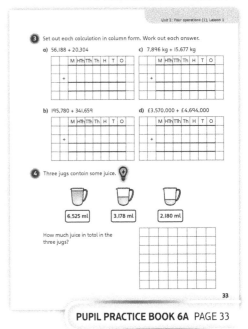

PUPIL PRACTICE BOOK 6A PAGE 33

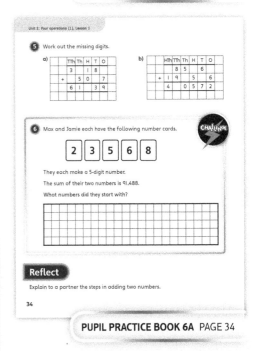

PUPIL PRACTICE BOOK 6A PAGE 34

Subtract integers

Learning focus

In this lesson, pupils rehearse and refine their use of the column method of subtraction for whole numbers.

Before you teach 🕚

- Can children use the column method to subtract 2- and 3-digit numbers?
- Have children had the opportunity to explore different contexts where subtractions are needed. For example:

  Take away: Ask: *I have 101 balloons but 25 of them burst. How many are left?*

  Find a part: Ask: *There are 101 balloons. 25 are red, the rest are yellow. How many are yellow?*

  Find the difference: Ask: *I have 101 balloons. You have 25 balloons. How many more balloons do I have?*

NATIONAL CURRICULUM LINKS

Year 6 Number – addition, subtraction, multiplication and division

Solve addition and subtraction multi-step problems in contexts, deciding which operations and methods to use and why.

ASSESSING MASTERY

Children can accurately carry out subtraction calculations using the column method for subtracting whole numbers.

COMMON MISCONCEPTIONS

Children may become confused when encountering calculations where exchanges are required. Ask:
- *What column are you thinking about? What digit are you subtracting?*
- *What will you need to do next?*

STRENGTHENING UNDERSTANDING

Provide place value equipment and encourage children to check and confirm their arithmetic and their answers.

GOING DEEPER

Give children opportunities to apply column subtraction methods to a range of problems from real-life or cross-curricular topics. For example, in history, ask: *How many years was the reign of …, or How many years between the birth of … and …?*

KEY LANGUAGE

In lesson: subtract, difference, column method, exchange

STRUCTURES AND REPRESENTATIONS

Column methods

RESOURCES

Mandatory: place value equipment

 In the eTextbook of this lesson, you will find interactive links to a selection of teaching tools.

Quick recap

Ask children to lay out and complete column subtractions of these 3-digit numbers:

879 – 213

925 – 816

602 – 185

Discover

Unit 2: Four operations (1), Lesson 2

Subtract integers

WAYS OF WORKING Pair work

ASK

- Question ① a): *What sort of calculation is needed here: addition, subtraction, multiplication or division?*
- Question ① b): *What different methods could you use to solve this calculation?*

IN FOCUS Children will need to recognise which operation is required in order to find the difference between two 4-digit numbers, in the context of historical dates. They should discuss and choose an appropriate method to find the answer.

PRACTICAL TIPS Represent the information in the question on a history timeline which can be displayed in the classroom.

ANSWERS

Question ① a) Find the difference using a method such as the column method of subtraction.

Question ① b) The difference between 1558 and 1952 is $1,952 - 1,558 = 394$.
The answer is 394 years.

Discover

① a) How can you work out the number of years between the beginning of Queen Elizabeth I's reign and the beginning of Queen Elizabeth II's reign?

   b) Work out the number of years between the beginnings of the two reigns.

48

PUPIL TEXTBOOK 6A PAGE 48

Share

WAYS OF WORKING Whole class teacher led

ASK

- Question ① a): *What does 'find the difference' mean?*
- Question ① a): *Can you explain why a subtraction is needed to solve this problem?*
- Question ① b): *What is the first step?*
- Question ① b): *What do the crossings out mean?*
- Question ① b): *Why are exchanges required?*

IN FOCUS In question ① a), children choose the operation that is most appropriate for solving problems that involve finding the difference. Listen out for children who do not identify that subtraction is the correct operation to use. In question ① b), children then correctly apply the column subtraction to find the answer. They will notice that exchanges are needed from the 10s to the 1s and from the 100s to the 10s.

PUPIL TEXTBOOK 6A PAGE 49

Think together

WAYS OF WORKING Whole class teacher led (I do, We do, You do)

ASK

- Question **1**: *Do you predict that any columns will need exchanges?*
- Question **2**: *How will you set out part b) correctly?*
- Question **3**: *Will you always end up making exchanges in problems like this?*

IN FOCUS Question **1** requires children to demonstrate the correct application of the column method of subtraction. The calculation is laid out for them. In question **2**, children will need to correctly lay out and apply the column method themselves, including when the numbers have a different number of digits. In question **3**, children investigate which calculations will require exchanges in relation to the digits in each number.

STRENGTHEN Provide place value equipment to support children's working, in particular to model any exchanges. Remind them to also check their answers.

DEEPEN Use question **3** as an opportunity to challenge children to explore similar puzzles, involving numbers with 3, 4, 5, 6 or 7 digits.

ASSESSMENT CHECKPOINT Question **2** assesses whether children can use the correct layout and application of the column method of subtraction, including when the numbers have a different number of digits.

ANSWERS

Question **1** a): Yes, she has set the question out correctly.

Question **1** b): 32,516 – 15,141 = 17,375

Question **2** a): 764,198 – 124,305 = 639,893

Question **2** b): 205,918 – 25,126 = 180,792

Question **2** c): 284,000 kg – 192,700 kg = 91,300 kg

Question **2** d): 200,000 – 176,365 = 23,635

Question **3** a): Answers depend on the numbers chosen. You will always need to exchange across two columns because in two pairs of digits the number being subtracted will be bigger and in the other two pairs of digits the number you are subtracting from will be bigger.

Question **3** b): Answers will vary. Look for children being able to use specialised examples to begin generalising about the pairs of digits in the calculations they are creating and solving.

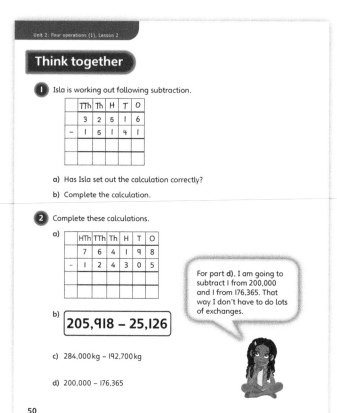

PUPIL TEXTBOOK 6A PAGE 50

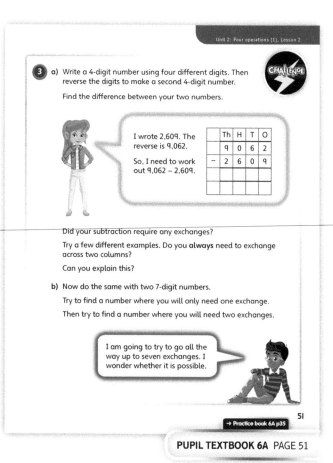

PUPIL TEXTBOOK 6A PAGE 51

Practice

WAYS OF WORKING

IN FOCUS In question ❶, children complete subtractions that are laid out in the column subtraction method. In question ❷ children will need to lay out and complete the subtractions themselves. Question ❸ requires children to find and correct a common error that can occur with this method. In question ❹, children lay out and complete subtractions where the numbers have a different number of digits. Question ❺ requires children to make an exchange across more than one column because there is a 0 digit in the larger number. In question ❻, children use what they have learned to calculate missing digits.

STRENGTHEN Ask children to model each exchange using place value equipment to keep track as numbers move between columns. Reinforce the importance of place value in questions like this.

DEEPEN In question ❼, children apply a flexible approach to problem solving in subtraction calculations. They explore how numbers can be adjusted slightly to remove the need for making exchanges. Can they explain why the same adjustment must be made to both numbers in the calculation?

THINK DIFFERENTLY Question ❻ provides a good opportunity for children to demonstrate a deeper understanding of the mathematics behind the process of the column subtraction method. Ask them to explain the steps they take to find each missing digit.

ASSESSMENT CHECKPOINT Question ❺ assesses whether children can exchange accurately.

ANSWERS Answers for the **Practice** part of the lesson can be found in the *Power Maths* online subscription.

Reflect

WAYS OF WORKING Independent thinking

IN FOCUS The **Reflect** part of the lesson prompts children to consider what they have learnt about the different subtraction methods.

ASSESSMENT CHECKPOINT Assess whether children can use what they have learned about column subtraction to generate calculations that will require more than one exchange.

ANSWERS Answers for the **Reflect** part of the lesson can be found in the *Power Maths* online subscription.

After the lesson ⏸

- How will you give children the opportunity to regularly rehearse column methods as part of continued procedural fluency practice?

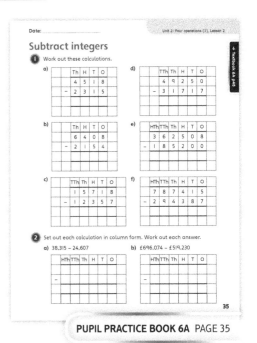

PUPIL PRACTICE BOOK 6A PAGE 35

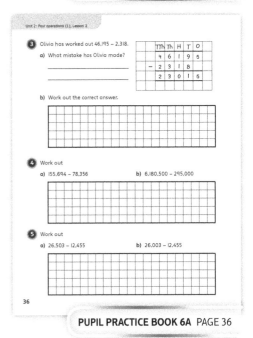

PUPIL PRACTICE BOOK 6A PAGE 36

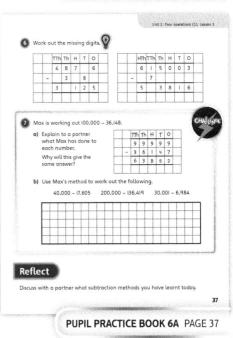

PUPIL PRACTICE BOOK 6A PAGE 37

Problem solving – addition and subtraction

Learning focus

In this lesson, children will develop their understanding of the columnar written methods of addition and subtraction where exchanges are sometimes necessary.

Before you teach

- How confident are children with written methods of addition and subtraction?
- Will you need to spend more teaching time on one method than on the others?
- What extra teaching strategies and experiences will you offer children to support each operation?

NATIONAL CURRICULUM LINKS

Year 6 Number – addition, subtraction, multiplication and division

Solve addition and subtraction multi-step problems in contexts, deciding which operations and methods to use and why.

ASSESSING MASTERY

Children can fluently and efficiently use columnar written methods to solve addition and subtraction problems, including those that involve exchanges. They can explain why and how these methods work and can represent them clearly.

COMMON MISCONCEPTIONS

Children may confuse the place value headings above the columns. Ask:
- *Can you show me the place value headings above each column?*
- *Does each number fit its column heading? Explain.*

STRENGTHENING UNDERSTANDING

Before the lesson, give children concrete opportunities to experience and revise addition and subtraction, such as building numbers with place value counters or base 10 equipment and adding or subtracting by adding or taking away resources, or using money in context through role playing or visiting a shop.

GOING DEEPER

Encourage children to create their own missing number calculations (for example 456,232 + ☐ = 563,213) and use them to challenge a partner.

KEY LANGUAGE

In lesson: addition, total, subtraction, method, column, calculate, calculation

Other language to be used by the teacher: difference, exchange

STRUCTURES AND REPRESENTATIONS

Column addition, column subtraction, bar model

RESOURCES

Optional: place value counters, printed place value grids, base 10 equipment

 In the eTextbook of this lesson, you will find interactive links to a selection of teaching tools.

Quick recap

Ask children to invent a story problem for each of these calculations:

25 + 75 = ?

200 − 150 = ?

Discover

WAYS OF WORKING Pair work

ASK

- Question ❶ a): *How will you find out how many runners actually completed the race?*
- Question ❶ a): *What would be the most efficient and accurate way of calculating this difference? Explain.*
- Question ❶ b): *What calculation is needed here? What do you notice about the numbers in the ones column?*

IN FOCUS Question ❶ a) offers children an opportunity to calculate the difference between two numbers. Encourage them to discuss and decide what would be the most efficient and accurate method for solving this problem.

Question ❶ b) involves an addition that requires an exchange from the ones to the tens. Encourage children to lay the addition out carefully with the digits in the correct columns, and ensure they know how to represent and carry out the exchange.

PRACTICAL TIPS: Children could be encouraged to build the numbers in the picture with place value counters or base 10 equipment to help scaffold their concrete understanding of addition. Discuss with children how they could organise their resources to make the calculations clear, moving towards organising them in columns according to place value.

ANSWERS

Question ❶ a): 2,145 runners completed the race.

Question ❶ b): 32,145 + 4,307 = 36,452 is the correct answer.
They started with 36,452 bottles of water.

Share

WAYS OF WORKING Whole class teacher led

ASK

- Question ❶ a): *Do any of the representations match how you would have solved the subtraction?*
- Question ❶ a): *Which method is more efficient? Explain.*
- Question ❶ b): *What calculation is needed here?*
- Question ❶ b): *How will you lay out your calculation?*

IN FOCUS It will be important to use the multiple representations and methods shown in question ❶ a) to scaffold children's revision of subtraction and to assess their current confidence and understanding. Children should be encouraged to use each of the methods and representations to help secure the links in their mathematical understanding.

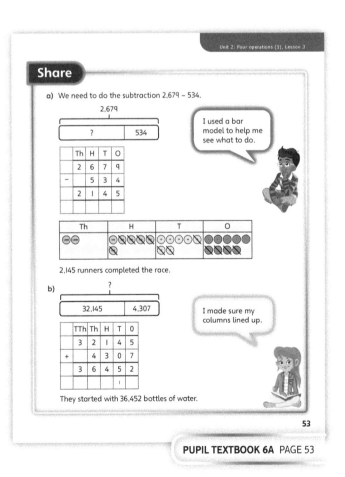

Problem solving – addition and subtraction

Discover

Mud Mayhem race!

We have had 2,679 runners this year, but 534 were unable to complete the race.

We have 32,145 empty water bottles. 4,307 bottles have not been used.

❶ a) How many runners completed the race?

b) How many bottles of water did they start with?

52

PUPIL TEXTBOOK 6A PAGE 52

Share

a) We need to do the subtraction 2,679 – 534.

I used a bar model to help me see what to do.

2,145 runners completed the race.

b)

I made sure my columns lined up.

They started with 36,452 bottles of water.

53

PUPIL TEXTBOOK 6A PAGE 53

Think together

Whole class teacher led (I do, We do, You do)

ASK

- Question ❶: *What operation is needed to solve this question? How do you know?*
- Question ❷: *Do all parts of this question require the same operation? Explain how you know.*
- Question ❷ b) and c): *Where will you find the information needed to solve this question?*
- Question ❸: *What will you do first? How does the bar model help you?*

IN FOCUS Question ❶ helps children with their conceptual understanding of addition which requires an exchange. They may benefit from having concrete resources available to them while they solve the question. Encourage them to make the calculation with place value counters while solving the abstract calculation and discuss what is the same and different about the representations. Question ❷ offers the opportunity to solve subtraction and addition calculations in context. Make sure children are aware that they need information from the **Discover** section of the lesson.

STRENGTHEN If children are struggling to decide what operation to use for each part of question ❷, ask: *Can you make the problem using resources? Does the question suggest you will take away from what you have or add to it? Explain how you know.*

DEEPEN Question ❸ is a multi-step problem. If children are quick to solve it encourage them to create a similar challenge for a partner. Ask: *Can you create one which is simple and one which is tricky? How are they the same and how are they different?*

ASSESSMENT CHECKPOINT Can children recognise different representations and use them to solve calculations? Do children recognise addition and subtraction calculations in the context of word problems? Do children understand how the columnar methods for addition and subtraction work, and can they use these methods with fluency?

ANSWERS

Question ❶: The marathon runners raised £43,937.

Question ❷ a): 1,222 runners finished the marathon.

Question ❷ b): 1,061 more runners started the Mud Mayhem race than started the marathon.

Question ❷ c): 3,367 runners finished both races.

Question ❸ a): The value of A is 700.
The value of B is 500.

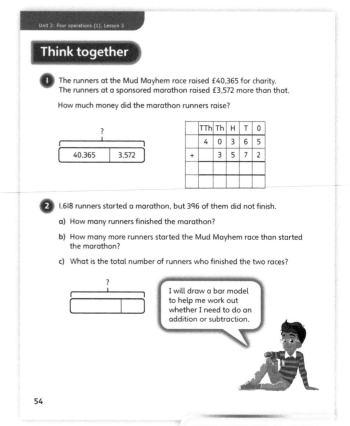

PUPIL TEXTBOOK 6A PAGE 54

PUPIL TEXTBOOK 6A PAGE 55

Practice

WAYS OF WORKING Independent thinking

IN FOCUS In questions **1**, **2** and **3**, children read and understand addition and subtraction in the context of word problems. Question **2** challenges children to choose the correct operation.

Question **3** is a multi-step word problem.

STRENGTHEN If children are struggling to decide how to solve the word problems in question **1**, ask: *Can you explain what is happening in the story of the question? Is something being added or is it being taken away? How do you know?*

DEEPEN Question **5** deepens children's fluency and problem solving when calculating with addition and subtraction. The question is written in a way that requires some 'untangling'. The bar model will support children with this.

THINK DIFFERENTLY Question **4** offers children the opportunity to think differently as they must interpret the numbers given on the number line in the context of the question, to reason about the value of the third number. They are required to find the sum of all numbers so they must remember to complete this final step of the problem.

ASSESSMENT CHECKPOINT Can children use a formal written method with fluency and link their understanding to pictorial representations? Do they draw out what they know and what they need to find to solve problems?

ANSWERS Answers for the **Practice** part of the lesson can be found in the *Power Maths* online subscription.

Reflect

WAYS OF WORKING Independent thinking and pair work

IN FOCUS This question will offer children the opportunity to explore meta-cognitive strategies for their own learning. Which question did they find most challenging and why? Can they make suggestions to help them feel more confident with this sort of question in future?

ASSESSMENT CHECKPOINT Assess children's confidence in identifying and carrying out the calculation needed for a given problem. Do they know which operation or operations are needed, and can they use visual representations to help them? Children should be able to show fluency with these types of calculation by demonstrating their ability to put the calculation into an appropriate context.

ANSWERS Answers for the **Reflect** part of the lesson can be found in the *Power Maths* online subscription.

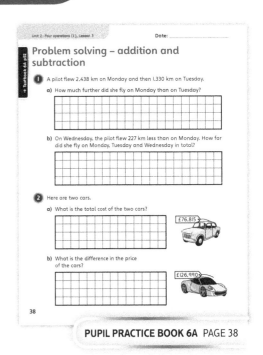

PUPIL PRACTICE BOOK 6A PAGE 38

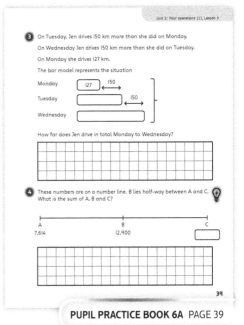

PUPIL PRACTICE BOOK 6A PAGE 39

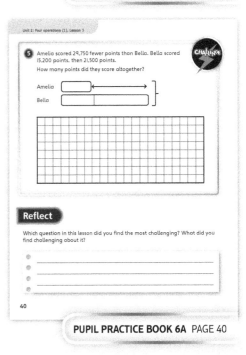

PUPIL PRACTICE BOOK 6A PAGE 40

After the lesson ⏸

- Are all children sufficiently confident with the columnar methods for both addition and subtraction?
- What support will you offer to children who are still struggling with one or both of the methods?
- How did this lesson develop children's use of mathematical vocabulary?

Common factors

Learning focus

In this lesson, children will develop their understanding of factors and how common factors link two or more numbers. They will use this understanding to find common factors.

NATIONAL CURRICULUM LINKS

Year 6 Number – addition, subtraction, multiplication and division

Identify common factors, common multiples and prime numbers.

ASSESSING MASTERY

Children can, given two or more numbers, find the common factors. They can use this ability to help solve mathematical problems and puzzles reliably, explaining their reasoning and understanding confidently.

COMMON MISCONCEPTIONS

Children may mistakenly look for common multiples of a number instead of factors. Ask:
- *What are factors? What operation can you use to find them?*

Get children to create an array for the number they are finding factors of. Ask:
- *What factors of [number] can you see in this array?*

STRENGTHENING UNDERSTANDING

Before this lesson, give children opportunities to rehearse multiplication facts. For example, use songs and chants, and activities such as 'follow me' cards (for example, *I am 6, who is 3 × 7? I am 21, who is 8 × 12?*).

GOING DEEPER

Encourage children to set challenges for each other, such as: *My two numbers share the common factors of 2, 3 and 5. What might my numbers be?*

KEY LANGUAGE

In lesson: factor, **common factor**, divide, remainder, multiplication, array

Other language to be used by the teacher: multiply, division

STRUCTURES AND REPRESENTATIONS

Arrays, sorting circles, tables

RESOURCES

Optional: 'follow me' cards, counters, multiplication grids

 In the eTextbook of this lesson, you will find interactive links to a selection of teaching tools.

Quick recap

Challenge children to find all the factors of 30. Discuss how you know if a number is a factor of another number or not. Ask children how they can make sure they have found all of the factors of a number.

Discover

Common factors

Discover

WAYS OF WORKING Pair work

ASK

- Question ❶ a): *What numbers will split equally into 4 groups?*
- Question ❶ b): *Can you prove your ideas?*

IN FOCUS Children are required to use knowledge of multiples or divisibility to find factors for given numbers.

PRACTICAL TIPS Children could solve the problem posed in the picture practically, with some children pretending to be adult helpers and others being the children. Change the challenge by varying the numbers used in the question.

ANSWERS

Question ❶ a): The adults can be split into 4 equal groups of 6 as $24 \div 4 = 6$.
The children cannot be split into 4 equal groups as $30 \div 4 = 7$ remainder 2.
The children cannot divide equally into 4 groups, because 4 is not a factor of 30.

Question ❶ b): Factors of 24 are 1, 2, 3, 4, 6, 8, 12 and 24.
Factors of 30 are 1, 2, 3, 5, 6, 10, 15 and 30.

We should divide into groups. We are 24 adults and 30 children.

Let's have 4 groups.

Make sure there are the same number of adults and children in each group.

❶ a) Show that the adults can be split into 4 equal groups but the children cannot.

b) Find the factors of each number.

56

PUPIL TEXTBOOK 6A PAGE 56

Share

WAYS OF WORKING Whole class teacher led

ASK

- Question ❶ a): *What other numbers are not factors of 30? How can you prove this?*
- Question ❶ b): *How did you show the groups that would be possible with those numbers?*
- Question ❶ b): *Can you see any numbers that are in both lists?*
- Question ❶ b): *Are the common factors of 24 and 30 also factors of any other numbers?*

IN FOCUS To ensure children's concrete understanding of this concept, it is important to link the idea with their experience of arrays. Children should be encouraged to build (for example, using counters) or draw arrays that prove the link between the common factors.

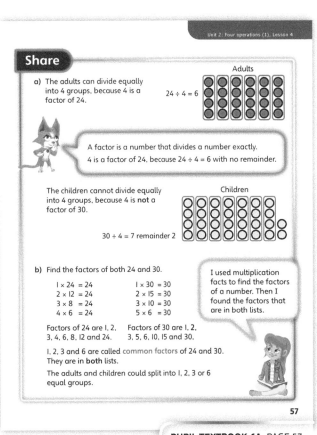

Share

a) The adults can divide equally into 4 groups, because 4 is a factor of 24.

$24 \div 4 = 6$

Adults

A factor is a number that divides a number exactly.
4 is a factor of 24, because $24 \div 4 = 6$ with no remainder.

The children cannot divide equally into 4 groups, because 4 is **not** a factor of 30.

Children

$30 \div 4 = 7$ remainder 2

b) Find the factors of both 24 and 30.

$1 \times 24 = 24$	$1 \times 30 = 30$
$2 \times 12 = 24$	$2 \times 15 = 30$
$3 \times 8 = 24$	$3 \times 10 = 30$
$4 \times 6 = 24$	$5 \times 6 = 30$

I used multiplication facts to find the factors of a number. Then I found the factors that are in both lists.

Factors of 24 are 1, 2, 3, 4, 6, 8, 12 and 24. Factors of 30 are 1, 2, 3, 5, 6, 10, 15 and 30.

1, 2, 3 and 6 are called common factors of 24 and 30. They are in **both** lists.

The adults and children could split into 1, 2, 3 or 6 equal groups.

57

PUPIL TEXTBOOK 6A PAGE 57

Think together

Whole class teacher led (I do, We do, You do)

ASK

- Question **1**: *How could you show the factors of each number?*
- Question **2**: *What times-tables knowledge will you need to use to solve this?*
- Question **3**: *How can you prove that you have found all the common factors?*

IN FOCUS Children should be encouraged to use their fluency with multiplication tables to help solve the problems listed in this part of the lesson. Develop children's ability to work systematically by referring back to Astrid's comment in the previous section. Question **1** can be used to link children's multiplicative understanding to their new understanding of factors by demonstrating the full multiplication calculations. It may be beneficial to also link them to the inverse division calculations.

STRENGTHEN If children are not yet fluent in the multiplication tables facts, offer multiplication grids for them to use. Ask: *How can you use this multiplication grid to help you? Is it possible to find the common factors using this?*

DEEPEN In question **3**, extend children's reasoning around common factors. Ask: *What other general statements can you make about types of numbers and their common factors? For example, what can you say about the common factors of all even numbers?*

ASSESSMENT CHECKPOINT At this point in the lesson, children should be able to explain what common factors are and how they link two or more numbers. Children should be confident when finding the common factors of two numbers and will be growing in confidence when finding common factors of more than two numbers. Question **3** will demonstrate these skills in particular; children should be able to explain the generalisations about multiplication facts that will help them find common factors.

ANSWERS

Question **1**: $1 \times 12 = 12$, $2 \times 6 = 12$, $3 \times 4 = 12$
$1 \times 15 = 15$, $3 \times 5 = 15$
Factors of 12 are 1, 2, 3, 4, 6 and 12.
Factors of 15 are 1, 3, 5 and 15.
The common factors of 12 and 15 are 1 and 3.

Question **2**:
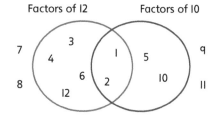
The common factors of 10 and 12 are 1 and 2. These appear in the overlap of the two circles.

Question **3** a): 1 and 5.

Question **3** b): 2 and 10.

Question **3** c): 3, 4, 15 and 20.

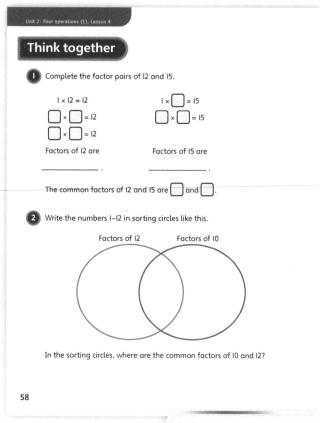

PUPIL TEXTBOOK 6A PAGE 58

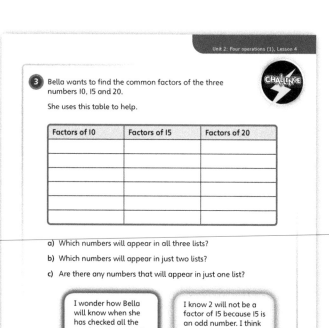

PUPIL TEXTBOOK 6A PAGE 59

Practice

WAYS OF WORKING Independent thinking

IN FOCUS Question ❶ encourages children to use arrays to explain their reasoning and prove their ideas. This will help them to link their concrete understanding with the more abstract ideas presented in the lesson. Children could be encouraged to build and manipulate these arrays to develop their reasoning and explanations. While solving question ❺, children should be encouraged to provide proof of their ideas through use of an array.

STRENGTHEN When solving question ❻ b), if children are having trouble getting started, ask: *How could you make finding the factors of each number easier? What resource might help you to get started finding the factors of each number? How could you write your findings?*

DEEPEN While solving question ❻ a), deepen children's thinking about the link between numbers by asking: *Does knowing the factors of 35 help you to find the factors of 70 more quickly? Does knowing the factors of 6 help you to find the factors of 60 more quickly? How about 600?*

ASSESSMENT CHECKPOINT Question ❺ assesses whether children are able to use sorting circles to confidently find common factors of two given numbers.

ANSWERS Answers for the **Practice** part of the lesson can be found in the *Power Maths* online subscription.

Reflect

WAYS OF WORKING Independent thinking

IN FOCUS This section assesses children's number sense. They should be able to recognise that, once they have found the factors of both numbers, up to the value of 15, it is not worthwhile going any further as any number bigger than 15 will not be a factor of it.

ASSESSMENT CHECKPOINT Look for children's understanding of the reasoning noted above.

ANSWERS Answers for the **Reflect** part of the lesson can be found in the *Power Maths* online subscription.

After the lesson

- Were children able to confidently use arrays and their multiplication fluency to clearly explain their reasoning?
- Were there any multiplication tables that children were less confident with? How will you develop their confidence with these in the future?

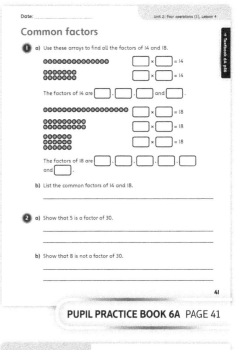

PUPIL PRACTICE BOOK 6A PAGE 41

PUPIL PRACTICE BOOK 6A PAGE 42

PUPIL PRACTICE BOOK 6A PAGE 43

Common multiples

Learning focus

In this lesson, children will develop their understanding of multiples and how common multiples link two or more numbers. They will use this understanding to find common multiples.

Before you teach

- Were there any multiplication tables that needed further input before this lesson?
- How will this influence your teaching?

NATIONAL CURRICULUM LINKS

Year 6 Number – addition, subtraction, multiplication and division

Identify common factors, common multiples and prime numbers.

ASSESSING MASTERY

Children can, given two or more numbers, find the common multiples. They can use this ability to help solve mathematical problems and puzzles reliably, explaining their reasoning and understanding confidently.

COMMON MISCONCEPTIONS

Children may mistakenly look for common factors of a number instead of multiples. Ask:
- *What are factors? What are multiples? How are they different?*

STRENGTHENING UNDERSTANDING

Children should be given opportunities to develop their fluency with multiplication tables, especially any of those highlighted as areas of development in the previous lesson. Again, give children the opportunity to recite rhymes, songs and chants and play multiplication games.

GOING DEEPER

Children could challenge each other, for example: *My two numbers share the common multiples 60 and 75. What numbers could they be?*

KEY LANGUAGE

In lesson: multiple, common multiple, common factor

Other language to be used by the teacher: multiplication

STRUCTURES AND REPRESENTATIONS

100 squares, sorting circles, bar models

RESOURCES

Optional: multiplication grids, hoops, bean bags

 In the eTextbook of this lesson, you will find interactive links to a selection of teaching tools.

Quick recap

Challenge children to find all the multiples of 12 between 100 and 200. Support them in working systematically.

Discover

Common multiples

WAYS OF WORKING Pair work

ASK

- Question ❶ a): *Can you predict the days that one of the jobs will be done?*
- Question ❶ a): *Can you predict the days that both jobs will be done together? Explain how you know.*
- Question ❶ b): *What is interesting about the days where more than one job is required?*

IN FOCUS Question ❶ a) introduces the concept of common multiples of two numbers. It is important for children to recognise that the common multiples are numbers that feature in both counts. This idea is further developed in question ❶ b) with the introduction of a third count.

PRACTICAL TIPS This concept could be introduced in small games outside. For example, a number of hoops could be laid on the ground and children challenged to throw bean bags into the hoops. Give restrictions, such as blue bean bags can only be thrown in every second hoop; red bean bags can only be thrown in every third hoop. Ask: *Which hoops will have both colours? Why?*

To engage children further, it would be introduced with the presentation of a real or imagined class pet, or perhaps a photograph of a real pet from a pupil or an adult at school. Give children the instructions as given in the picture and use them to help solve questions ❶ a) and b).

ANSWERS

Question ❶ a): Lexi will need to do both jobs on day 15 and day 30. (Days that are common multiples of 3 and 5.)

Question ❶ b): Lexi will need to do all three jobs on day 30. (Days that are common multiples of 2, 3 and 5.)

Share

WAYS OF WORKING Whole class teacher led

ASK

- Question ❶ a): *What numbers are multiples of 3? What numbers are multiples of 5?*
- Question ❶ a): *Can you find multiples of 5 in the multiples of 3?*
- Question ❶ b): *Can you use what you have found to predict all the common multiples?*
- Question ❶ b): *If a common multiple is 30, does that mean 300 will be a common multiple too? How about 3,000? Why?*

IN FOCUS In question ❶ a) children look at multiples of two numbers and then identify the common multiples between them. A third number is introduced in question ❶ b) and children find a common multiple of all three numbers. For both questions, children should be encouraged to use 100 squares to help them identify and write the multiples of each number. Discuss patterns children spot and how these can be used to help predict further common multiples.

Discover

❶ a) On which days will Lexi need to change the bedding and give carrots?

b) On which days will Lexi need to do all three jobs?

60

PUPIL TEXTBOOK 6A PAGE 60

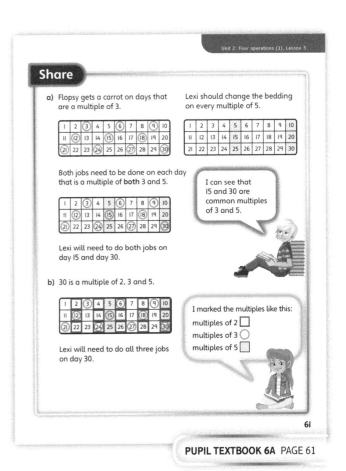

PUPIL TEXTBOOK 6A PAGE 61

Think together

WAYS OF WORKING Whole class teacher led (I do, We do, You do)

ASK

- Question ❶: *How could you write all the multiples of each number?*
- Question ❷: *How will you find the common multiples?*
- Question ❸: *Are you able to find all the common multiples? Explain your ideas.*

IN FOCUS Questions ❶, ❷ and ❸ offer children opportunities to find common multiples of different numbers in different contexts. Children should be encouraged to justify their solutions with evidence.

STRENGTHEN To help children begin finding common multiples of the numbers in each question, it may be helpful to give them 100 squares they can colour or write on. Ask: *How can you use this to help you? What patterns can you spot?*

DEEPEN Children could deepen their understanding of common multiples by investigating what happens if they look for common multiples of more numbers. Ask: *What happens if you look for the common multiples of three numbers? Will there be more or fewer common multiples? Why?*

Children could also be encouraged to investigate how much of a difference is made if one of the numbers is increased or decreased by 1. For example, for question ❶ ask: *Would there be a big change in common multiples of 5 and 6 instead of 4 and 6? Explain your ideas.*

ASSESSMENT CHECKPOINT By the end of question ❸, children should be able to find common multiples of two numbers. Use question ❸ to assess whether children are able to apply their learning in a given context.

ANSWERS

Question ❶ a):

I	2	3	4	5	⑥	7	8	9	10
II	⑫	13	14	15	16	17	⑱	19	20
21	22	23	㉔	25	26	27	28	29	㉚
31	32	33	34	35	㊱	37	38	39	40
41	㊷	43	44	45	46	47	㊽	49	50

Question ❶ b): Common multiples of 4 and 6 are 12, 24, 36 and 48.

Question ❷: The common multiples of:
   a) 6 and 9 are: 18, 36, 54, 72, 90
      (and all following multiples of 18)
   b) 5 and 6 are: 30, 60, 90
      (and all following multiples of 30)
   c) 20 and 100 are: 100, 200, 300
      (and all following multiples of 100).

Question ❸: The towers could be 36 cm, 72 cm, 108 cm and all subsequent multiples of 36.

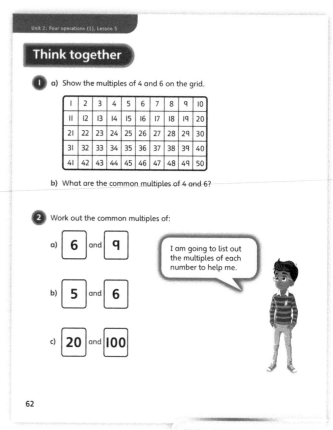

Unit 2: Four operations (1), Lesson 5

Think together

❶ a) Show the multiples of 4 and 6 on the grid.

I	2	3	4	5	6	7	8	9	10
II	12	13	14	15	16	17	18	19	20
21	22	23	24	25	26	27	28	29	30
31	32	33	34	35	36	37	38	39	40
41	42	43	44	45	46	47	48	49	50

b) What are the common multiples of 4 and 6?

❷ Work out the common multiples of:

a) **6** and **9**

b) **5** and **6**

c) **20** and **100**

I am going to list out the multiples of each number to help me.

62

PUPIL TEXTBOOK 6A PAGE 62

Unit 2: Four operations (1), Lesson 5

❸ Emma has cubes that are 9 cm tall.

Ambika has cubes that are 12 cm tall.

CHALLENGE

12 cm

9 cm

They each make a tower. The two towers are equal in height.

How tall could the towers be?

I wonder if there is an easy way to work out the heights of the possible towers.

I will think about common multiples. I think there is more than one solution.

63

→ Practice book 6A p44

PUPIL TEXTBOOK 6A PAGE 63

Practice

WAYS OF WORKING Independent thinking

IN FOCUS Question ③ assesses whether children can find common multiples of two numbers. It also gives an opportunity to observe whether children are able to make generalisations about the numbers they are finding common multiples of. Ask: *Do your generalisations apply to other numbers? Prove it.*

Question ⑤ challenges the assumption that merely multiplying the two given numbers together will give the first common multiple. Children should be encouraged to explain why this is not the case.

STRENGTHEN If children are struggling to place the numbers into the sorting diagram in question ③, ask: *What resource can you use to begin finding multiples of 5? How can you use it to find common multiples of 4 and 5?*

DEEPEN Building on question ③, children could be challenged to create a sorting diagram with three circles, each looking for multiples of a different number, less than 10. Can they find the common multiples for these numbers? Can they make any generalisations about the common multiples they find for all three?

ASSESSMENT CHECKPOINT Children should be able to confidently find common multiples of any given numbers. Question ② assesses this skill and also gives you the opportunity to assess children's ability to begin making generalisations about multiples. Look for children linking their multiplicative understanding to patterns they find in common multiples.

ANSWERS Answers for the **Practice** part of the lesson can be found in the *Power Maths* online subscription.

Reflect

WAYS OF WORKING Independent thinking

IN FOCUS This question provides a final opportunity to assess whether children can confidently find common multiples of numbers, demonstrating their understanding through the explanation of their reasoning. Ask children to independently find the three common multiples the question asks for. Once they have been given time to do this, they could share their findings with their partner. Ask: *Did you both find the same multiples? Why might they be different?*

ASSESSMENT CHECKPOINT Children should be able to find the common multiples and also recognise that there are an infinite number of possible multiples to pick from.

ANSWERS Answers for the **Reflect** part of the lesson can be found in the *Power Maths* online subscription.

After the lesson ⏸

- Could children recognise and explain the difference between common multiples and common factors?
- Were children able to understand and explain why there are infinite common multiples?
- How could you have made this lesson more practical?

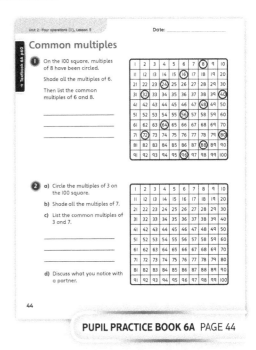

PUPIL PRACTICE BOOK 6A PAGE 44

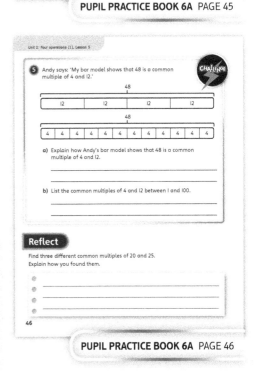

PUPIL PRACTICE BOOK 6A PAGE 45

PUPIL PRACTICE BOOK 6A PAGE 46

Rules of divisibility

Learning focus

In this lesson, children learn to recognise the properties of numbers that can be used to deduce their divisibility by 2, 3, 5 and other factors.

Before you teach

- Do children recognise multiples of 10 and 100?
- Have children had opportunities to discuss odd and even numbers between 0 and 100?
- Can children count up in 5s from 0, or from any other multiple of 5?

NATIONAL CURRICULUM LINKS

Year 6 Number – addition, subtraction, multiplication and division

Identify common factors, common multiples and prime numbers.

Use their knowledge of the order of operations to carry out calculations involving the four operations.

ASSESSING MASTERY

Children can use rules of divisibility to identify multiples of 2, 3 and 5.

COMMON MISCONCEPTIONS

Children may think that the ones digit is always the indicator of divisibility, for example, that any number that ends in 3 is a multiple of 3. Ask:
- *When is the ones digit helpful? When is it not helpful?*

STRENGTHENING UNDERSTANDING

Start by asking children to recognise multiples of 10 and to describe the patterns that they see in this multiplication.

GOING DEEPER

Children can begin to explore the properties of numbers that are multiples of 6 and 9.

KEY LANGUAGE

In lesson: multiple, divisible

Other language to be used by the teacher: digit, rule

RESOURCES

Optional: number tracks

 In the eTextbook of this lesson, you will find interactive links to a selection of teaching tools.

Quick recap

Ask children to make a list of multiples of 10 that are greater than 1,000. Then ask them to make a list of multiples of 100 that are greater than 1,000.

Discover

Unit 2: Four operations (1), Lesson 6

Rules of divisibility

Discover

WAYS OF WORKING Pair work

ASK

- Question **1** a): *Can you say the number in full?*
- Question **1** a): *What do you notice about the number?*
- *What do you wonder about it?*

IN FOCUS Children are introduced to rules that can be used to recognise which numbers are multiples of 3 and 5, including very big numbers.

PRACTICAL TIPS Ask children to count up in 5s together from 900. Support them in continuing the count beyond 1,000, What do they notice?

ANSWERS

Question **1** a): 2,370,165 is a multiple of 5 because the last digit is a 5.

Question **1** b): If you add all the digits in 2,370,165 together you get

$2 + 3 + 7 + 0 + 1 + 6 + 5 = 24$

24 is a multiple of 3, so 2,370,165 is a multiple of 3.

1 a) How do you know 2,370,165 is a multiple of 5 just by looking at the number?

b) Is the teacher correct to say 2,370,165 is a multiple of 3? How can you check without dividing the number by 3?

64

PUPIL TEXTBOOK 6A PAGE 64

Share

WAYS OF WORKING Whole class teacher led

ASK

- Question **1** a): *What do you notice about the pattern of counting in 5s?*
- Question **1** a): *How far does this pattern continue?*
- Question **1** a): *What rule can we find for recognising multiples of 5?*
- Question **1** b): *What is the rule used to check for divisibility by 3?*
- Question **1** b): *How could you test this rule?*

IN FOCUS In question **1** a), children familiarise themselves with the rule that can be used to identify multiples of 5. In question **1** b), they explore a rule for identifying multiples of 3.

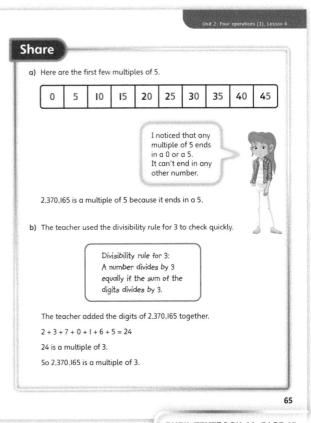

Share

a) Here are the first few multiples of 5.

0	5	10	15	20	25	30	35	40	45

I noticed that any multiple of 5 ends in a 0 or a 5. It can't end in any other number.

2,370,165 is a multiple of 5 because it ends in a 5.

b) The teacher used the divisibility rule for 3 to check quickly.

Divisibility rule for 3:
A number divides by 3 equally if the sum of the digits divides by 3.

The teacher added the digits of 2,370,165 together.

$2 + 3 + 7 + 0 + 1 + 6 + 5 = 24$

24 is a multiple of 3.

So 2,370,165 is a multiple of 3.

65

PUPIL TEXTBOOK 6A PAGE 65

Think together

Unit 2: Four operations (1), Lesson 6

WAYS OF WORKING Whole class teacher led (I do, We do, You do)

ASK

- Question ❶: *Which column is easiest to check? Why?*
- Question ❷: *Could an odd number be a multiple of 4? Why not?*
- Question ❸: *Can a number be a multiple of both 2 and 5 at the same time? Can a number be a multiple of both 2 and 3 at the same time?*

IN FOCUS In question ❶, children apply the rules of divisibility for multiples of 2, 3 and 5. In question ❷ they are introduced to the rule of divisibility by 4 and use this to identify multiples of 4. Question ❸ gives an overview of divisibility rules and requires children to apply what they have learned to suggest numbers that they know to be divisible by 2, 3, 5 or 6.

STRENGTHEN Build children's confidence by first exploring rules of divisibility by 10, 2 and 5.

DEEPEN Challenge children to explore and explain why an even multiple of 5 is also a multiple of 10.

ASSESSMENT CHECKPOINT Use question ❶ to assess whether children can recognise key divisibility properties with multiples of 2, 3 and 5.

ANSWERS

Question ❶:

Number	Divisible by 2	Divisible by 3	Divisible by 5
124	✓	✗	✗
405	✗	✓	✓
166	✓	✗	✗
216	✓	✓	✗
176	✓	✗	✗

Question ❷: 924, 6,320 and 26,352 are divisible by 4. 1,514 is not divisible by 4.

Question ❸ a): 0, 2, 4, 6 or 8

Question ❸ b): 0 or 5

Question ❸ c): 2, 5 or 8 (digits then add to a multiple of 3)

Question ❸ d): 2 or 8 (an even multiple of 3)

Think together

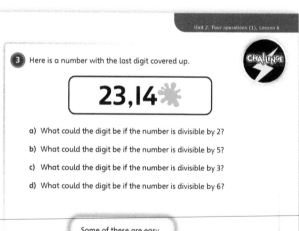

❶ Complete the table.

The first row has been done for you.

Number	Divisible by 2	Divisible by 3	Divisible by 5
124	✓	✗	✗
405			
166			
216			
176			

Discuss with your class how you decided.

I wonder if I can tell which of these numbers divide by 6 easily.

❷ Use the divisibility rule for 4 to decide whether the numbers are divisible by 4.

924   6,320

1,514   26,352

Divisibility rule for 4:
Do the last two digits of the number divide by 4?
If so, then the number divides by 4.

66

PUPIL TEXTBOOK 6A PAGE 66

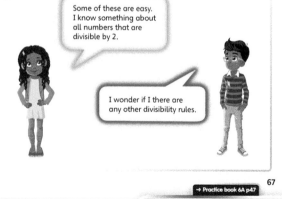

Unit 2: Four operations (1), Lesson 6

❸ Here is a number with the last digit covered up.

23,14✹

a) What could the digit be if the number is divisible by 2?

b) What could the digit be if the number is divisible by 5?

c) What could the digit be if the number is divisible by 3?

d) What could the digit be if the number is divisible by 6?

Some of these are easy. I know something about all numbers that are divisible by 2.

I wonder if I there are any other divisibility rules.

67

→ Practice book 6A p47

PUPIL TEXTBOOK 6A PAGE 67

Practice

WAYS OF WORKING Independent thinking

IN FOCUS In question ①, children apply the rules of divisibility to multiples of 2 and 5, before moving on to identify multiples of 3 in question ②. In question ③, children find numbers which are divisible by 2 and/or 3. By implication, this will also enable them to identify divisibility by 6. In question ④, children apply the rule of divisibility by 4. In question ⑤, they take an overview of the rules of divisibility in order to find missing digits that meet given criteria.

STRENGTHEN Focus on asking children to work carefully through questions ① to ③ in order to consolidate the rules of divisibility for 2, 3, 5 and 6.

DEEPEN Challenge children to explore and explain why an even multiple of 3 is also a multiple of 6.

ASSESSMENT CHECKPOINT Question ② will allow you to assess whether children are correctly applying the rule of divisibility to identify which numbers are multiples of 3, including very big numbers.

ANSWERS Answers for the **Practice** part of the lesson can be found in the *Power Maths* online subscription.

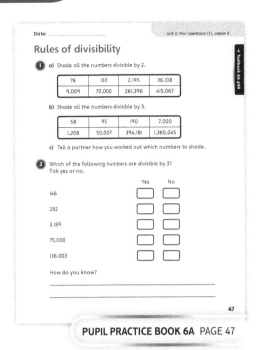

PUPIL PRACTICE BOOK 6A PAGE 47

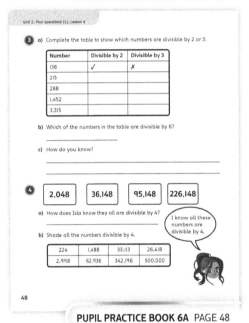

PUPIL PRACTICE BOOK 6A PAGE 48

Reflect

WAYS OF WORKING Pair work

IN FOCUS The **Reflect** part of the lesson prompts children to discuss the multiples of 2 and 3 and what their relationship is to numbers that are divisible by 6.

ASSESSMENT CHECKPOINT Assess whether children can apply reasoning about tests for divisibility by 2 and 3.

ANSWERS Answers for the **Reflect** part of the lesson can be found in the *Power Maths* online subscription.

PUPIL PRACTICE BOOK 6A PAGE 49

After the lesson

- How will you ensure children are getting regular practice in recognising multiples of 2, 3 and 5?
- What opportunities can you give children to rehearse recognising multiples of 2, 3 and 6?

Primes to 100

Learning focus

In this lesson, children will learn to recognise and identify prime numbers. They will explore how these numbers are different from other numbers.

Before you teach

- How will you ensure children are given practical opportunities to investigate prime numbers in this lesson?
- How will you ensure that children's learning from the previous two lessons is drawn on explicitly in this one?

NATIONAL CURRICULUM LINKS

Year 6 Number – addition, subtraction, multiplication and division

Identify common factors, common multiples and prime numbers.

ASSESSING MASTERY

Children can recognise and identify prime numbers, fluently explaining their unique properties. They can explain why 2 is the only even prime number and why 1 is not a prime number.

COMMON MISCONCEPTIONS

Children may assume that 1 is a prime number as it is only divisible by 1 and, therefore, itself. Ask:
- *How many factors does a prime number have?*
- *How many factors does 1 have?*

STRENGTHENING UNDERSTANDING

Arrays will be a powerful tool to help strengthen children's understanding in this lesson. Ask: *How many arrays can you make for this number? How many arrays are possible if the number is prime?*

GOING DEEPER

Children could be encouraged to write each number in question ❸ of the **Think together** section as a product of factors with at least one of the factors as a prime number.

KEY LANGUAGE

In lesson: prime, array, remainder, divide, factor, composite number, reasoning

Other language to be used by the teacher: multiple, multiply, multiplication, division

STRUCTURES AND REPRESENTATIONS

Arrays, tables

RESOURCES

Optional: counters, 100 square

 In the eTextbook of this lesson, you will find interactive links to a selection of teaching tools.

Quick recap 🔁

Ask children which number between 5 and 10 they think has the most factors. Challenge them to make a prediction and then to explore it.

Discover

PUPIL TEXTBOOK 6A PAGE 68

WAYS OF WORKING Pair work

ASK

- Question ❶ a): *How many factors does 16 have?*
- Question ❶ a): *What happens when one more counter is added? How many factors can you find for your new number?*
- Question ❶ b): *How many factors does 13 have? How is this number similar to 17?*
- Question ❶ b): *Can you find any other numbers that have similar properties?*

IN FOCUS The arrays in the picture show that 16 has several factors. By adding one more counter to the arrays, children should quickly spot that the only arrays without remainders are 1 × 17 and 17 × 1. Children should be encouraged to begin generalising about the numbers mentioned. Can they find any other numbers that have similar properties? Ensuring this is done practically will help secure children's understanding.

PRACTICAL TIPS Children should be given ample opportunities to create the arrays linked to the numbers they are investigating. The scenario shown in the picture can be recreated in the classroom practically using counters. Once these numbers have been explored, can children find other examples where adding one counter creates a number with fewer factors?

ANSWERS

Question ❶ a): Only two different arrays are possible using 17 counters: 1 row of 17 because 17 ÷ 1 = 17 and 17 rows of 1 because 17 ÷ 17 = 1. Isla cannot make more arrays using Aki's counter.

Question ❶ b): 13 and 19 are both prime numbers so you can only make two arrays for each.

Share

WAYS OF WORKING Whole class teacher led

ASK

- Question ❶ a): *How many ways did you try to make an array for 17? How many were successful?*
- Question ❶ b): *What was the same about all the prime numbers? What was different?*

IN FOCUS When looking at arrays for 17, 13 and 19, it is important to ensure that children understand the property that any prime number has exactly two factors: 1 and itself.

DEEPEN It may be that, at this point, children have only found odd numbered primes and so generalise that all primes must be odd. This should be used as an interesting learning point. Ask: *Are there any even primes?*

Primes to 100

Discover

1 a) Can Isla make more arrays if she uses 17 counters?

b) How many arrays can you make using 13 or 19 counters?

68

PUPIL TEXTBOOK 6A PAGE 68

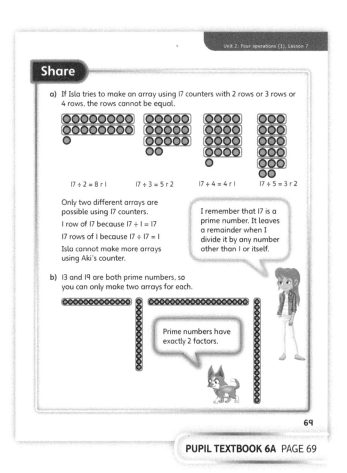

Share

a) If Isla tries to make an array using 17 counters with 2 rows or 3 rows or 4 rows, the rows cannot be equal.

17 ÷ 2 = 8 r 1   17 ÷ 3 = 5 r 2   17 ÷ 4 = 4 r 1   17 ÷ 5 = 3 r 2

Only two different arrays are possible using 17 counters.
1 row of 17 because 17 ÷ 1 = 17
17 rows of 1 because 17 ÷ 1 = 1
Isla cannot make more arrays using Aki's counter.

I remember that 17 is a prime number. It leaves a remainder when I divide it by any number other than 1 or itself.

b) 13 and 19 are both prime numbers, so you can only make two arrays for each.

Prime numbers have exactly 2 factors.

69

PUPIL TEXTBOOK 6A PAGE 69

Think together

WAYS OF WORKING Whole class teacher led (I do, We do, You do)

ASK

- Question **1**: *Does finding just the factors 1 and the number itself prove that a number is definitely prime?*
- Question **2**: *How could you check that a number is definitely prime?*
- Question **3**: *Are there any ways to quickly judge if a number is prime or not?*

IN FOCUS Question **1** challenges the assumption that if children can prove that a number has the factors 1 and itself, then it must be prime. Be sure to discuss the importance of gathering enough evidence to prove their ideas.

STRENGTHEN For question **3** b), children can be helped to more efficiently identify numbers that are not prime by ensuring they are aware that 2 is the only even prime number. Ask: *Are prime numbers more likely to be even or odd? Explain. How will this help you solve this question more efficiently?*

DEEPEN Question **3** b) provides a good opportunity to make generalisations about numbers with different properties to help children identify primes more efficiently. Children should be encouraged to spot patterns, for example multiples of 5 (except 5 itself) and multiples of 10 are never prime and are easy to identify. Ask: *Can you create some rules for identifying prime numbers?*

ASSESSMENT CHECKPOINT Children should be able to identify the properties of prime numbers and prove a number is prime using resources and arrays. They should be beginning to identify and explain generalisations to help find prime numbers more efficiently.

ANSWERS

Question **1**: Disagree. Mo has not proved that the numbers are definitely prime. While 11 is prime, 21 has the factors 1, 3, 7 and 21.

Question **2**: Alex has circled 39, which is not prime. She has missed 41.

Question **3** a): Bella's method will find out whether or not a number is prime. She can stop at 10, because 10 × 10 is 100 and 100 > 97. So when she gets to 10 she will have found any factor pairs, each of which must contain a number smaller than 10. She will not find any, because 97 is prime.

Question **3** b): Prime numbers: 71, 79

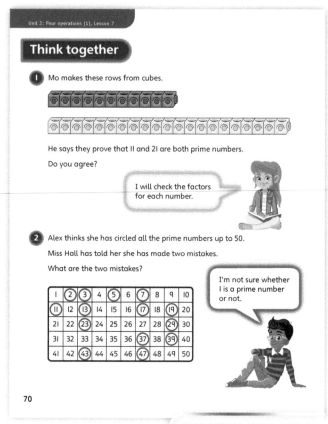

Think together

1. Mo makes these rows from cubes.

He says they prove that 11 and 21 are both prime numbers.
Do you agree?

I will check the factors for each number.

2. Alex thinks she has circled all the prime numbers up to 50.
Miss Hall has told her she has made two mistakes.
What are the two mistakes?

I'm not sure whether 1 is a prime number or not.

1	2	3	4	5	6	7	8	9	10
11	12	13	14	15	16	17	18	19	20
21	22	23	24	25	26	27	28	29	30
31	32	33	34	35	36	37	38	39	40
41	42	43	44	45	46	47	48	49	50

70

PUPIL TEXTBOOK 6A PAGE 70

3. Bella and Richard are discussing how to find out if 97 is a prime number.

CHALLENGE

I can check if a number is prime by dividing by 2, then 3, then 4, to see if there is always a remainder.

But how do you know when to stop checking?

Bella

Richard

a) Will Bella's method work?
When should she stop checking?

b) Look at these numbers. Which numbers are prime?

56   79
99
170   77   1,200   355
71   348   321

I can tell just by looking at some of these that they are not prime numbers.

Numbers that are not prime are called composite numbers.

71

→ Practice book 6A p50

PUPIL TEXTBOOK 6A PAGE 71

Practice

WAYS OF WORKING Independent thinking

IN FOCUS Throughout this part of the lesson, it is important for children to be able to build or draw the arrays to represent the numbers they are dealing with, where possible. Questions **1** and **2** encourage children to look for rules to identify which numbers are not prime, as well as finding which numbers are prime.

STRENGTHEN If children are struggling to find the prime numbers in question **4**, it may be beneficial to recap the generalisations made about the properties of numbers (for example, multiples of 10 are never prime) and display these somewhere prominent in the classroom.

DEEPEN Question **7** offers children an opportunity to share their reasoning. This activity could be deepened by asking children to investigate if there are any patterns of numbers (for example, 3, 13, 23, 33, 43, 53) where every number is prime. Ask: *Can you find a regular sequence of numbers that are all prime?* Question **5** requires children to identify prime numbers as factors of other numbers. This introduces the concept of prime factors.

ASSESSMENT CHECKPOINT Children should now be more confident with the properties of numbers and how these can help them identify whether a number is prime or not. They should be able to share their reasoning confidently, using the lesson vocabulary accurately.

ANSWERS Answers for the **Practice** part of the lesson can be found in the *Power Maths* online subscription.

Reflect

WAYS OF WORKING Independent thinking

IN FOCUS This question will demonstrate whether children are able to explain how to find out whether a number is prime or not. Children should be given time to formulate and write their proof, which they can then share and discuss with their partner. Ask: *Have you investigated the numbers in the same way? Is one method more efficient than the other? Explain how.*

ASSESSMENT CHECKPOINT Children should be using the concrete or pictorial representations of arrays, coupled with their number knowledge, to identify whether the two numbers are prime.

ANSWERS Answers for the **Reflect** part of the lesson can be found in the *Power Maths* online subscription.

After the lesson ⏸

- How was children's ability to generalise developed in this lesson?
- How could you continue to develop children's use of prime numbers in other areas of the curriculum? (For example, possible team groupings in PE – there are 17 people, what equal teams can you make?)

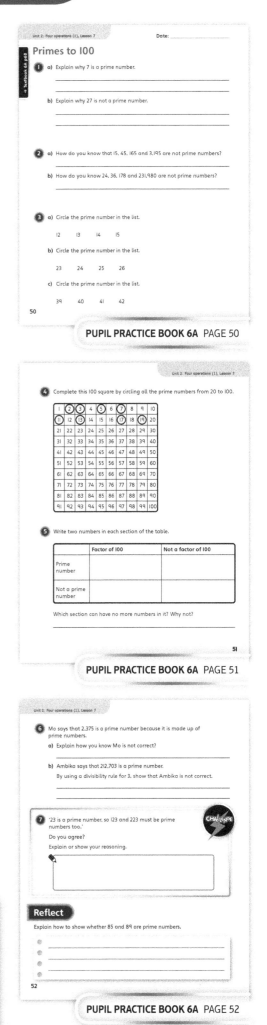

PUPIL PRACTICE BOOK 6A PAGE 50

PUPIL PRACTICE BOOK 6A PAGE 51

PUPIL PRACTICE BOOK 6A PAGE 52

Squares and cubes

Learning focus

In this lesson, children will learn to recognise and identify square and cube numbers. They will explore how these numbers are different from others.

Before you teach

- Are children confident with finding the area of a square and the volume of a cube?
- What resources will you provide to make this link to the properties of shapes explicit?

NATIONAL CURRICULUM LINKS

Year 5 Number – multiplication and division

Recognise and use square numbers and cube numbers, and the notation for squared (^{2}) and cubed (^{3}).

ASSESSING MASTERY

Children can recognise, identify and calculate square and cube numbers, fluently explaining their unique properties. They are able to recognise and use the mathematical notation for squared (^{2}) and cubed (^{3}) and can confidently explain their reasoning using their mathematical understanding.

COMMON MISCONCEPTIONS

Children may mistakenly assume that x^2 means x multiplied by 2, and similarly, x^3 means x multiplied by 3. Ensure children link the terminology of 'squared' and 'cubed' to the properties of their namesake shapes. Ask:
- *To find the area of a square with a side length of 4, would you calculate 4 × 2 or 4 × 4? Why?*
- *How does this relate to 4^2?*

STRENGTHENING UNDERSTANDING

Before beginning this lesson, children could be reminded of their work on area and volume. Give children variously-sized squares and cubes for them to measure and find the area and volume of.

GOING DEEPER

Children could be set 'Always, Sometimes, Never' statements to investigate. For example: *When you square an even number, the result is divisible by 4.*

KEY LANGUAGE

In lesson: square, cube, multiplication, multiply (×), array, prime

Other language to be used by the teacher: squared, cubed, multiplied

STRUCTURES AND REPRESENTATIONS

Array, multiplication grid, sorting circles

RESOURCES

Mandatory: counters, multilink cubes, 2D square, 3D cube

Optional: multiplication grids, 100 square

 In the eTextbook of this lesson, you will find interactive links to a selection of teaching tools.

Quick recap

Ask children to make an array using exactly 8 counters. Then ask them to make an array using exactly 9 counters. Ask: *Does everyone's array look the same?*

Discover

Squares and cubes

WAYS OF WORKING Pair work

ASK

- Question ① a): *Can you make a solid cube with 16 small cubes? What regular shape can you make?*
- Question ① b): *Could you make one layer then another?*
- Question ① b): *What other amounts make solid cubes?*

IN FOCUS Question ① a) requires children to understand the difference between square numbers and cube numbers, and to recognise the common misconception of mistaking one for the other. Children will explore and explain based on the arrangement of 2D square arrays and 3D cube representations. The focus of question ① b) is to explore the numerical value of different cube numbers.

PRACTICAL TIPS Children should be encouraged to follow Lee's line of enquiry practically in class. If children are given the opportunity to build the shape being described by Lee, they should be able to explain his mistake more easily.

ANSWERS

Question ① a): Lee is incorrect. He cannot make a large solid cube with all 16 cubes.

Question ① b): The largest solid cube Lee can make is a 2×2×2 cube using 8 small cubes.
Lee would need another 11 small cubes to make a 3×3×3 large solid cube (the next largest cube).

Discover

① a) Is Lee correct? Can he make a large solid cube using all 16 small cubes?

b) What is the largest solid cube Lee can make? How many more cubes would he need to make the next largest cube?

72

PUPIL TEXTBOOK 6A PAGE 72

Share

WAYS OF WORKING Whole class teacher led

ASK

- Question ① b): *What numbers of small cubes can you make a square with? How are they similar and how are they different?*
- Question ① b): *What numbers of small cubes can you make large solid cubes with? How are they similar and how are they different?*
- Question ① b): *Can you find any patterns in the square or cube numbers?*

IN FOCUS In this section, it is important to make explicit the link between the dimensions of the 2D and 3D shapes and finding square and cube numbers.

DEEPEN Give children other numbers to investigate using multilink cubes. Children could begin recording their findings in a systematic way. Their findings could be kept and displayed prominently in the classroom as a learning aid for the rest of the lesson. Make sure to list the numbers with a picture of the corresponding square or cube.

Share

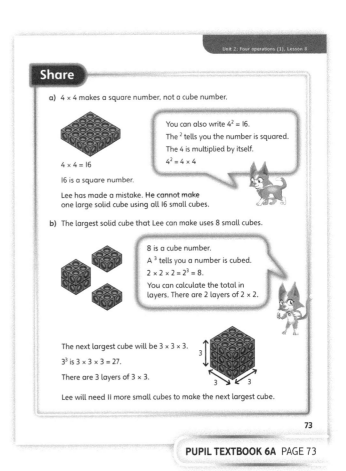

a) 4×4 makes a square number, not a cube number.

$4 \times 4 = 16$

16 is a square number.

Lee has made a mistake. He cannot make one large solid cube using all 16 small cubes.

You can also write $4^2 = 16$.
The ^{2} tells you the number is squared.
The 4 is multiplied by itself.
$4^2 = 4 \times 4$

b) The largest solid cube that Lee can make uses 8 small cubes.

8 is a cube number.
A ^{3} tells you a number is cubed.
$2 \times 2 \times 2 = 2^3 = 8$.
You can calculate the total in layers. There are 2 layers of 2×2.

The next largest cube will be $3 \times 3 \times 3$.
3^3 is $3 \times 3 \times 3 = 27$.
There are 3 layers of 3×3.

Lee will need 11 more small cubes to make the next largest cube.

73

PUPIL TEXTBOOK 6A PAGE 73

Think together

WAYS OF WORKING Whole class teacher led (I do, We do, You do)

ASK

- Question **1**: *How do you know if you need to square or cube a number?*
- Question **1**: *What do ^{2} ('squared') and ^{3} ('cubed') mean?*
- Question **2**: *What is different about x^2 and x multiplied by 2?*

IN FOCUS Question **2** approaches the misconception of reading the square sign as '× 2' and the cube sign as '× 3'. Children should be encouraged to show evidence that this is a misconception by building the shapes to match. When working on question **3**, it is important to discuss why a pattern for cube numbers is not evident on a multiplication grid. Ask: *Can you see any cube numbers on the multiplication grid? What is the same and what is different about square and cube numbers? Why are there so few cube numbers on the grid?*

STRENGTHEN For question **1**, if children are struggling to match the pictures to the calculations, provide them with the resources necessary to build their own versions. Ask: *What multiplication does this array represent? How do you know? Can you write the multiplication it is representing? Can you find it in the list?*

DEEPEN For question **3**, children could be given a 100 square to investigate whether cube numbers create any patterns on that type of grid. Ask: *Do you predict a pattern will be visible on this type of grid? Explain your prediction.*

ASSESSMENT CHECKPOINT At this point, children should be able to explain what the square and cube signs mean and what calculation they will need to solve when they meet them. Question **1** assesses whether they are able to link their understanding to the concrete representations of a square and a cube.

ANSWERS

Question **1** a): $5^2 = 25$

Question **1** b): $4^3 = 64$

Question **1** c): $10^2 = 100$

Question **1** d): $5^3 = 125$

Question **2**: Luis has misunderstood the square and cube signs. He has mistaken their meaning as '× 2' and '× 3'. The correct working is $4^3 = 4 × 4 × 4 = 64$ and $6^2 = 6 × 6 = 36$. Therefore 6^2 is not greater than 4^3.

Question **3**: The square numbers appear diagonally downwards from the top left (1–144).

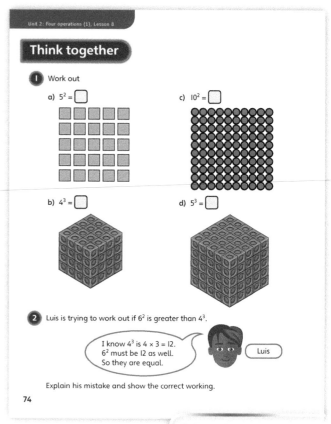

Practice

WAYS OF WORKING Independent thinking

IN FOCUS Question **1** links the pictorial representations with the abstract representations of squared and cubed numbers. If a class record of square and cube numbers has been placed in the classroom from the **Discover** section of the lesson, it would be beneficial to remind children of it for their independent activities.

STRENGTHEN If children are struggling with the pictorial and abstract representation in question **1**, they should be encouraged to make concrete versions of the numbers, for example, using multilink cubes. Likewise, if children are struggling to visualise the cube structure in question **4**, offer them a concrete version of the problem for them to manipulate. This will help them to explain their reasoning more clearly.

DEEPEN In question **6**, if children complete the sorting circles/diagram, they could be challenged to find all numbers that are both squares and cubes. An example of this is 64, which can be found by squaring 8 or cubing 4.

THINK DIFFERENTLY Question **5** challenges children's assumptions about how the square and cube signs represent numbers that increase alongside the numbers they follow. Children are likely to assume that $30^2 = 90$ is correct, forgetting that the square sign now represents '× 30' not '× 3' as in $3^2 = 9$.

ASSESSMENT CHECKPOINT Children should be able to fluently understand and interpret the square and cube signs, linking them to pictorial and concrete representations of numbers. Questions **1** and **2** assess how accurate children are when finding square and cube numbers. Question **3** assesses whether children can find the result of squaring and cubing numbers and also the number that would need to be squared or cubed to find a particular result.

ANSWERS Answers for the **Practice** part of the lesson can be found in the *Power Maths* online subscription.

Reflect

WAYS OF WORKING Pair work

IN FOCUS This question assesses whether children understand the two concepts well enough to apply them to previous mathematical knowledge. Children should work with their partner to explain how the mathematical notation has been misused or misinterpreted and devise advice to give Danny. The question highlights three misconceptions; if children are unable to explain where these are, it may indicate that they are liable to make the same mistakes.

ASSESSMENT CHECKPOINT Children should be able to confidently diagnose the misconception and explain where the student has gone wrong. Using their knowledge and understanding of the lesson's concepts, they should be able to give advice on how to correctly solve the problem.

ANSWERS Answers for the **Reflect** part of the lesson can be found in the *Power Maths* online subscription.

After the lesson

- Were children equally confident with both mathematical concepts?
- If they were weaker in one than the other, how will you support their understanding moving forward?
- How can these concepts be brought into other areas of the curriculum?

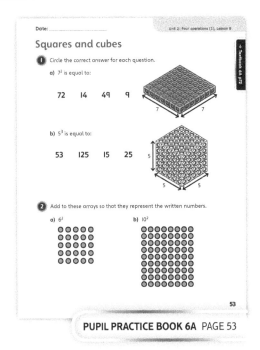

PUPIL PRACTICE BOOK 6A PAGE 53

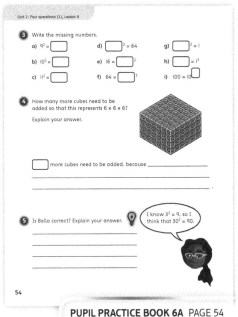

PUPIL PRACTICE BOOK 6A PAGE 54

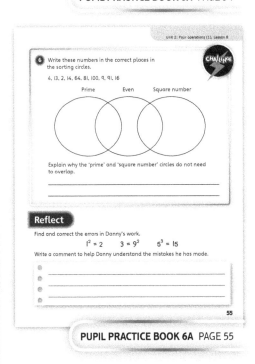

PUPIL PRACTICE BOOK 6A PAGE 55

End of unit check

> **Don't forget the unit assessment grid in your *Power Maths* online subscription.**

WAYS OF WORKING Group work teacher led

WAYS OF WORKING Group work teacher led

IN FOCUS

- Question ❶ assesses children's ability to apply the correct use of the columnar method of written addition.
- Question ❷ assesses children's ability to interpret a subtraction calculation and to use the columnar written method.
- Question ❸ assesses children's understanding of square numbers and how they can be represented.
- Question ❹ assesses children's understanding of and ability to recognise multiples, factors, prime numbers and square numbers.
- Question ❺ assesses children's understanding of cube numbers and how they can be represented.
- Question ❻ assesses children's ability to recognise and find common factors of given numbers.
- Check that children have fluency when working with number facts and using them to check calculations or solve a missing number problem.

ANSWERS AND COMMENTARY

Children will be able to efficiently and accurately solve addition and subtraction problems using a columnar method. Children who have mastered the concepts in this unit will recognise what a common factor is. They will be able to identify how prime numbers differ from other numbers. They can confidently cube a number, recognising how square and cube numbers can be represented pictorially.

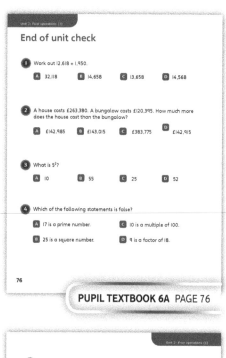

PUPIL TEXTBOOK 6A PAGE 76

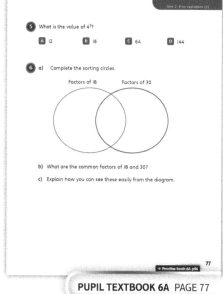

PUPIL TEXTBOOK 6A PAGE 77

Q	A	WRONG ANSWERS AND MISCONCEPTIONS	STRENGTHENING UNDERSTANDING
1	D	A suggests children have made the place value error of misaligning the digits in the two numbers.	For squaring and cubing numbers, ask: *If you were going to find the area or volume of this shape, what calculation would you use?*
2	A	C suggests children have added instead of subtracting.	Build confidence with factors by asking children to explore the following rules and explain their findings clearly:
3	C	A suggests children have mistaken ^{2} for × 2.	– Prime numbers have exactly two factors.
4	C	D may suggest that children do not understand the difference between a factor and a multiple.	– A square number has an odd number of factors, including one factor that is multiplied by itself.
5	C	B suggests children are confusing squared and cubed.	
6		a) Left circle: 9, 18; right circle: 5, 10, 15, 30; middle: 1, 2, 3, 6	
		b) The common factors of 18 and 30 are 1, 2, 3, 6	
		c) They are all in the middle section.	

My journal

WAYS OF WORKING Independent thinking

ANSWERS AND COMMENTARY

Encourage children to use digit cards, or to make their own by writing the digits on small scraps of paper or sticky notes.

They should try out different combinations and explore the effects, then apply some reasoning to describe and justify any properties they are looking for.

Possible answers include:

a square number: 4, 16, 25, 36, 64

a cubed number: 64

a 3-digit prime number: 131, 251, 563

a 3-digit number that has 3 and 5 as common factors: 345, 435, 315, 135, 465, 645

Power check

WAYS OF WORKING Independent thinking

ASK

- How confident are you that you could identify common factors and common multiples of two or more numbers?
- Do you understand what square numbers and cube numbers are? Can you describe how they might relate to area and volume?
- Do you think you could solve any given word problem? Explain how you know.

Power puzzle

WAYS OF WORKING Independent thinking

IN FOCUS Use this **Power puzzle** to assess children's ability to accurately subtract a 3-digit number from a 4-digit number. To deepen children's investigation in this **Power puzzle**, ask:
- What is the longest chain you can make? What is the shortest?
- Is there any way of making sure you end up with a long chain of calculations?

ANSWERS AND COMMENTARY Children should find that, whatever numbers they begin with, they eventually find themselves 'stuck', constantly using and reusing the digits 6, 1, 4, 7. This chain gives 7,173 6,354 3,087 8,352 **6,174 6,174**. If children do not find this it suggests they need more practice with subtraction.

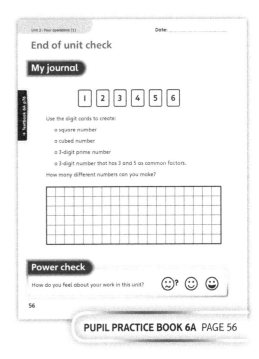

PUPIL PRACTICE BOOK 6A PAGE 56

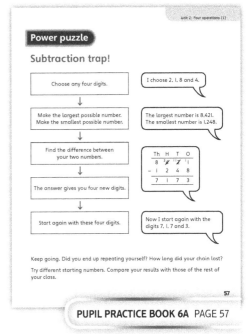

PUPIL PRACTICE BOOK 6A PAGE 57

After the unit ⏸

- Were children more confident with one method than with others? How could you tell?
- If this was the case, how will you continue to offer support in the future to secure their understanding of the other written methods?

Strengthen and **Deepen** activities for this unit can be found in the *Power Maths* online subscription.

Unit 3
Four operations ②

WHY THIS UNIT IS IMPORTANT

This unit develops children's understanding of how the four operations can be used to manipulate numbers and solve problems. Children will deepen their understanding of the columnar method for multiplication of 4-digit numbers by 1- and 2-digit numbers and develop an understanding of written methods for division. Children will make links to methods they have met before and apply new learning to contextual word problems.

After this, children learn about the order of operations, investigating its effect on calculations and considering why it is important to have an agreed order. They then learn how brackets can affect the order of operations. Using these concepts, they complete calculations, solve problems and diagnose mistakes in calculations.

Finally, children learn methods to solve mental calculations with small and large numbers. They consider where mental methods are appropriate and where written methods are appropriate. They also use number facts they already know to solve problems involving related number facts.

WHERE THIS UNIT FITS

→ Unit 2: Four operations (1)
→ **Unit 3: Four operations (2)**
→ Unit 4: Fractions (1)

This unit builds on children's knowledge of using formal columnar written methods for multiplication and division, including with real-life contexts. They learn about the order of operations and mental methods, before moving on to work with fractions in Unit 4.

Before they start this unit, it is expected that children:
- are fluent in their multiplication tables
- understand the terms, and are able to find, factors and multiples
- understand and can use the four operations
- are able to interpret and solve word problems.

ASSESSING MASTERY

Children will efficiently and accurately solve multiplication of a 4-digit number by a 1- or 2-digit number using a columnar method. They will be able to use multiple methods to solve division calculations, including short and long division, explaining how their understanding of multiples and factors can help them. Children will be able to fluently adhere to the correct order of operations, demonstrating and explaining how brackets can affect this. Finally, they will be able to solve mathematical problems, using efficient mental methods and explaining where written methods are more appropriate.

COMMON MISCONCEPTIONS	STRENGTHENING UNDERSTANDING	GOING DEEPER
Children may multiply incorrectly when using short multiplication, for example, attempting to solve 2,345 × 6 by calculating '5 × 6', then '45 × 6', then '345 × 6' and finally '2,345 × 6'.	Show the numbers in a place value grid and ask what number each digit represents.	Children could write their own story problems for each of the four different operations.
Children may muddle the order of operations or neglect to remember how brackets influence how to solve a calculation.	To help children remember the order of operations, they could be encouraged to create a class rhyme or song.	Give children 4–5 random 1-digit numbers. Can they use their numbers and understanding of the order of operations and brackets to create another given number? If not, how close can they get?

UNIT STARTER PAGES

Use these pages to introduce the focus to children. You can use the characters to explore different ways of working too!

STRUCTURES AND REPRESENTATIONS

Array: Arrays are a visual representation of multiplication and division. They are an excellent tool for showing equal groups within a number.

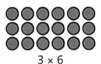

3 × 6

Bar model: Bar models enable children to more easily represent a problem. In the context of this unit, they are used to show different types of calculations.

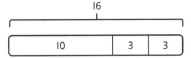

Number line: A number line is a more abstract representation of a sequence of numbers. It is used in this unit to represent different calculations or lists of multiples.

Part-whole model: Part-whole models help to clearly show the different ways a number can be partitioned.

KEY LANGUAGE

There is some key language that children will need to know as a part of the learning in this unit.

→ method, column, columnar

→ multiply, multiplication, product, approximation

→ divide, division, short division, long division

→ factor, multiple, divisor, dividend, remainder

→ order of operations, brackets

→ inverse operation

PUPIL TEXTBOOK 6A PAGE 78

PUPIL TEXTBOOK 6A PAGE 79

Multiply by a 1-digit number

Learning focus

In this lesson, children will develop their understanding of the multiplication of 4-digit numbers by 1-digit numbers. They will use multiple representations and methods to solve these calculations.

Before you teach

- What methods for solving multiplication calculations are children fluent with already?
- How will you use children's prior learning to engage them and secure progress?

NATIONAL CURRICULUM LINKS

Year 6 Number – addition, subtraction, multiplication and division

Multiply multi-digit numbers up to 4 digits by a 2-digit whole number using the formal written method of long multiplication.

ASSESSING MASTERY

Children can fluently and reliably multiply a 4-digit number by a 1-digit number. They can demonstrate their thinking and understanding through multiple representations and written methods.

COMMON MISCONCEPTIONS

Children may multiply incorrectly when using short multiplication, for example, attempting to solve 2,345 × 6 by calculating '5 × 6', then '45 × 6', then '345 × 6' and finally '2,345 × 6'. Ask:
- *What do you have if you recombine 5, 45, 345, and 2,345? Is that different from the original calculation?*
- *What number does the digit '4' represent? How do you know?*

STRENGTHENING UNDERSTANDING

Give children opportunities to recap written methods from previous units and years. Show a multiplication and ask: *What is this calculation asking you to do? Can you represent it using resources or with a picture? What written method would you use to solve it? Show me.*

GOING DEEPER

Encourage children to write word problems for each other that require them to multiply 4-digit numbers by 1-digit numbers.

KEY LANGUAGE

In lesson: multiplication, total, digit, estimate, rounding

Other language to be used by the teacher: repeated addition, multiply, times, product

STRUCTURES AND REPRESENTATIONS

Short multiplication, grid method, place value counters

RESOURCES

Optional: place value counters

 In the eTextbook of this lesson, you will find interactive links to a selection of teaching tools.

Quick recap

Ask children to practise chanting each of the times-tables, up to 9 × 9.

Discover

Multiply by a 1-digit number

WAYS OF WORKING Pair work

ASK

- Question ① a): *What number facts could you use to help you estimate the solution to this problem?*
- Question ① b): *How will you go about solving this calculation? What operation could be used?*
- Question ① b): *Can you represent the problem with resources?*

IN FOCUS Question ① a) gives children the opportunity to begin investigating how to solve problems that require multiplication. Be aware that neither of the given totals is correct, which may baffle children; encourage them to think about which estimate is more sensible.

Question ① b) gives children their first opportunity to calculate a 4-digit number multiplied by a 1-digit number.

PRACTICAL TIPS Children could role-play a travel agency scenario. Give a list of holiday destinations with prices. Children, working in small groups, investigate how much it would cost their group to go on each holiday. What would the price be if two groups went? (Be careful not to let the number of children go over 9.)

ANSWERS

Question ① a): Neither of the totals is correct, but rounding shows that £12,905 is more likely to be correct.

Question ① b): £3,225 × 4 = £12,900. The trip will cost £12,900 for four people.

Discover

① a) Without calculating, how can you tell which total is more likely to be correct, £128,820 or £12,905?

b) How much will the trip actually cost for four people?

80

PUPIL TEXTBOOK 6A PAGE 80

Share

WAYS OF WORKING Whole class teacher led

ASK

- Question ① a): *How have Astrid and Dexter shown that both totals are wrong?*
- Question ① a): *How did you prove they are wrong? Can you make a model or draw a picture to show this?*
- Question ① b): *Can you explain the two different methods shown by the place value counters?*
- Question ① b): *Did you use a written method? Which one?*
- Question ① b): *Can you explain how your method works?*

IN FOCUS When looking at question ① a), discuss why it is important to use the mental methods and number facts demonstrated in Dexter and Astrid's comments when solving any calculation, as they can help children determine whether their solutions are correct or not. Use the pictures of place value counters in question ① b) to develop children's fluency with number and to build their awareness of the commutative nature of multiplication.

Also use question ① b) to show that multiplication is far more efficient than repeated addition.

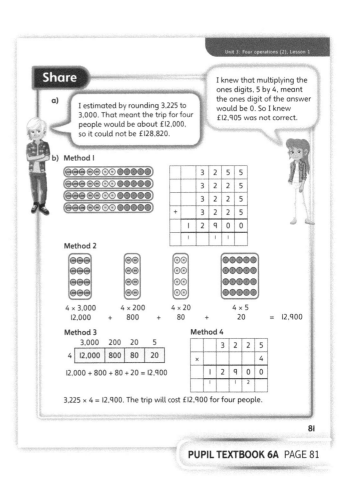

PUPIL TEXTBOOK 6A PAGE 81

Think together

Whole class teacher led (I do, We do, You do)

ASK

- Question ❶: *Can you explain how the place value grid demonstrates the multiplication?*
- Question ❷: *How can you use the methods shown to help you find the solution?*
- Question ❸: *What will you look for to know which digits go where in the calculation?*

IN FOCUS Question ❶ demonstrates the use of place value counters on a grid to represent multiplication with a 4-digit number. Question ❷ requires children to think about two different multiplication methods and to consider which is more efficient.

STRENGTHEN If children are struggling with the methods shown in questions ❶ and ❷, ask: *Look back at the models in **Share**. Can they help you? How will you use them to help you solve these calculations?*

DEEPEN Give children 5 digit cards with the numbers 2, 3, 4, 5 and 6. Challenge them to create a set of multiplications that:

- have an odd answer
- have an answer with '0' in the ones place
- have an answer which has '8' in the ones place and is between 15,000 and 20,000.

Can children find 3 different solutions to each challenge?

ASSESSMENT CHECKPOINT Do children recognise different multiplication methods, and can they make links with previous written methods? Can children use written multiplication methods with fluency to solve problems?

ANSWERS

Question ❶ a): Each row shows 2,345. There are four rows.

Question ❶ b): 9,380

Question ❶ c): Children will use their own methods.

Question ❷ a): 1,718 × 3 = 5,154

Question ❷ b): 2,536 × 7 = 17,752

Question ❸ a):

TTh	Th	H	T	O
	1	4	5	3
×				2
	2	9	0	6
			1	

Question ❸ b): A 4- or 5-digit answer can be made. For example:

TTh	Th	H	T	O
	3	2	1	4
×				5
1	6	0	7	0
	1	1	2	

Question ❸ c): The greatest answer you can make is:

TTh	Th	H	T	O
	4	3	2	1
×				5
2	1	6	0	5
	2	1	1	

It is not possible to make a 6-digit answer with the digit cards available.

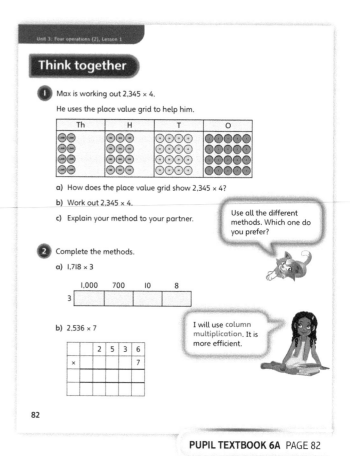

Think together

❶ Max is working out 2,345 × 4.

He uses the place value grid to help him.

Th	H	T	O

a) How does the place value grid show 2,345 × 4?

b) Work out 2,345 × 4.

c) Explain your method to your partner.

Use all the different methods. Which one do you prefer?

❷ Complete the methods.

a) 1,718 × 3

	1,000	700	10	8
3				

b) 2,536 × 7

		2	5	3	6
×					7

I will use column multiplication. It is more efficient.

82

PUPIL TEXTBOOK 6A PAGE 82

---

❸ a) Use each card once to complete the multiplication.

CHALLENGE

☐ ☐ ☐ ☐
× ☐
‾‾‾‾‾‾‾‾
2 9 0 6
1

| 1 | 2 | 3 |
| 4 | 5 | |

I think there is more than one possible answer.

b) Rearrange the cards and use each card once to make a new calculation with a different answer.

☐ ☐ ☐ ☐
× ☐
‾‾‾‾‾‾‾‾

| 1 | 2 | 3 |
| 4 | 5 | |

c) Can you make a calculation with a 5-digit answer?
Can you make a calculation with a 6-digit answer?

I am going to try to find the largest answer.

83

→ Practice book 6A p58

PUPIL TEXTBOOK 6A PAGE 83

Practice

WAYS OF WORKING Independent thinking

IN FOCUS Questions ❶ and ❸ give children the opportunity to practise the short multiplication method. Children should independently link their understanding of the grid method to short multiplication. Question ❷ links children's understanding of multiplication along a number line with the short multiplication written method.

STRENGTHEN It may help children's understanding in question ❹ if they are able to use resources to create the questions and the calculations. Ask: *How could you use resources to represent the problems? How could you use resources or draw pictures to represent the calculations you need to solve?*

DEEPEN When working on question ❺, deepen children's generalisations about the calculations by asking: *What is the same/different about the calculations that make the greatest and smallest product? Is there a way to be sure of making the greatest or smallest product, no matter what digits you use?*

ASSESSMENT CHECKPOINT Can children recognise and interpret different representations of multiplication and link these with the short multiplication written method? Can children solve multiplications using short multiplication with fluency?

ANSWERS Answers for the **Practice** part of the lesson can be found in the *Power Maths* online subscription.

Reflect

WAYS OF WORKING Independent thinking

IN FOCUS This question will give you the opportunity to assess whether children have been able to identify links between a multiplication method they have learnt before and the method they are currently using. These links will scaffold and cement their understanding of short multiplication so it is important they see and understand them.

ASSESSMENT CHECKPOINT Look for children recognising that short multiplication, like the grid method, uses partitioning when multiplying a number by another number, even though this is less visually obvious than in the grid method.

ANSWERS Answers for the **Reflect** part of the lesson can be found in the *Power Maths* online subscription.

After the lesson ⏸

- How clearly did your lesson make the links between methods of multiplication?
- What would you do differently next time you teach this, to cement those links even more effectively?
- What percentage of your class is confident and fluent in the use of short multiplication?

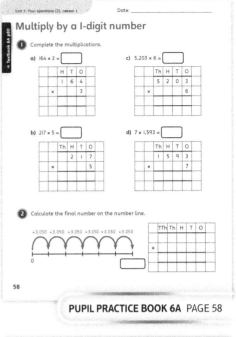

PUPIL PRACTICE BOOK 6A PAGE 58

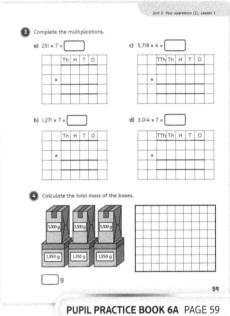

PUPIL PRACTICE BOOK 6A PAGE 59

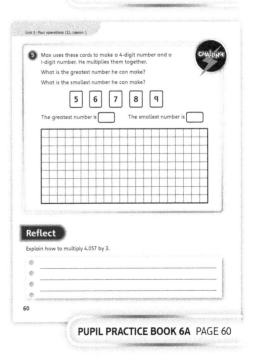

PUPIL PRACTICE BOOK 6A PAGE 60

Multiply up to a 4-digit number by a 2-digit number

Learning focus

In this lesson, children will develop their understanding of the multiplication of numbers with up to four digits by 2-digit numbers. They will use multiple representations and methods to solve these calculations.

Before you teach

- How confident are children with multiplying by a 1-digit number?
- Were there any misconceptions in the previous lesson that could impact on this lesson? How will you plan for these?

NATIONAL CURRICULUM LINKS

Year 6 Number – addition, subtraction, multiplication and division

Multiply multi-digit numbers up to four digits by a 2-digit whole number using the formal written method of long multiplication.

ASSESSING MASTERY

Children can fluently and reliably multiply a number with up to four digits by a 2-digit number. They can demonstrate their thinking and understanding through multiple representations and written methods.

COMMON MISCONCEPTIONS

Children may multiply incorrectly when using the column method for multiplication, for example, calculating 345 × 26 as 5 × 6, 40 × 6, 300 × 6, followed by 5 × 2, 40 × 2 and 300 × 2. Ask:
- *What is the place value of each digit in both of the numbers in the calculation?*
- *Is the value of the digit 2 two or twenty? How do you know?*

STRENGTHENING UNDERSTANDING

If children are not confident in using short multiplication to multiply by a 1-digit number, offer opportunities to find the product of a 4-digit and a 1-digit number, using the grid method to support conceptual understanding. Ask: *How do the grid method and short multiplication show the multiplication of the 1s × 1s? How do they both show the multiplication of the 10s × 1s? How are the methods the same and different? How do the similarities help you to use short multiplication?*

GOING DEEPER

Children could write a short multiplication, showing the product of a 4-digit and a 1-digit number. Children could then challenge their partner by hiding one or two of the digits in the calculation. Ask: *How can you make the challenge trickier or easier? Explain how it is trickier/easier now.*

KEY LANGUAGE

In lesson: column multiplication, long multiplication, area, grid, digit

Other language to be used by the teacher: repeated addition, times, multiply, product, calculate, double

STRUCTURES AND REPRESENTATIONS

Grid method, long multiplication, short multiplication

RESOURCES

Optional: place value grids and counters, place value equipment

 In the eTextbook of this lesson, you will find interactive links to a selection of teaching tools.

Quick recap

Ask children to complete the following multiplication facts:

10 × 10 =   10 × 11 =   10 × 12 =

11 × 11 =   11 × 12 =

12 × 12 =

Discuss whether they could use any of these answers to work out the answer to 13 × 11.

Discover

Multiply up to a 4-digit number by a 2-digit number

Discover

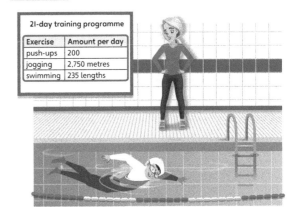

WAYS OF WORKING Pair work

ASK

- Question ① a): *What do you notice about the numbers in the training programme?*
- Question ① a): *What other calculations can you think of to do with the numbers in this context?*

IN FOCUS This section gives children the opportunity to recognise the use of multiplication calculations with bigger numbers in a real-life context. Discuss the two methods mentioned in Questions ① a) and ① b). Do children recall how to lay out the grid method? Do they recall how to line up the columns in the column method, and how to use both methods?

PRACTICAL TIPS Ask children to represent the number of lengths, 235, using place value equipment. Discuss the place value of each digit and how this will help with each of the methods given.

ANSWERS

Question ① a): 4,935 (using the grid method)

Question ① b): 4,935 (using column multiplication)

21-day training programme	
Exercise	Amount per day
push-ups	200
jogging	2,750 metres
swimming	235 lengths

① a) How many lengths do the athletes swim altogether in the 21 days? Use an area/grid method to work this out.

b) Use column multiplication to work out the same calculation.

84

PUPIL TEXTBOOK 6A PAGE 84

Share

WAYS OF WORKING Whole class teacher led

ASK

- Question ① a): *Can you explain how each part of this diagram is related to a different part of the calculation?*
- Question ① b): *How is this method related to the method in question ① a)?*
- Question ① b): *What mistakes could someone make in this calculation?*

IN FOCUS Question ① a) models the use of the area model, also called the grid method, to multiply a 3-digit number by a 2-digit number. Question ① b) demonstrates the written column multiplication method which is used to solve the same multiplication calculation. Encourage children to discuss which method they found quicker or easier, and to explain why this is.

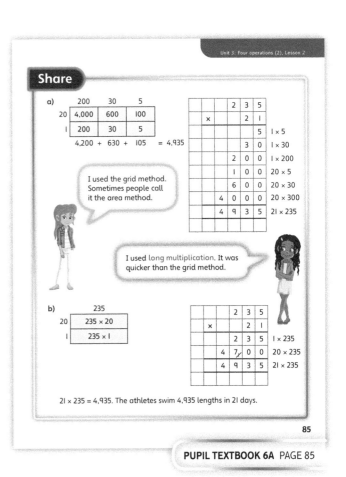

PUPIL TEXTBOOK 6A PAGE 85

Think together

WAYS OF WORKING **WAYS OF WORKING** Whole class teacher led (I do, We do, You do)

ASK

- Question **1**: *Can you see how the number 418 has been partitioned? How does this help you to work out the answer?*
- Question **2**: *Where is the best place to start?*
- Question **3**: *What is the same and what is different in each method?*

IN FOCUS Question **1** gives children the opportunity to use the grid method or area model to solve a multiplication. In question **2**, children are required to use the written column method. For question **3**, children should demonstrate a flexible approach to calculations, exploring the use of several different multiplication methods for one given calculation, and then deciding which method they prefer.

STRENGTHEN Build children's confidence by first asking them to work with these two methods for multiplications of a 2-digit number by another 2-digit number.

DEEPEN Challenge children to explore the use of various methods to solve other calculations, as shown in question **3**.

ASSESSMENT CHECKPOINT Use question **2** to assess whether children can correctly lay out and use the written column method to multiply 2- and 3-digit numbers by a 2-digit number.

ANSWERS

Question **1**:

	400	10	8
10	4,000	100	80
6	2,400	60	48

418 × 16 = 4,000 + 2,400 + 100 + 60 + 80 + 48
= 6,688

Question **2** a):

		7	3	2
×			3	7
	5	1₂	2₁	4
2	1	9	6	0
2	7	0	8	4
	✗			

Question **2** b):

	2	1	0	8
×			2	4
	8	4	3₃	2
4	2	1₁	6	0
5	0	5	9	2
✗				

Question **3**: All of Olivia's methods produce the same answer: 130,000.

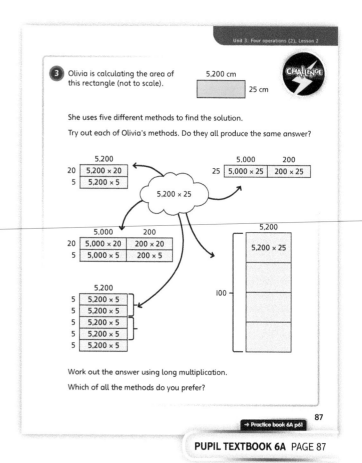

Think together

1 Use the grid method to work out 418 × 16.

	400	10	8
10			
6			

2 Work out the multiplications.

a) 732 × 37

b) 2,108 × 24

I will check each answer by estimating what it should be close to.

86

PUPIL TEXTBOOK 6A PAGE 86

3 Olivia is calculating the area of this rectangle (not to scale).

5,200 cm
25 cm

CHALLENGE

She uses five different methods to find the solution.

Try out each of Olivia's methods. Do they all produce the same answer?

Work out the answer using long multiplication.

Which of all the methods do you prefer?

87

→ Practice book 6A p61

PUPIL TEXTBOOK 6A PAGE 87

Practice

WAYS OF WORKING Independent thinking

IN FOCUS In question ❶, children use the grid method or area model to solve multiplications involving multiplying 3-digit numbers by 2-digit numbers. In question ❷, they use the column method for multiplying 3- and 4-digit numbers by 2-digit numbers. In question ❸, children select their own method. Ask them to explain why they have chosen their preferred method.

Question ❹ is a problem-solving question, in the context of capacity. In question ❺ children solve a problem finding missing digits.

STRENGTHEN Children can first try solving the calculations in question ❷ using the grid method to support their conceptual understanding.

DEEPEN Children can explore question ❻ in more depth. Ask: *How could you represent this differently to make it clearer? Could this be solved through trial and improvement? Is there a more efficient way? Why?*

THINK DIFFERENTLY Question ❹ requires children to apply their understanding of multiplication to a problem-solving situation. Through rounding and estimating, they should recognise that there is not enough water to fill the pool. They should also be encouraged to use a written method to find out exactly how much water is in the pool. Ask: *How many more buckets are needed? Can you explain how you know it must be more than 100 extra buckets?*

ASSESSMENT CHECKPOINT Do children make links with previous understanding to fluently and accurately use column multiplication? Are they flexible and fluent in their use of partitioning, when using the grid method, to make multiplications simpler?

ANSWERS Answers for the **Practice** part of the lesson can be found in the *Power Maths* online subscription.

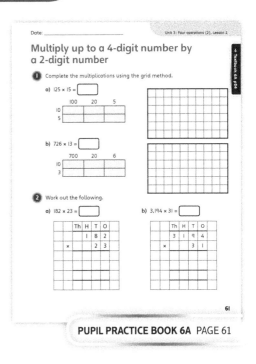

PUPIL PRACTICE BOOK 6A PAGE 61

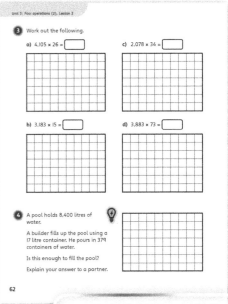

PUPIL PRACTICE BOOK 6A PAGE 62

Reflect

WAYS OF WORKING Independent thinking

IN FOCUS Children may solve this using any of the methods from this lesson. To assess their ability to solve long multiplications, ask: *Can you represent this with two different methods, one column multiplication and one of your choice? Explain how the two methods are linked.*

ASSESSMENT CHECKPOINT Look for children accurately explaining which product is greater, justifying their solution. Children should find that column multiplication is the most efficient method.

ANSWERS Answers for the **Reflect** part of the lesson can be found in the *Power Maths* online subscription.

After the lesson ⏸

- Could children fluently and accurately explain the links between the methods shown in this lesson?
- How will you continue to develop the links in children's learning in future lessons?

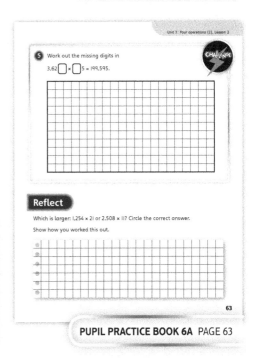

PUPIL PRACTICE BOOK 6A PAGE 63

Short division

Learning focus

In this lesson, children learn to use a written short division method.

Before you teach

- How can you use children's understanding of multiplication to deepen their understanding of division?
- How confident are children with division already?
- What misconceptions will you need to plan for?

NATIONAL CURRICULUM LINKS

Year 6 Number – addition, subtraction, multiplication and division

Divide numbers up to four digits by a 2-digit number using the formal written method of short division where appropriate, interpreting remainders according to the context.

ASSESSING MASTERY

Children can use the written method of short division to solve division calculations, including those with remainders.

COMMON MISCONCEPTIONS

Children may have inadequate conceptual understanding of short division, even if they do find the correct answer. For example, thinking that $132 \div 6$ is solved as '6 into 1 goes 0, 6 into 13 goes 2, 6 into 12 goes 2' without an awareness of the place values. Ask:
- *Why is your first digit 0 when there are lots of 6s in 100?*

You could explain that the first digit is the hundreds column, so it represents how many hundreds of 6s there are in 132. Since there are fewer than 100, the first digit is 0.

STRENGTHENING UNDERSTANDING

Encourage children to use place value equipment to explore the place value of the digits in each division calculation.

GOING DEEPER

Encourage children to write real-life division word problems to challenge their partner. Ask: *Can you write a problem that has more than one step?*

KEY LANGUAGE

In lesson: divide, share, groups of, remainder

STRUCTURES AND REPRESENTATIONS

Place value grid, short division

RESOURCES

Mandatory: place value equipment

 In the eTextbook of this lesson, you will find interactive links to a selection of teaching tools.

Quick recap

Ask children to complete the following missing number calculations:

$36 \div ? = 9$	$28 \div ? = 4$	$54 \div ? = 6$
$? \div 6 = 7$	$? \div 8 = 6$	$? \div 9 = 5$

Discover

WAYS OF WORKING Pair work

ASK

- Question ① a): *What operation do you need to solve this question? Explain how you know.*
- Question ① a): *How can you represent your thinking and solution? Can you use more than one method?*
- Question ① b): *How is this question similar to question ① a)? How is it different?*
- Question ① b): *How will you represent your working for this question? How is it similar to your working in question ① a)? How is it different?*

IN FOCUS In question ① a), children divide a 3-digit number by a 1-digit number. Encourage them to show their working using short division.

Question ① b) develops children's thinking by extending the calculation to a 3-digit number divided by a 2-digit number. Discuss the ways in which their approach is the same as, and is different from, the one used in question ① a).

PRACTICAL TIPS Children can use place value equipment to model each of the divisions, including any exchanges that are required.

ANSWERS

Question ① a): 22 days

Question ① b): 23 tubes

Short division

Discover

We have 276 tubes of fruit purée to last 12 weeks.

We each need to drink 6 bottles of water a day.

① a) Each astronaut has 132 bottles of water for their stay in the space station. How many days will this last for one astronaut?

b) The astronauts eat the same amount of fruit purée each week. How many tubes of fruit purée will they eat each week?

88

PUPIL TEXTBOOK 6A PAGE 88

Share

WAYS OF WORKING Whole class teacher led

ASK

- Question ① a): *Where has 132 been written in the short division calculation? Where has 6 been written? What is the first step?*
- Question ① a): *Can you see the exchange of 1 hundred?*
- Question ① a): *Which method do you prefer to use: the written method or the place value equipment? Why?*

IN FOCUS When introducing the short division method, ensure children understand what they are actually calculating. This will help them to realise how and why place value fits into the written method and will help secure their conceptual understanding.

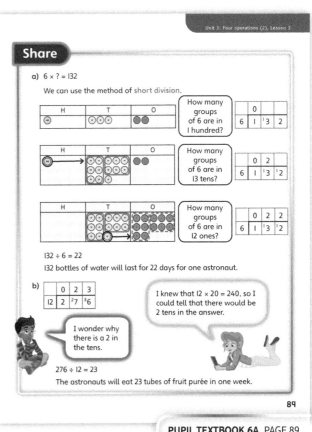

Share

a) $6 \times ? = 132$

We can use the method of short division.

$132 \div 6 = 22$

132 bottles of water will last for 22 days for one astronaut.

b) I knew that $12 \times 20 = 240$, so I could tell that there would be 2 tens in the answer.

I wonder why there is a 2 in the tens.

$276 \div 12 = 23$

The astronauts will eat 23 tubes of fruit purée in one week.

89

PUPIL TEXTBOOK 6A PAGE 89

Think together

WAYS OF WORKING Whole class teacher led (I do, We do, You do)

ASK

- Question ❶: *What is the first calculation you need to do? Where do you write the remainder for each step?*
- Question ❷: *How will you decide the calculations for each step?*
- Question ❸: *What numbers give a remainder when you divide by 5?*

IN FOCUS In questions ❶ and ❷, children use the written method of short division. Question ❸ presents children with division problems that will result in remainders.

STRENGTHEN Ask children to use place value equipment to model each of the divisions and any exchanges that are required.

DEEPEN Challenge children to investigate the following problems about remainders: *How can you tell if a number will leave a remainder of 1 when you divide by 2? How can you tell if a number will leave a remainder of 1 when you divide by 5? How can you tell if a number will leave a remainder of 1 when you divide by 4?*

ASSESSMENT CHECKPOINT Question ❷ can be used to assess whether children can neatly and accurately use the written method of short division, demonstrating an understanding of place value.

ANSWERS

Question ❶: 67 days

Question ❷ a): 238

Question ❷ b): 1,073

Question ❷ c): 246

Question ❷ d): 466

Question ❸ a): For the first division, the number must end in 0 or 5 to not have a remainder. For the second division, the last two digits must divide by 4 to not have a remainder.

Question ❸ b): 159 r 2 and 445 r 3

Question ❸ c): 789

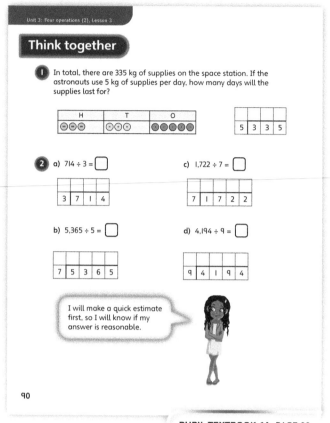

PUPIL TEXTBOOK 6A PAGE 90

PUPIL TEXTBOOK 6A PAGE 91

Practice

WAYS OF WORKING Independent thinking

IN FOCUS In question **1**, children divide 3-digit numbers by 1-digit numbers, using place value counters to help them. Question **2** requires children to complete divisions using the written method. Question **3** is a problem-solving question where children use division in context.

Question **4** gives children an opportunity to independently use their knowledge of number to solve division calculations. It will be important to discuss with children how, after solving 468 ÷ 9, they can use their knowledge of that calculation to more easily solve 4,689 ÷ 9.

In question **5**, children need to reason about when remainders will occur. Question **7** requires children to solve divisions where the divisors are 11 or 12.

STRENGTHEN Build children's confidence by first asking them to work on divisions without remainders, for example 366 ÷ 3, 996 ÷ 3, 8,488 ÷ 2, 6,648 ÷ 2.

DEEPEN Ask children to write other short divisions with 4-digit numbers divided by teen numbers, including those that will have remainders, for a partner to solve. How will they check each other's answers?

ASSESSMENT CHECKPOINT Can children use short division to solve division calculations with fluency? Can children solve divisions of up to 4-digit numbers by 1- and 2-digit numbers, recognising where their knowledge of number facts can help them to solve a calculation more efficiently?

ANSWERS Answers for the **Practice** part of the lesson can be found in the *Power Maths* online subscription.

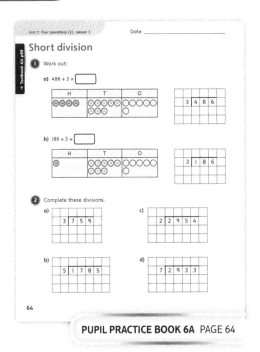

PUPIL PRACTICE BOOK 6A PAGE 64

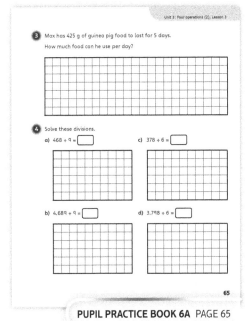

PUPIL PRACTICE BOOK 6A PAGE 65

Reflect

WAYS OF WORKING Independent thinking

IN FOCUS The **Reflect** part of the lesson prompts children to explore potential misconceptions or errors that are likely to occur when using the written method of short division.

ASSESSMENT CHECKPOINT Assess whether children can explain how to identify and subsequently avoid making mistakes when using this method.

ANSWERS Answers for the **Reflect** part of the lesson can be found in the *Power Maths* online subscription.

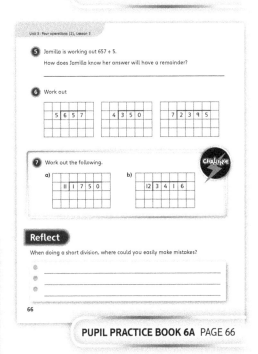

PUPIL PRACTICE BOOK 6A PAGE 66

After the lesson ⏸

- How will you make sure children can regularly complete divisions as part of their ongoing calculation fluency work?
- How can you offer children opportunities to use this method in other areas of the curriculum?

Division using factors

Learning focus

In this lesson, children use factors of the divisor in order to divide efficiently.

Before you teach

- Can children identify the factors of 2-digit numbers?
- Can children accurately divide by a 1-digit number?
- Can children explain why different methods sometimes work to solve different calculations efficiently?

NATIONAL CURRICULUM LINKS

Year 6 Number – addition, subtraction, multiplication and division

Divide numbers up to four digits by a 2-digit number using the formal written method of short division where appropriate, interpreting remainders according to the context.

Identify common factors, common multiples and prime numbers.

ASSESSING MASTERY

Children can divide using factors of the dividend for certain divisions where this will make them easier to solve.

COMMON MISCONCEPTIONS

Children may assume all divisions can be completed using this method. Ask:
- *What can you say about the number you are dividing by?*
- *What will happen if there is a remainder?*

STRENGTHENING UNDERSTANDING

Help children to explore how they can divide 2-digit numbers by 4 by halving, and then halving again.

GOING DEEPER

Challenge children to explore how to use division by factors to divide by 4, 8, 16 and 32. They should find that they can use repeated halving.

KEY LANGUAGE

In lesson: divide, factors

STRUCTURES AND REPRESENTATIONS

Bar model/area model (grid method)

RESOURCES

Optional: small objects, such as counters, to make arrays

 In the eTextbook of this lesson, you will find interactive links to a selection of teaching tools.

Quick recap

Ask children to use short division to solve these calculations:

884 ÷ 4   2,868 ÷ 2   9,639 ÷ 3   8,088 ÷ 4

(Note, these divisions have been chosen so that there will be no remainders at any stage.)

Discover

WAYS OF WORKING Pair work

ASK

- Question **1** a): *Why is 750 divided by 15? Where is 15 in the picture?*
- Question **1** b): *What do you notice about the arrangement of the 15 people in the picture?*
- Question **1** b) *How could this help you with the division?*

IN FOCUS Children can see an array displayed in the context of people arranged on seats in a boat. They use this to help them recognise the role of factors in divisions.

PRACTICAL TIPS Ask children to arrange small objects such as counters into a 5 × 3 array to represent the people in the boat. Discuss how this shows that 3 and 5 are factors of 15.

ANSWERS

Question **1** a): The 750 represents the total number of people who rode the log flume that day. The 15 represents the number of people in each boat. So to find the number of times the log flume ran, you need to divide the total number of people (750) by the number of people in each boat (15).

Question **1** b): $750 \div 3 = 250$, $250 \div 5 = 50$. The log flume ran 50 times that day.

Division using factors

Discover

750 people rode in the log flume boat today. It was full every time.

1 a) You need to know how many times the log flume boat ran today. Explain why you need to work out $750 \div 15$.

b) Work out $750 \div 15$ by first dividing by 3 and then by 5.

92

PUPIL TEXTBOOK 6A PAGE 92

Share

WAYS OF WORKING Whole class teacher led

ASK

- Question **1** a): *What does the 750 mean in this context? What does the 15 mean in this context?*
- Question **1** b): *How does this method work? Why not divide by 10 and then 5, to divide by 15?*

IN FOCUS In question **1**, children will recognise how divisions can be used in problem-solving contexts and what each of the numbers in the word problem represents in the corresponding division calculation. In question **1** b), children then solve the problem by using repeating division with the factors of the divisor.

Share

a) and b)

There are 15 seats on the log flume boat. I worked out $750 \div 15$.

I saw a way to make this easier. I know that 15 is 3×5 so I divided by using factors.

$750 \div 15 = 750 \div 3 \div 5$

The log flume boat ran 50 times today.

93

PUPIL TEXTBOOK 6A PAGE 93

Think together

Unit 3: Four operations (2), Lesson 4

Think together

WAYS OF WORKING Whole class teacher led (I do, We do, You do)

ASK

- Question **1**: *What is the first step? What is the second step?*
- Question **2**: *What is the same and what is different about the two approaches?*
- Question **3**: *Which order is most efficient for you? Which makes the calculation more manageable?*

IN FOCUS In question **1**, children further explore division by using factors in a real-life context. In question **2**, they explore a potential misconception that can occur around the use of factors and partitioning with division. For question **3**, children select the factors and the best order to use for given calculations in order for the method to be as efficient as possible.

STRENGTHEN Work with children to explore dividing numbers by 4 by halving, and then halving again.

DEEPEN Challenge children to explore various ways in which to divide 1,764 by 36 using factors. (Note that 1,764 has been chosen here as it is a multiple of 36. The same exercise could also be repeated with other multiples of 36.)

ASSESSMENT CHECKPOINT Use question **2** to assess whether children can correctly identify the factors of a given divisor and use these to efficiently divide a 4-digit number where there will be no remainder.

ANSWERS

Question **1**: 1,260 ÷ 2 = 630
630 ÷ 7 = 90
90 people per day

Question **2**: ÷ 6 then ÷ 3
£5,490 ÷ 18 = 305
305 tickets were sold

Question **3** a): Yes, they all give the same answer. Children should discuss the method they used.

Question **3** b): Possible answers include:
÷ 6 → ÷ 4,
÷ 3 → ÷ 2 → ÷ 4
÷ 3 → ÷ 2 → ÷ 2 → ÷ 2

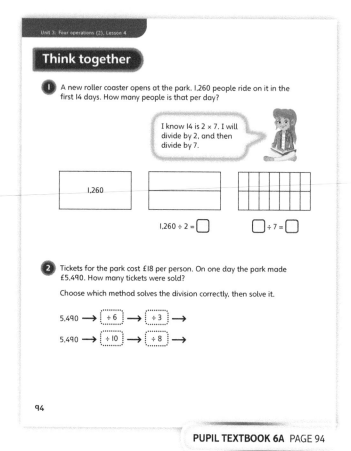

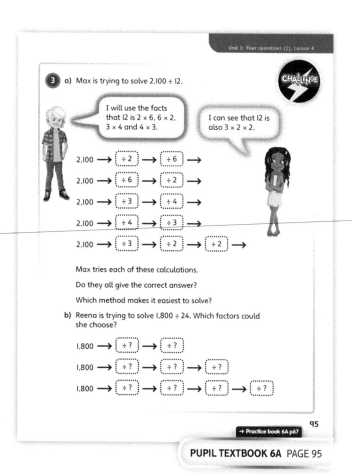

PUPIL TEXTBOOK 6A PAGE 94

PUPIL TEXTBOOK 6A PAGE 95

Practice

WAYS OF WORKING Independent thinking

IN FOCUS Question ① a) links children's understanding of the bar model with their new learning in this lesson. Children should be able to use their understanding of the factors of 14 from earlier in the lesson to scaffold their independent work. Question ① b) moves on from the bar model to a grid representation to represent division calculations.

Question ② moves on to abstract written calculations. The calculations support children's thinking by giving the factors they need to use, but this is withdrawn in question ③.

STRENGTHEN If children are struggling to identify the factors to use to solve the calculations in question ③, give them a multiplication square and ask: *How can you use this to help you find the factors you can use?*

DEEPEN Question ④ gives children a good opportunity to begin generalising about the calculations and the properties of the numbers they are working with. Deepen children's reasoning by asking: *How can you represent your idea with a picture? Can you use more than one type of picture to support your thinking? Explain how your pictures demonstrate your ideas.*

ASSESSMENT CHECKPOINT Use question ③ to assess whether children can choose which factors to use in order to divide efficiently.

ANSWERS Answers for the **Practice** part of the lesson can be found in the *Power Maths* online subscription.

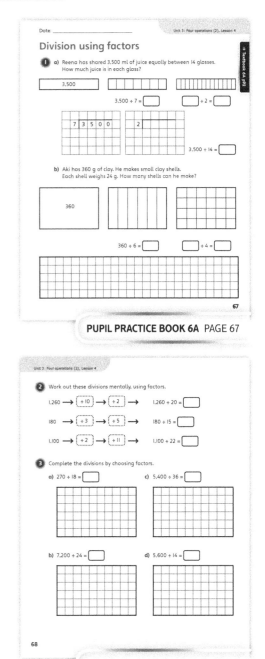

PUPIL PRACTICE BOOK 6A PAGE 67

PUPIL PRACTICE BOOK 6A PAGE 68

Reflect

WAYS OF WORKING Independent thinking

IN FOCUS The **Reflect** part of the lesson prompts children to consider different possible methods of division that can be carried out by using factors.

ASSESSMENT CHECKPOINT Assess whether children can show that two different methods of division with factors can both lead to the same correct calculation and answer.

ANSWERS Answers for the **Reflect** part of the lesson can be found in the *Power Maths* online subscription.

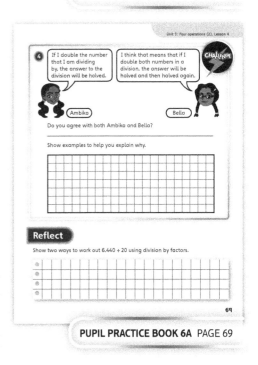

PUPIL PRACTICE BOOK 6A PAGE 69

After the lesson ⏸

- How fluently were children able to identify the factors they could use to solve the divisions efficiently?
- Do children understand why this method only applies when there will be no remainders?

Divide a 3-digit number by a 2-digit number (long division)

Learning focus

In this lesson, children will learn to recognise and identify square and cube numbers. They will explore how these numbers are different from others.

Before you teach

- How will you show the links between the method in today's lesson and those from previous lessons?
- Are children confident using multiples? Will they need more support or practice to ensure sufficient progress in this lesson?

NATIONAL CURRICULUM LINKS

Year 6 Number – addition, subtraction, multiplication and division

Divide numbers up to four digits by a 2-digit whole number using the formal written method of long division, and interpret remainders as whole number remainders, fractions, or by rounding, as appropriate for the context.

ASSESSING MASTERY

Children can use their fluency with multiples of 2-digit numbers to accurately divide a number with up to 4 digits by a 2-digit number. They can represent this understanding using an efficient written method.

COMMON MISCONCEPTIONS

When using long division, children may subtract the factor as well as the multiple from the dividend. Ask:
- *What does each of these numbers represent?*
- *Which number should be subtracted from the number you are dividing? Explain how you know.*

STRENGTHENING UNDERSTANDING

Children who find it difficult to secure conceptual understanding of long division may benefit from practising with a 2-digit number divided by a 1-digit number, for example, 28 ÷ 4. Link long division to an array, discussing the multiples children could use to solve the division more quickly. Ask: *Can you organise 28 counters into the array that shows this division?*

GOING DEEPER

Children could investigate the statement and say whether it is always, sometimes or never true. When calculating a division where a number is divided by a 2-digit prime number, short division won't work.

KEY LANGUAGE

In lesson: group, division, divide, method

Other language to be used by the teacher: multiple, divisor, inverse grid method, approximation, short division, long division, area model, factor

STRUCTURES AND REPRESENTATIONS

Grid method, long division

RESOURCES

Optional: place value counters

 In the eTextbook of this lesson, you will find interactive links to a selection of teaching tools.

Quick recap

Ask children to complete these subtraction calculations:

2,358 – 1,200   9,205 – 4,101
3,575 – 1,455   2,991 – 598

Discover

Divide a 3-digit number by a 2-digit number (long division)

WAYS OF WORKING Pair work

ASK

- Question ① a): *How could you use rounding to help you approximate an answer?*
- Question ① a): *Which method will you use to solve this calculation?*
- Question ① b): *What multiplication will you try? Why?*

IN FOCUS Question ① a) recaps children's previous learning about division and gives them an opportunity to investigate whether the methods they already understand work effectively in this situation. Children may begin to recognise that not all methods are suitable for this calculation. Question ① b) reminds children about the use of the inverse operation to check calculations.

PRACTICAL TIPS When solving question ① a), encourage children to investigate which of the methods they have learnt about previously works best to solve this calculation. Children should recognise that the methods they already know are less efficient and could result in errors being made.

ANSWERS

Question ① a): There are 29 security officers in each group.

Question ① b): 29 × 13 = 377
So the answer 29 is correct.

Discover

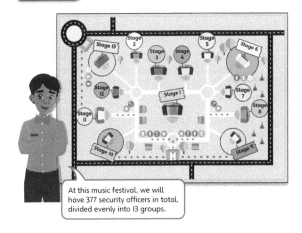

At this music festival, we will have 377 security officers in total, divided evenly into 13 groups.

① a) How many security officers will there be in each group?

b) What multiplication can you do to check your answer is correct?

96

PUPIL TEXTBOOK 6A PAGE 96

Share

WAYS OF WORKING Whole class teacher led

ASK

- Question ① a): *How is long division different from short division? Which method is needed here?*
- Question ① a): *How do the bar models help with the long division?*
- Question ① a): *Why is knowing how to find multiples important when using this method?*
- Question ① b): *Did you get back to the number you started with? What does this tell you?*

IN FOCUS At this point in the lesson, it will be important to make the link between the inverse grid method and long division as explicit as possible. Long division is more abstract in its presentation but identical in the thinking that goes on to solve it. Children should be made aware of this to secure their understanding. During class conversation, encourage children to identify the differences between long division and short division.

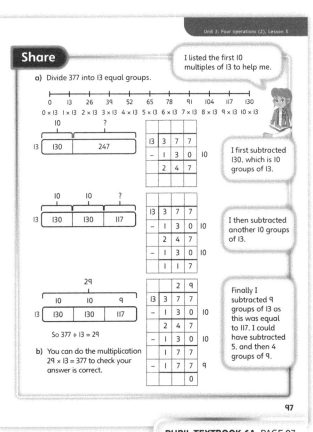

Share

I listed the first 10 multiples of 13 to help me.

a) Divide 377 into 13 equal groups.

I first subtracted 130, which is 10 groups of 13.

I then subtracted another 10 groups of 13.

Finally I subtracted 9 groups of 13 as this was equal to 117. I could have subtracted 5, and then 4 groups of 9.

So 377 ÷ 13 = 29

b) You can do the multiplication 29 × 13 = 377 to check your answer is correct.

97

PUPIL TEXTBOOK 6A PAGE 97

Think together

Whole class teacher led (I do, We do, You do)

ASK

- Question ❶: *Describe and explain how the grid method and the long division calculation are linked.*
- Question ❶: *What numbers are missing from the grid method? Explain how you know.*
- Question ❷: *Can you use the grid method to help you complete the divisions?*
- Question ❸: *Can you divide 799 by any factors of 17? What numbers multiply together to make 17?*
- Question ❸: *What method can you use when the divisor isn't a neat multiple of other digits?*

IN FOCUS Question ❶ scaffolds children's understanding of the link between the grid method to represent their multiplicative reasoning and the newly learnt method of long division. Discuss these links before moving on to question ❷ where children are required to apply methods independently.

STRENGTHEN If children do not know how to approach question ❷, ask them to look at the divisor, 31. Ask: *Do you recognise this number from your times-tables? What does that tell you? What method will be needed here? What do the 6 and the 8 mean? How will you work out how many 31s go into 682? How will you write this down?*

DEEPEN When solving question ❸, encourage children to generalise the type of question Emma is solving, then make up a similar problem for their partner. Ask: *What is different about Emma's problem compared with Reena's? Can you think of other divisors that would require long division? How could you make up a similar question for your partner, with no remainder? Could you start with the inverse calculation?*

ASSESSMENT CHECKPOINT Do children understand the processes involved when solving a division calculation using long division? Can they explain how to correctly set out the written calculation? Can children compare all the methods they have learnt, recognising where each one is most and least useful?

ANSWERS

Question ❶: There will be 19 balloons in each group.

Question ❷ a): 22

Question ❷ b): 46

Question ❸: $588 \div 28 = 588 \div 4 \div 7$
$= 147 \div 7$
$= 21$

$799 \div 17 = 47$ (using long division)

Look for children who suggest the appropriate method to solve Emma's problem, i.e. long division, and who can explain that, as 17 is a prime number, they cannot use factors of 17 to help.

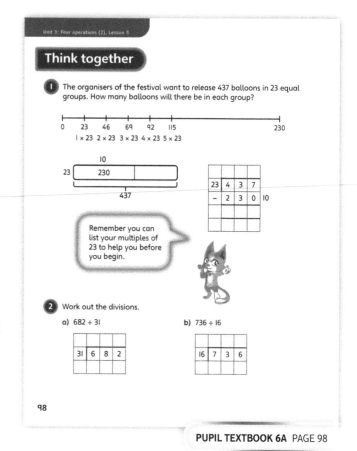

PUPIL TEXTBOOK 6A PAGE 98

PUPIL TEXTBOOK 6A PAGE 99

Practice

WAYS OF WORKING Independent thinking

IN FOCUS Question ① links children's prior learning with their newly learnt written method of long division. Where possible, they should use multiples of 10 and 100 first to solve calculations efficiently.

STRENGTHEN For questions ②, ③ and ④, children should develop their efficiency by identifying where it is possible to use multiples of 10 to quickly make progress through the division calculations. Ask: *Can you fit ten 13s into 364? How about 20? How can this fact help you?*

Question ⑤ provides a real-life context in which to practise division.

DEEPEN Question ⑥ requires children to find the missing divisor, rather than the quotient. Children will need to recognise that they can rearrange the numbers given so that they can still carry out a division in the style of other calculations in this section.

ASSESSMENT CHECKPOINT Do children know how to use multiples of 10 to efficiently solve a long division calculation? Can they use their knowledge of multiples, factors and number facts to efficiently and confidently solve calculations? Do children recognise how an understanding of doubling and halving can help them to solve calculations quickly?

ANSWERS Answers for the **Practice** part of the lesson can be found in the *Power Maths* online subscription.

Reflect

WAYS OF WORKING Independent thinking and pair work

IN FOCUS This question offers an opportunity to assess whether children can confidently use the inverse operation to check the result of a division. They may choose to carry out the actual division using a method of their choice. Children should demonstrate confidence in the method they choose. When children have finished calculating, give them an opportunity to compare their methods. Ask: *Have you used the same method to check? Can you explain which method is more efficient and why?*

ASSESSMENT CHECKPOINT Can children confidently use the methods they have been taught? Can they recognise the most efficient ways of using these methods or the inverse to solve a problem?

ANSWERS Answers for the **Reflect** part of the lesson can be found in the *Power Maths* online subscription.

After the lesson ⏸

- What percentage of the class has mastered long division?
- Are children as confident with this method as they are with short division? How can you tell?

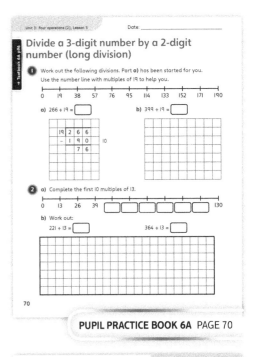

PUPIL PRACTICE BOOK 6A PAGE 70

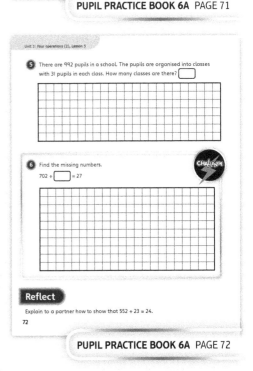

PUPIL PRACTICE BOOK 6A PAGE 71

PUPIL PRACTICE BOOK 6A PAGE 72

Divide a 4-digit number by a 2-digit number (long division)

Learning focus

In this lesson, children use long division to divide by 2-digit numbers.

Before you teach

- Can children use a formal written method for dividing by a 1-digit number?
- Can children subtract efficiently?
- Can children talk about the remainders in different stages of a calculation?
- Can children write out the multiples of a given 2-digit number?

NATIONAL CURRICULUM LINKS

Year 6 Number – addition, subtraction, multiplication and division

Divide numbers up to four digits by a 2-digit whole number using the formal written method of long division, and interpret remainders as whole number remainders, fractions, or by rounding, as appropriate for the context.

Divide numbers up to four digits by a 2-digit number using the formal written method of short division where appropriate, interpreting remainders according to the context.

ASSESSING MASTERY

Children can use the long division method to complete divisions which involve dividing by 2-digit numbers.

COMMON MISCONCEPTIONS

Children may struggle with the fact that the calculation has several different stages. Ask:
- *Do you need to pause and take a breath? What is the first step? What will the second step be?*

STRENGTHENING UNDERSTANDING

Refer back to previous lessons so that children can work with this method for 1-digit numbers, in order to build their confidence.

GOING DEEPER

Challenge children to explore, describe and compare the various methods that they know for solving divisions, including mental methods.

KEY LANGUAGE

In lesson: divide, division, long division, multiple, remainder

STRUCTURES AND REPRESENTATIONS

Long division

RESOURCES

Optional: empty number lines

 In the eTextbook of this lesson, you will find interactive links to a selection of teaching tools.

Quick recap

Ask children to list the multiples of 7 from 0 to 70.

Then ask them to list the multiples of 11 from 0 to 110. Then ask them to list the multiples of 15 from 15 to 150.

Discuss the mental methods you can use to complete the lists of multiples

Discover

Divide a 4-digit number by a 2-digit number (long division)

Discover

Our rescue centre received a donation of 2,478 tins of cat food.

WAYS OF WORKING Pair work

ASK

- Question ① a): *How many digits are there in the number of tins?*
- Question ① a): *How could this be part of a division lesson?*
- Question ① b): *What would be an easier division to complete in this context?*
- Question ① b): *What would be a more challenging calculation?*

IN FOCUS This section gives children the opportunity to recognise the use of division in a real-life context. They are first encouraged to list the multiples of a given 2-digit number and then apply this to solving a division calculation.

PRACTICAL TIPS Work together and discuss how to write out the multiples of 2-digit numbers efficiently and accurately. Consider providing empty number lines on which children jot the multiples.

ANSWERS

Question ① a): 21, 42, 63, 84, 105, 126, 147, 168, 189, 210

Question ① b): 118 days

① a) The cat rescue centre uses 21 tins of food each day. List the multiples of 21 up to 10 × 21.

   b) How many days will the food last for?

100

Share

WAYS OF WORKING Whole class teacher led

ASK

- Question ① a): *What method will you use to write out the multiples? Would you work from left to right in order? Or would you use known patterns or relationships to find multiples? How can you check you have found all the multiples?*
- Question ① b): *How are the visual and the written methods related? Can you see the parts of the division in the diagram? How do the multiples from question ① a) help you?*

IN FOCUS In question ① a), children work on writing out the multiples of the given divisor. This could be generated as a list or represented on a number line. In question ① b), children then select the relevant multiples to complete a division calculation using the long division method. Ensure children understand why each multiple is being used and where each one has been placed in the visual model and the long division layout.

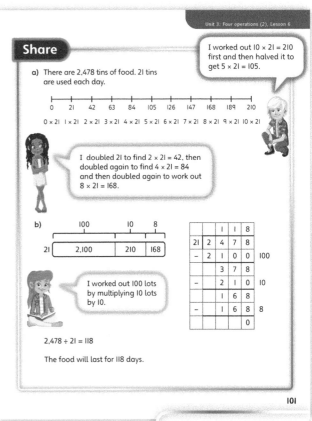

Think together

WAYS OF WORKING Whole class teacher led (I do, We do, You do)

ASK

- Question ❶: *How do the multiples help you with this method?*
- Question ❷: *What step has been completed? What will you do next?*
- Question ❸: *What will you look for in each calculation to decide which method will work best?*
- Question ❸: *Does one method work for more types of calculation than the others?*

IN FOCUS In question ❶, children complete a long division calculation with scaffolding to support them with each of the steps. In question ❷, children complete a division using the long division method where the first step has been done for them. Question ❸ requires children to choose a different division method for several different given calculations.

STRENGTHEN Give children the opportunity to practise the long division method with calculations where 3-digit numbers are being divided by 1-digit numbers, in order to build their confidence.

DEEPEN Question ❸ deepens children's reasoning with the methods they have learnt. Encourage them to find a checking strategy for these calculations. They will need to fluently identify the properties of the calculations that will lead to the use of different checking strategies.

ASSESSMENT CHECKPOINT Assess whether children can work through the long division process step-by-step, and show an understanding of what happens at each stage.

ANSWERS

Question ❶ a): 23, 46, 69, 92, 115, 138, 161, 184, 207, 230

Question ❶ b): Use 23 × 5 and 23 × 6 and multiply these answers by 10.

Question ❶ c): 193

Question ❷: 54

Question ❸: Children choose various methods and explain why they chose each method. They should aim to choose the most efficient method for each calculation, for example, using factors to divide or using long division.

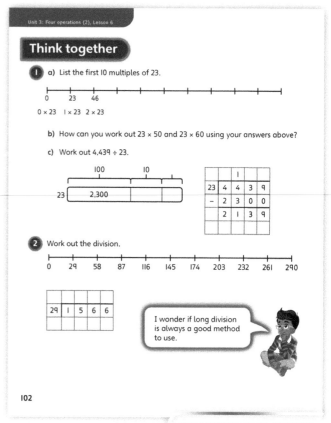

PUPIL TEXTBOOK 6A PAGE 102

PUPIL TEXTBOOK 6A PAGE 103

Practice

WAYS OF WORKING Independent thinking

IN FOCUS Question **1** provides practice in the factor method of division.

Questions **2** and **3** require children to complete given divisions accurately, using a formal written method.

Question **4** presents a real-life problem with space for children to apply their chosen method independently. Encourage children to justify the method they choose to use.

Question **5** reminds children that it is useful to list the first 10 multiples of a divisor and use knowledge of place value when doing long division.

STRENGTHEN To help decide which method is best in questions **4** and **5**, children can look back at the textbook to help them. Ask: *Will the calculations needed be similar to any you have done before? How? Can you use the way you solved earlier divisions to help you?*

DEEPEN Challenge children to explore the different methods that they could use to complete divisions such as 5,775 ÷ 25. Can they explain which method they prefer and why?

ASSESSMENT CHECKPOINT Do children select an appropriate method and use it fluently? Can they use appropriate strategies to find approximate answers and check calculations?

ANSWERS Answers for the **Practice** part of the lesson can be found in the *Power Maths* online subscription.

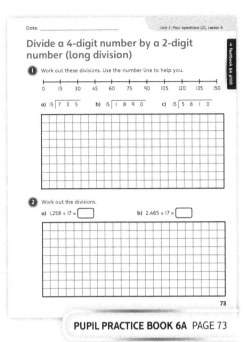

PUPIL PRACTICE BOOK 6A PAGE 73

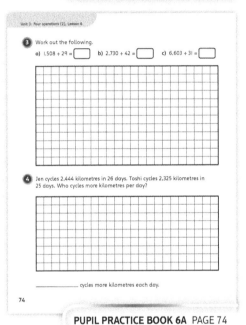

PUPIL PRACTICE BOOK 6A PAGE 74

Reflect

WAYS OF WORKING Independent thinking

IN FOCUS This question allows children to recognise and demonstrate where the factor method can and cannot be used.

It will give a final opportunity to assess children's understanding of when the use of long division is appropriate.

ASSESSMENT CHECKPOINT Children should be able to fluently recognise types of calculation where the factor method is, and is not, appropriate. Look for children using knowledge of prime numbers to help them make decisions.

ANSWERS Answers for the **Reflect** part of the lesson can be found in the *Power Maths* online subscription.

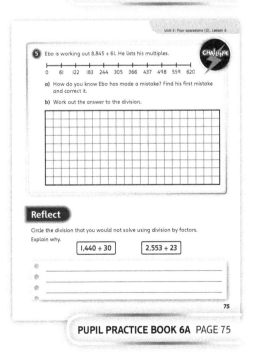

PUPIL PRACTICE BOOK 6A PAGE 75

After the lesson

- Are children fluent in determining which method is appropriate for which situation?
- What proportion of the class can reliably select an appropriate method for each calculation they approach?

Long division with remainders

Learning focus

In this lesson, children will develop their understanding of division with remainders. They will learn how the written methods for division they have learnt can represent and solve a division calculation that has a remainder. They will also learn that representing a remainder as a fraction can give a more accurate answer.

Before you teach

- Have children already recognised or found divisions that have remainders?
- How will you ensure children's learning is deepened further in this lesson?

NATIONAL CURRICULUM LINKS

Year 6 Number – addition, subtraction, multiplication and division

Divide numbers up to four digits by a 2-digit whole number using the formal written method of long division, and interpret remainders as whole number remainders, fractions, or by rounding, as appropriate for the context.

Divide numbers up to four digits by a 2-digit number using the formal written method of short division where appropriate, interpreting remainders according to the context.

ASSESSING MASTERY

Children can use their fluency with multiples of 2-digit numbers to divide a number up to four digits by a 2-digit number and find a remainder if there is one. They can use an efficient written method to represent this.

COMMON MISCONCEPTIONS

When solving a division with a remainder, children may add the remainder to the factors or multiples in their calculation. Ask:
- *What is the problem with what you have found in your running total?*
- *What do we do with the remainder?*

Children may say that a division 'doesn't work' if they find there is a remainder. Ask:
- *Can you say how many full groups of the divisor fit into the dividend?*
- *What do we call the number that is left over?*

STRENGTHENING UNDERSTANDING

To help children with the idea of remainders, it may be useful to give them a simpler division calculation that they can make with counters or blocks. Using 25 ÷ 4 as an example, ask: *Can you find 25 counters and share them into groups of 4? What do you notice about the groups? Is it possible to do this? What is the leftover counter called?*

GOING DEEPER

Children can write division word problems that have remainders. To deepen this further, children could write a problem where it is necessary to round the remainder up, round it down or leave it as it is.

KEY LANGUAGE

In lesson: divide, division, remainder, subtract, fraction, multiple

STRUCTURES AND REPRESENTATIONS

Long division

RESOURCES

Optional: empty number lines

 In the eTextbook of this lesson, you will find interactive links to a selection of teaching tools.

Quick recap

Ask children to complete these calculations using long division. (These are straightforward divisions with no remainders.)

195 ÷ 13   6,996 ÷ 11

Discover

Long division with remainders

Discover

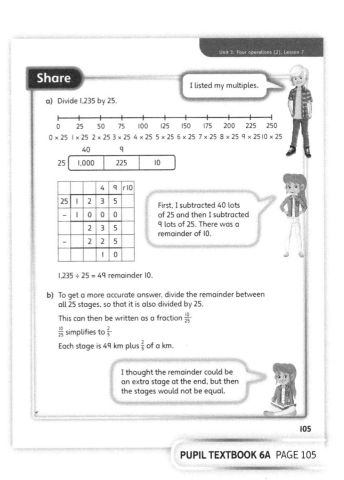

Vintage car race, 1,235 km

Stage 1 of 25

1 a) Divide 1,235 by 25.

b) The race is split into 25 equal stages.
   How long is each stage?

104

PUPIL TEXTBOOK 6A PAGE 104

WAYS OF WORKING Pair work

ASK

- Question 1 a): *Does this calculation result in a remainder? Can you prove it?*
- Question 1 a): *How many different ways could you represent this calculation?*
- Question 1 a): *Are there any methods you know that don't work with this calculation?*
- Question 1 b): *Does the way in which we record remainders work for this question? How?*
- Question 1 b): *What do we need to do to the remainder to make the solution make sense with the question?*

IN FOCUS To ensure quick progress in the lesson, it is important to recap children's understanding of the methods they have learnt in this unit. Question 1 b) offers an opportunity to discuss how a remainder can be used, being rounded up, rounded down or left as it is. Recording the remainder as 'remainder 10' would not make sense in the context of the question. Children should discuss how the remainder can be used to solve this problem.

PRACTICAL TIPS Encourage children to make a list of the multiples before they start to work through the calculation. Children could jot the multiples onto empty number lines if they find this helpful.

ANSWERS

Question 1 a): 49 r 10

Question 1 b): $49\frac{2}{5}$ km

Share

WAYS OF WORKING Whole class teacher led

ASK

- Question 1 a): *How does the model help you?*
- Question 1 a): *Which method did you use to solve the question?*
- Question 1 b): *Can you explain Astrid's comment?*
- Question 1 b): *What does the fraction represent?*
- Question 1 b): *Can it be simplified further than $\frac{2}{5}$?*

IN FOCUS At this point in the lesson, it will be important to recap and secure children's understanding that a fraction is a representation of a division and that $\frac{10}{25}$ is the same as 10 ÷ 25. Making sure children understand this will ensure that their conceptual understanding is secured throughout the lesson.

Share

I listed my multiples.

a) Divide 1,235 by 25.

```
0    25   50   75  100  125  150  175  200  225  250
0×25 1×25 2×25 3×25 4×25 5×25 6×25 7×25 8×25 9×25 10×25
```

	40	9	
25	1,000	225	10

			4	9	r 10
25	1	2	3	5	
−	1	0	0	0	
		2	3	5	
−		2	2	5	
			1	0	

First, I subtracted 40 lots of 25 and then I subtracted 9 lots of 25. There was a remainder of 10.

1,235 ÷ 25 = 49 remainder 10.

b) To get a more accurate answer, divide the remainder between all 25 stages, so that it is also divided by 25.

This can then be written as a fraction $\frac{10}{25}$.

$\frac{10}{25}$ simplifies to $\frac{2}{5}$.

Each stage is 49 km plus $\frac{2}{5}$ of a km.

I thought the remainder could be an extra stage at the end, but then the stages would not be equal.

105

PUPIL TEXTBOOK 6A PAGE 105

Think together

Whole class teacher led (I do, We do, You do)

ASK

- Question **1**: *How does the list of multiples help with the calculation?*
- Question **2**: *Can you first estimate the answers to these divisions?*
- Question **3**: *What could the number be if there was no remainder after dividing by 29?*

IN FOCUS In question **1**, children work through and complete the long division method, where each step is scaffolded and a list of multiples of the divisor is provided. Question **2** requires children to set out and complete long divisions by themselves. In question **3**, children explore the inverse operation to find the dividend when the divisor, quotient and remainder are known.

STRENGTHEN Give children opportunities to rehearse writing out the multiples of the divisor accurately and efficiently. Jotting them onto empty number lines may help children to organise their thinking.

DEEPEN Challenge children to investigate this question: The answer to a division is 99 r 98. What could the question have been? Ask them to explain how they solved it and what they learned about remainders with division.

ASSESSMENT CHECKPOINT Use question **2** to assess whether children can accurately lay out the steps of the long division method and use it to solve division calculations where scaffolding is not provided and where they may be a remainder.

ANSWERS

Question **1**: 36 r 5

Question **2** a): 42 r 2

Question **2** b): £87

Question **3**: 1,573

PUPIL TEXTBOOK 6A PAGE 106

PUPIL TEXTBOOK 6A PAGE 107

Practice

WAYS OF WORKING Independent thinking

IN FOCUS Question ❶ provides children with practice in setting out and completing long divisions that all have the same divisor. Questions ❶ and ❸ give children the opportunity to solve division calculations using the methods they have learnt. This extends and challenges their understanding as they select methods to apply independently.

Questions ❷ and ❺ challenge children to apply their understanding of remainders in different contexts. In question ❷, it is appropriate to record the remainder as a whole number that is 'left over'.

STRENGTHEN If children are convinced that only 13 bags of bird seed are needed in question ❺, ask: *Will they be able to make all the bird feeders with only 13 bags? Why? How many bags should they buy to get the extra 20 kg they need? What did we have to do with the remainder and why?*

DEEPEN Question ❷ helps deepen children's understanding of remainders. They should reason that: When 1,000 is divided by 16, the remainder (8) is less than 16. So, when 1,001 is divided by 16, you will still get the same whole number of 16s, but the remainder will increase by 1. Hence, when 1,001 is divided by 16 the remainder will be 9.

THINK DIFFERENTLY In Question ❹, children need to use given information to work out what the remainder will be in a linked division calculation. They explain how the information they are given helps them to know the remainder.

ASSESSMENT CHECKPOINT Do children demonstrate fluency and reasoning when using known number facts and linking them with division? Can children accurately and fluently solve division calculations that include remainders in multiple representations, including abstract calculations and real-life written problems?

ANSWERS Answers for the **Practice** part of the lesson can be found in the *Power Maths* online subscription.

Reflect

WAYS OF WORKING Pair work

IN FOCUS Once children have written their story problem, they can challenge a partner to solve it, to check it has all the required features.

ASSESSMENT CHECKPOINT Look for children having written a contextual problem requiring a division that fits the parameters. Children may want to write the abstract calculation first, before putting it into a story.

ANSWERS Answers for the **Reflect** part of the lesson can be found in the *Power Maths* online subscription.

After the lesson ⏸

- How confident were children at using written methods to solve division calculations with remainders?
- Were children able to link the context of the written problems to rounding a remainder up or down?
- Could children clearly explain how the fractions are linked to the remainders they find?
- How could you implement this skill in other areas of the curriculum?

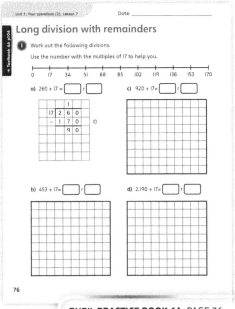

PUPIL PRACTICE BOOK 6A PAGE 76

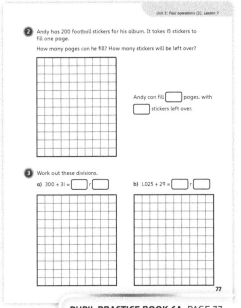

PUPIL PRACTICE BOOK 6A PAGE 77

PUPIL PRACTICE BOOK 6A PAGE 78

Order of operations

Learning focus

In this lesson, children will learn the correct order of operations and use this to help solve multi-step calculations.

NATIONAL CURRICULUM LINKS

Year 6 Number – addition, subtraction, multiplication and division

Use their knowledge of the order of operations to carry out calculations involving the four operations.

ASSESSING MASTERY

Children can recognise and explain the correct order of operations. They can explain why the order of operations is important and can identify where the order has not been followed.

COMMON MISCONCEPTIONS

Children may be inclined to calculate from left to right, ignoring the order of operations. Have the correct order displayed prominently in the classroom. Ask:
- *Show me how you solved the calculation.*
- *Did you follow the order on the display?*

STRENGTHENING UNDERSTANDING

To help children remember the order of operations, it may be helpful and fun to create a class rhyme or song. It is too early to use 'BIDMAS' (**B**rackets, **I**ndices, **D**ivision and **M**ultiplication, **A**ddition and **S**ubtraction) as children have not yet learnt about brackets.

GOING DEEPER

Children could be given a selection of five numbers and four operations. Ask: *How many different solutions can you find using these numbers and operations?*

KEY LANGUAGE

In lesson: order of operations, calculation, addition, subtraction, multiplication, division

Other language to be used by the teacher: calculate, add, subtract, multiply, divide

RESOURCES

Optional: ten frames, counters, multilink cubes, bead strings

 In the eTextbook of this lesson, you will find interactive links to a selection of teaching tools.

Quick recap

Play a game of 'Follow my calculation steps'.
For example:

- Start with the number 5
- Add 10
- Now subtract 11
- Multiply by 10
- Divide by 2
- Add 1

Ask: *What number do you have now?*
Repeat with different chains of operations.

Discover

Order of operations

Discover

WAYS OF WORKING Pair work

ASK

- Question ❶ a): *Can a single calculation have two solutions?*
- Question ❶ a): *How has each child come to their solution?*
- Question ❶ a): *How does each child's model demonstrate their thinking?*

IN FOCUS Looking at the picture, children are likely to assume that Ebo is correct as they will be used to solving calculations reading from left to right. It is important at this stage not to give the game away and let children discuss their ideas without any help from you.

PRACTICAL TIPS Children could be encouraged to make their own model of the multi-step calculation using tens frames and counters, multilink cubes or bead strings.

ANSWERS

Question ❶ a): Ebo has solved the calculation as (3 + 5) × 2. Lexi has solved the calculation as 3 + (5 × 2).

Question ❶ b): Lexi is correct.

❶ a) Explain why Ebo and Lexi have produced different answers.
   b) Who is correct?

108

PUPIL TEXTBOOK 6A PAGE 108

Share

WAYS OF WORKING Whole class teacher led

ASK

- Question ❶ a): *Who do you think is correct and why?*
- Question ❶ b): *Why do you think it is important to have an agreed order of operations?*
- Question ❶ b): *What might happen if you did not agree on the order you solve operations?*

IN FOCUS At this point in the lesson, children are only learning about multiplication and addition. Before moving on, make sure they understand that multiplication is done first, then addition. If children ask about division and subtraction, it would be an interesting opportunity for them to predict where those operations will feature in the order based on what they already know, but these will be covered properly in the next section.

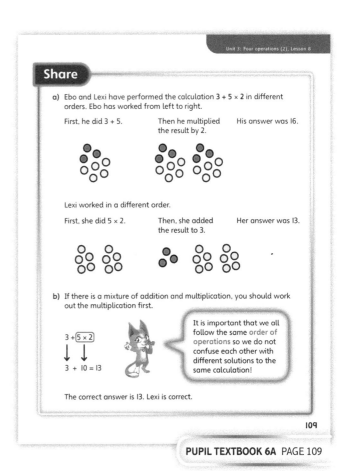

Share

a) Ebo and Lexi have performed the calculation **3 + 5 × 2** in different orders. Ebo has worked from left to right.

First, he did 3 + 5.   Then he multiplied the result by 2.   His answer was 16.

Lexi worked in a different order.

First, she did 5 × 2.   Then, she added the result to 3.   Her answer was 13.

b) If there is a mixture of addition and multiplication, you should work out the multiplication first.

3 + 5 × 2
↓   ↓
3 + 10 = 13

It is important that we all follow the same order of operations so we do not confuse each other with different solutions to the same calculation!

The correct answer is 13. Lexi is correct.

109

PUPIL TEXTBOOK 6A PAGE 109

Think together

Think together

WAYS OF WORKING Whole class teacher led (I do, We do, You do)

ASK

- Question **1**: *At what point in the order of operations do you think you should solve subtractions?*
- Question **2** b): *What happens if there are more than two operations?*
- Question **3** b): *What is different about these calculations from those you looked at before?*

IN FOCUS In this part of the lesson, children are introduced to the position of subtraction and division within the order of operations. Question **1** introduces subtraction, while question **3** introduces division. It is important to make sure the position of these calculations is made clear in your teaching, that is, division and multiplication first, then addition and subtraction. Link the additive and multiplicative operations to help children understand why the operations are in the order they are in.

In question **3** b), when children consider Ash's comments, they may conclude that the answer is always the same, regardless of the order in which they work out the multiplication and division. While this is true when the multiplication precedes the division in a written calculation, it is not true if the division precedes the multiplication, as in $10 \div 5 \times 2$. Use this to illustrate that children should work through multiplications and divisions in the order they appear.

STRENGTHEN For question **2** b), if children are struggling to know which multiplication to solve first in the three-part calculation, ask: *What happens if you solve the first multiplication first? What happens if you solve the second multiplication first?*

DEEPEN Children could be given calculations that include all four operations. They could also be given a sequence of three or four numbers (for example, 3 4 5 = 23) and asked to find the missing operations.

ASSESSMENT CHECKPOINT At this point in the lesson, children should be able to explain that the operations of multiplication and division are carried out before the operations of addition and subtraction. Children should be able to solve calculations that involve up to three operations and should be able to explain why people may find more than one solution. Question **2** gives you the opportunity to assess children's recognition of the order of multiplication and addition or subtraction operations. Children should recognise in both calculations that the multiplications should be done first.

ANSWERS

Question **1**: $(3 \times 5) - 2 = 13$ is correct

Question **2** a): Solve 25×2 first, then subtract from 100, giving an answer of 50.

Question **2** b): Solve 11×2 and 3×11 first, then add the two results, giving an answer of 55.

Question **3** a): $25 + 100 \div 4 = 50$
$45 = 500 \div 10 - 5$

Question **3** b): Both ways of solving the calculation result in the same solution (10).

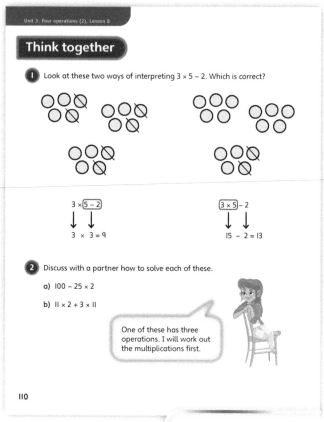

PUPIL TEXTBOOK 6A PAGE 110

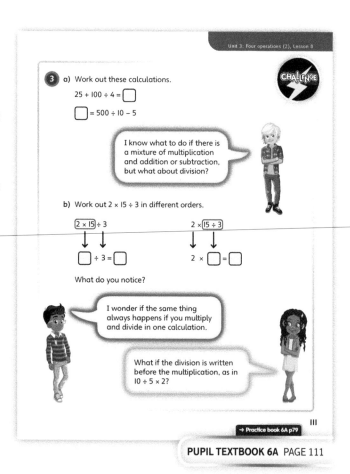

PUPIL TEXTBOOK 6A PAGE 111

Practice

WAYS OF WORKING Independent thinking

IN FOCUS Question ① scaffolds children's use of the order of operations by linking the abstract to pictures of the concrete representations. This link could be reinforced by offering children the resources pictured to make the calculations themselves. Question ② requires children to identify a mistake that has been made by not following the correct order of operations. While solving questions ④ a) and b), it would be beneficial to encourage children to discuss how the calculation and solutions change as the operations are altered. They should notice that though the calculations look different, the order in which the operations need to be carried out remains the same, and therefore so does the answer.

STRENGTHEN In question ③, children are shown how to draw a box around the operation that needs to be done first. Encourage children to do this for every problem that has more than one operation, to help them organise their thinking.

DEEPEN Question ④ can be deepened by asking children to suggest other pairs of linked calculations where the same numbers are arranged differently, but the order of operations means that the answer will still be the same.

THINK DIFFERENTLY The calculations in question ⑤ each have three operations. Listen carefully to children's explanations and reasoning about the order in which they carried out the three steps, and how they knew to do it that way.

ASSESSMENT CHECKPOINT Children should be able to confidently solve a calculation with more than one operation. Question ③ is a valuable opportunity to assess children's ability to recognise, explain and follow the correct order of operations reliably. Question ⑥ assesses whether children can recognise that, by knowing the order of operations, they can work backwards from a number to complete a missing number calculation. Look for children's clarity and confidence when giving explanations linked to their learning earlier in the lesson.

ANSWERS Answers for the **Practice** part of the lesson can be found in the *Power Maths* online subscription.

Reflect

WAYS OF WORKING Independent thinking, pair work

IN FOCUS Writing their own calculation requires children to demonstrate their grasp of the order of operations without prompts. The activity could be made into a challenge that children set their partner. Once children have designed their calculation, they could share it with a partner. Can they identify all the possible solutions and explain which is the correct one and why? Peer-to-peer feedback will further reinforce children's understanding.

ASSESSMENT CHECKPOINT Look for children's ability to design their own calculations and explain how to solve them correctly. Children should be able to confidently and fluently explain how to use the order of operations to solve their calculations.

ANSWERS Answers for the **Reflect** part of the lesson can be found in the *Power Maths* online subscription.

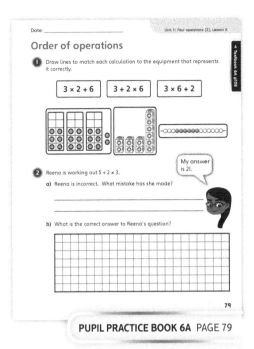

PUPIL PRACTICE BOOK 6A PAGE 79

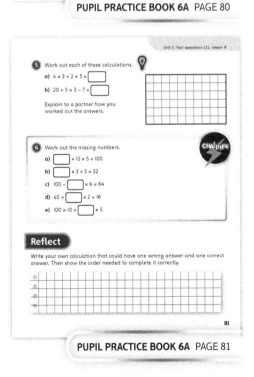

PUPIL PRACTICE BOOK 6A PAGE 80

After the lesson ⏸

- How confident were children with remembering the order of operations by the end of the lesson?
- Could this lesson have been made more practical? How will you facilitate this next time you teach it?

PUPIL PRACTICE BOOK 6A PAGE 81

Brackets

Learning focus

In this lesson, children will extend their understanding of the order of operations by investigating what effect brackets can have on a calculation.

Before you teach ⏸

- Were there any misconceptions from the previous lesson that will need to be overcome to ensure progress in this lesson?
- How will you integrate this teaching into today's lesson?

NATIONAL CURRICULUM LINKS

Year 6 Number – addition, subtraction, multiplication and division

Use their knowledge of the order of operations to carry out calculations involving the four operations.

ASSESSING MASTERY

Children can recognise and explain the effect brackets can have on the order of operations. They are able to confidently solve calculations that include brackets by solving what is inside the brackets first.

COMMON MISCONCEPTIONS

Children may ignore the brackets in a calculation and solve it either using the order of operations or ignoring that as well and solving it in the order it is presented. Ask:
- *What was the order of operations you learnt in the last lesson?*
- *What effect do brackets have within a calculation?*

STRENGTHENING UNDERSTANDING

When children are solving questions that require them to add brackets to an already written calculation, it may help to have the calculation written on a large piece of paper and have brackets on cards that can be placed in and around the calculation. This may help children to be more flexible in their approach, as they will be able to easily change the position of brackets without having multiple recorded mistakes across one calculation.

GOING DEEPER

Children could be given a selection of numbers and asked to find a calculation that equals a given (random) number. Ask: *Using these numbers – a, b, c, d, e, f – can you write a calculation that equals x?*

KEY LANGUAGE

In lesson: brackets, multiply, calculation, order of operations

Other language to be used by the teacher: calculate, divide, add, subtract

STRUCTURES AND REPRESENTATIONS

Bar models

RESOURCES

Optional: large paper, card, base 10 equipment, counters, multilink cubes, bead strings

 In the eTextbook of this lesson, you will find interactive links to a selection of teaching tools.

Quick recap

Ask children to complete these calculations with the correct order of operations:

$2 \times 5 \times 5$ $2 \times 5 + 5$ $2 + 5 \times 5$ $2 \times 5 \div 5$

Discover

Brackets

Discover

WAYS OF WORKING Pair work

ASK

- Question ① a): *Do you agree with the total the mechanic has come to?*
- Question ① a): *What do you notice about the calculation she has written?*
- Question ① a): *What total should the mechanic have found, using the calculation she has written?*
- *Can you show both calculations from questions ① a) and ① b) using a bar model? What is similar and what is different?*

IN FOCUS Question ① a) recaps children's understanding of the previous lesson. While the mechanic has noted the correct number of tyres, the calculation she has written does not equal 160 when following the order of operations. This will prompt children to see that something else may be needed to show when a calculation should be solved in a different order.

PRACTICAL TIPS Children could create their own versions of the scenario posed in the picture, using toy cars, lorries and trains. Children could be encouraged to see how the calculations stay the same and how they differ, depending on the vehicles used.

ANSWERS

Question ① a): The mechanic's written calculation is incorrect. It gives an answer of 100.

Question ① b):

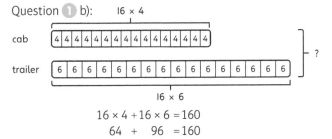

$$16 \times 4 + 16 \times 6 = 160$$
$$64 \ + \ 96 \ = 160$$

Share

WAYS OF WORKING Whole class teacher led

ASK

- Question ① a): *What was the problem with the context and the calculation representing it?*
- Question ① a): *How do the brackets help you make the calculation fit the problem?*
- Question ① b): *Can you write another calculation that uses brackets to change the order of operations?*

IN FOCUS In question ① a), it is important to discuss how the brackets have enabled the calculation to reflect the context of the problem. Children should be encouraged to recognise how this can help them to be more efficient, solving what would have otherwise needed to be two separate multiplications in one.

I have to check the tyres on 16 lorries today. Each cab has 4 wheels. Each trailer has 6 wheels.

① a) The mechanic wants to work out how many tyres she needs to check.

She writes down $4 + 6 \times 16 = 160$.

Is her written calculation correct?

b) Another mechanic works out how many tyres are on 16 cabs and then how many tyres are on 16 trailers. He adds the two answers.

Show his method using a bar model and write the calculation for each step.

112

PUPIL TEXTBOOK 6A PAGE 112

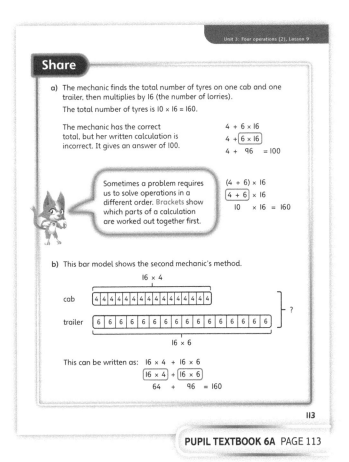

Share

a) The mechanic finds the total number of tyres on one cab and one trailer, then multiplies by 16 (the number of lorries).

The total number of tyres is $10 \times 16 = 160$.

The mechanic has the correct total, but her written calculation is incorrect. It gives an answer of 100.

$4 + 6 \times 16$
$4 + \boxed{6 \times 16}$
$4 + \ 96 \ = 100$

Sometimes a problem requires us to solve operations in a different order. **Brackets** show which parts of a calculation are worked out together first.

$(4 + 6) \times 16$
$\boxed{4 + 6} \times 16$
$10 \ \times 16 = 160$

b) This bar model shows the second mechanic's method.

This can be written as: $16 \times 4 + 16 \times 6$
$\boxed{16 \times 4} + \boxed{16 \times 6}$
$64 \ + \ 96 = 160$

113

PUPIL TEXTBOOK 6A PAGE 113

Think together

Whole class teacher led (I do, We do, You do)

ASK

- Question **1**: *What do you need to work out first?*
- Question **1**: *Does the calculation you need to do first come first in the order of operations? If not, what do you need to do?*
- Question **2**: *What do the brackets mean in the calculations?*
- Question **3** a): *How many places could brackets be put into the calculation? How many solutions could be found?*

IN FOCUS For question **1**, encourage children to try different possible solutions. Ask: *What is the 'story' of your calculation? Does it match the story in the problem?*

After question **1**, contexts are removed so that children think in the abstract.

STRENGTHEN It may help to link the calculations with concrete representations, using base 10 equipment, place value counters, multi-link cubes or bead strings. Ask: *What does the concrete representation show? How would you write this as a calculation?*

If children struggle with question **3**, it is important to direct them to Astrid's comment as up until this point they have only experienced brackets around two numbers. This could potentially lead to the misconception that that is the maximum amount seen in any example of brackets. Ask: *What is Astrid suggesting? Do you think she is able to do that? Can you try doing what she suggests? What happens?*

DEEPEN If children solve question **3**, they could be encouraged to continue with Ash's line of questioning. Ask *Do you predict it is possible to make calculations for all numbers from 1 to 20 using just four 4s? Explain. Show me how many you can make.*

ASSESSMENT CHECKPOINT Children should now be able to recognise the function of brackets within a calculation and know that whatever calculations are bracketed should be solved first. They should be able to confidently solve calculations with brackets and be more confident at finding where brackets need to be in a calculation, using trial and error. Question **2** assesses children's understanding of how the use of brackets influences their calculations.

ANSWERS

Question **1**: $4 \times (£7{\cdot}50 + £3{\cdot}50) = £44{\cdot}00$
$4 \times £11{\cdot}00 = £44{\cdot}00$

Question **2** a): $(15 - 5) \times 3 = 30$
$15 - (5 \times 3) = 0$

Question **2** b): $(15 + 5) \times (15 - 5) = 200$
$15 + (5 \times 15) - 5 = 85$

Question **3** a): $(4 + 4) \times (4 \div 4) = 8$
$4 + (4 \times 4 \div 4) = 8$
$(4 + 4 \times 4) \div 4 = 5$

Question **3** b): $(4 \div 4) + (4 \div 4) = 2$
or $(4 \times 4) \div (4 + 4) = 2$
$4 \times (4 + 4) - 4 = 28$

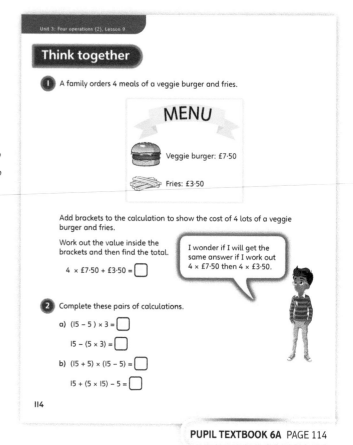

Think together

1. A family orders 4 meals of a veggie burger and fries.

MENU

Veggie burger: £7·50

Fries: £3·50

Add brackets to the calculation to show the cost of 4 lots of a veggie burger and fries.

Work out the value inside the brackets and then find the total.

$4 \times £7{\cdot}50 + £3{\cdot}50 = \Box$

I wonder if I will get the same answer if I work out $4 \times £7{\cdot}50$ then $4 \times £3{\cdot}50$.

2. Complete these pairs of calculations.

a) $(15 - 5) \times 3 = \Box$
$15 - (5 \times 3) = \Box$

b) $(15 + 5) \times (15 - 5) = \Box$
$15 + (5 \times 15) - 5 = \Box$

114

Unit 3: Four operations (2), Lesson 9

3. a) Add brackets to make each of these calculations correct. **CHALLENGE**

$4 + 4 \times 4 \div 4 = 8$

$4 + 4 \times 4 \div 4 = 5$

I will try putting brackets around three of the 4s.

b) Choose operations to make these calculations correct.

$(4 \bigcirc 4) \bigcirc (4 \bigcirc 4) = 2$

$4 \bigcirc (4 \bigcirc 4) \bigcirc 4 = 28$

I'll try and make all the numbers from 1 to 20 using just four 4s.

Sometimes I work systematically from 1 to 20, or I might try a different order.

115

→ Practice book 6A p82

Practice

WAYS OF WORKING Independent thinking

IN FOCUS Question ❶ links calculations to concrete and pictorial representations. This helps to secure children's understanding of the abstract calculations. If children used equipment earlier in the lesson, it may help them to do so again, before moving on to the abstract calculations in Question ❷. Question ❹ moves on to word problems, providing a good opportunity for children to apply their learning in context. Children could act out each problem with concrete resources to help them understand the 'story' of the calculation before turning it into a written calculation.

STRENGTHEN In question ❸, children underline the part of each calculation that needs to be done first, based on their knowledge of order of operations and brackets. Encourage them to do this for every multi-step problem and to describe their reasoning each time.

DEEPEN In question ❼, children are identifying both missing operations and missing brackets from given calculations. Ask: *What operations could you use? How can you record the different ways you've already tried? Is one operation more likely to work than another? Why?*

THINK DIFFERENTLY In question ❺, children insert brackets into multi-step calculations where the answer is known. Encourage them to justify their thinking each time. Children should notice that, without brackets, the calculations in questions ❺ d) and e) seem the same but give two different answers. This will reinforce the importance of brackets when working through a calculation in the correct order.

ASSESSMENT CHECKPOINT Children should be able to fluently create and solve calculations with brackets. They should be able to fluently reason and problem solve, completing partially finished calculations or identifying where mistakes have been made. Question ❹ is particularly useful for assessing whether children can recognise how a calculation, taken from a contextual problem, would be presented. Look for children linking their understanding of the problem's 'story' with the order of operations.

ANSWERS Answers for the **Practice** part of the lesson can be found in the *Power Maths* online subscription.

Reflect

WAYS OF WORKING Independent thinking

IN FOCUS This question offers an opportunity to assess whether children can recognise and describe the role of brackets. Can they explain why a calculation can give different answers depending on whether brackets have been used or not?

ASSESSMENT CHECKPOINT Children should recognise that the order of operations mean the multiplication would be done first, but that by putting brackets round the addition, that step will now be done first, and so the answer will change.

ANSWERS Answers for the **Reflect** part of the lesson can be found in the *Power Maths* online subscription.

After the lesson ⏸

- How flexible were children in their use of brackets?
- Could they use trial and error confidently to achieve a desired result?
- Were children still able to follow the order of operations outside of any brackets or does this need revisiting?

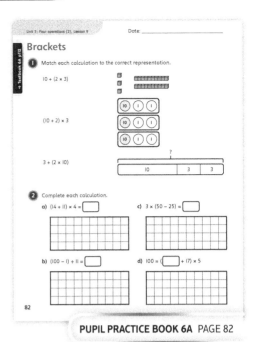

PUPIL PRACTICE BOOK 6A PAGE 82

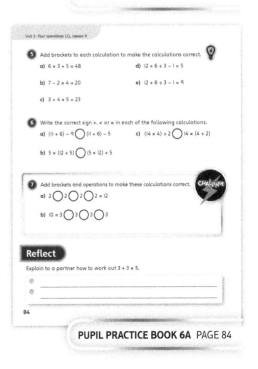

PUPIL PRACTICE BOOK 6A PAGE 83

PUPIL PRACTICE BOOK 6A PAGE 84

Mental calculations

Learning focus

In this lesson, children will learn efficient mental methods for solving calculations with smaller numbers, including decimals.

Before you teach 🔢

- What mental methods do children already know and use in the classroom?
- Are they more confident with one operation than another?
- How will you support those operations that need more practice?

NATIONAL CURRICULUM LINKS

Year 6 Number – addition, subtraction, multiplication and division
Perform mental calculations, including with mixed operations and large numbers.

ASSESSING MASTERY

Children can use efficient mental methods to confidently and fluently solve calculations with smaller numbers, including decimals. They can use these mental methods to help them solve number problems and puzzles, and can describe the methods they used, explaining how they were helpful.

COMMON MISCONCEPTIONS

When children are adding or subtracting by compensating, they may add or take away an insufficient amount at the end of the calculation. For example, when solving 0·99 + 5·98, they may calculate 1 + 6, but then either subtract the wrong amount at the end or not subtract anything at all. Ask:
- *What did you do to the numbers before you solved this mentally?*
- *Have you added back what you subtracted? Have you subtracted what you added?*

STRENGTHENING UNDERSTANDING

Children may benefit from having real-life practice of the concepts covered in this lesson, for example, by running a cake sale. Calculating the amount of ingredients needed to bake multiple cakes and dealing with money in a real context will provide a good opportunity for children to develop confidence with mental methods.

GOING DEEPER

Children could be challenged to investigate how the methods taught in this lesson might transfer to slightly larger numbers, prior to working with 1,000s and 1,000,000s in the next lesson. For example, ask: *If you can multiply by 9 easily, can you use the same method to multiply by 90 or 900? How about 999? Explain.*

KEY LANGUAGE

In lesson: mental method, written method, calculation, mentally

Other language to be used by the teacher: add, subtract, multiply

STRUCTURES AND REPRESENTATIONS

Tables, number lines, column additions, bar models

RESOURCES

Optional: base 10 equipment

 In the eTextbook of this lesson, you will find interactive links to a selection of teaching tools.

Quick recap

Ask children how many pence there are in £1.
Then ask them how many pence there are in £2.
Then ask: *What is 25p less than £5?*

Discover

Unit 3: Four operations (2), Lesson 10

Mental calculations ❶

Discover

WAYS OF WORKING Pair work

ASK

- Question ❶ a): *How much does bread cost?*
- Question ❶ b): *How much does each item cost?*
- Question ❶ b): *How much does each person spend? How did you work it out?*
- Question ❶ b): *Is there an easier way to find each total?*

IN FOCUS This part of the lesson provides a good opportunity to assess children's ability to calculate mentally. Make sure children are given time to feed back their ideas into the class discussion to allow you to judge the current class confidence level.

PRACTICAL TIPS Shopping role-play or, if the opportunity is available, a trip to a local shop would give children many chances to work out totals and differences in prices mentally.

ANSWERS

Question ❶ a): Holly receives 5p change.

Question ❶ b): Toshi spent £7·97.

❶ a) Holly buys 5 loaves of bread. She pays with £5.
   How much change does she receive?

b) Toshi buys yoghurt, bread and cereal.
   How much did Toshi spend?

116

PUPIL TEXTBOOK 6A PAGE 116

Share

WAYS OF WORKING Whole class teacher led

ASK

- Question ❶ a): *Why is it easier to round 0·99 up to 1 when calculating mentally?*
- Question ❶ a): *What is important to remember when using this method?*
- Question ❶ b): *What are the possible mistakes someone might make? Explain.*
- Question ❶ b): *Could this method be used for other numbers? Explain.*

IN FOCUS While discussing question ❶ a), make sure to point out the pattern in the numbers. Discuss how this can help children at the end to quickly and reliably compensate. Children could be encouraged to discuss how that pattern may change if they compensated numbers such as 0·98 or 0·97.

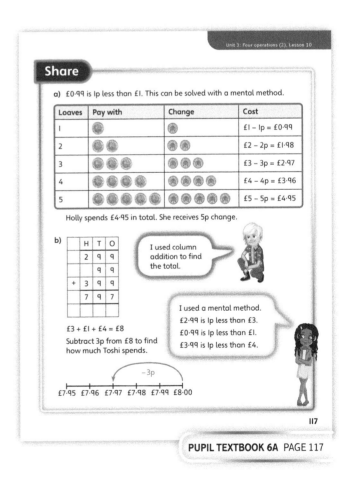

Share

a) £0·99 is 1p less than £1. This can be solved with a mental method.

Loaves	Pay with	Change	Cost
1			£1 – 1p = £0·99
2			£2 – 2p = £1·98
3			£3 – 3p = £2·97
4			£4 – 4p = £3·96
5			£5 – 5p = £4·95

Holly spends £4·95 in total. She receives 5p change.

b)

H	T	O
2	9	9
	9	9
+ 3	9	9
7	9	7

I used column addition to find the total.

I used a mental method.
£2·99 is 1p less than £3.
£0·99 is 1p less than £1.
£3·99 is 1p less than £4.

£3 + £1 + £4 = £8
Subtract 3p from £8 to find how much Toshi spends.

−3p

£7·95 £7·96 £7·97 £7·98 £7·99 £8·00

117

PUPIL TEXTBOOK 6A PAGE 117

Think together

WAYS OF WORKING Whole class teacher led (I do, We do, You do)

ASK

- Question **1**: *How could you multiply by 9 quickly?*
- Question **1**: *What easier number is 9 close to?*
- Question **2** b): *How do the brackets help you?*

IN FOCUS Question **1** looks at compensating by multiplying by 10 or 100, then subtracting 1 'group' or number to multiply by 99 or 9. Discuss with children how this can help them to multiply by other numbers. For example, multiplying by 4 could be achieved by multiplying by 5 and subtracting. Questions **2** and **3** develop children's flexibility and fluency with mental methods by asking them to consider which method is best suited for each calculation. Encourage children to share their reasoning about the methods they choose.

STRENGTHEN If children are struggling to mentally compensate, it may help to provide them with printed 'parts' of a bar model that would represent the problem they are solving. Children could, for example, lay down 10 parts to represent the simple multiplication, then take one of the parts away to represent the compensation.

DEEPEN While children are solving question **3**, ask: *What is similar and what is different about each calculation? How will you use that to help you? Can you write some more groups of questions like this where one number fact can help you solve several calculations?*

ASSESSMENT CHECKPOINT At this point in the lesson, children should be able to recognise compensation as an efficient mental method. They should be able to use this when solving addition, subtraction and multiplication calculations. Question **2** offers an opportunity to assess children's use of both mental methods taught in the lesson. Ask children to explain their method to ensure accurate assessment.

ANSWERS

Question **1** a): £4·95

Question **1** b): £4·05

Question **2** a): 19p + 29p + 39p should be solved as
(20p + 30p + 40p) − 3p = 87p

Question **2** b): £10 − (3 × £0·99) should be solved as
£10 − (3 × £1) + 3p = £7·03

Question **3**: Jamie can use a known fact such as:
100 × 24 = 2,400 or 50 × 25 = 1,250
to work out all the other calculations:
50 × 24 = 1,200
49 × 24 = 1,176
(23 × 24) + (27 × 24) = 1,200

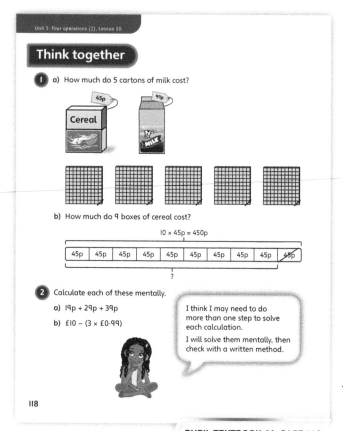

PUPIL TEXTBOOK 6A PAGE 118

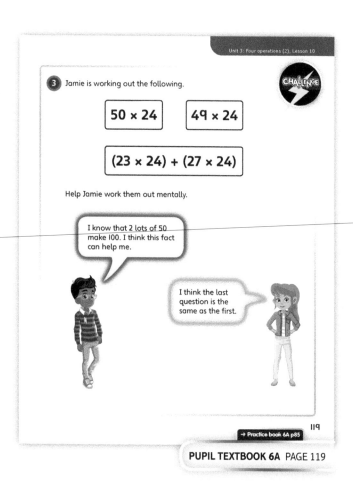

PUPIL TEXTBOOK 6A PAGE 119

Practice

WAYS OF WORKING Independent thinking

IN FOCUS It is important to focus on question ❶ as it links the mental method to the concrete and pictorial representations children will be familiar with. This will help them to visualise each problem and the method they need to use. In question ❸, watch for children who are tempted to use a written method. If you see any child doing this, ask: *How would you have solved this if a pencil and paper were not available? What is similar about your written method and your mental method and what is different?*

STRENGTHEN If children are struggling to solve question ❷, ask: *How could you represent each problem? Could you make finding the totals easier? How will you make sure you find the correct solution?*

DEEPEN Deepen children's reasoning when solving question ❻ by asking them to write an explanation, in a couple of sentences, about how they approached each calculation. Ask: *Could you have approached the question in a more efficient way?*

THINK DIFFERENTLY Question ❺ encourages children to recognise that the method of compensation can work for numbers other than 9. It also approaches the potential misconception where children add or subtract 1, regardless of the number they are dealing with and its difference between it and the next 10.

ASSESSMENT CHECKPOINT Children should be able to confidently solve mental calculations, using compensation to help them. They should be able to link this to concrete and pictorial representations of their mental methods and use this understanding to help them explain where use of mental methods is appropriate or where written methods are more so.

ANSWERS Answers for the **Practice** part of the lesson can be found in the *Power Maths* online subscription.

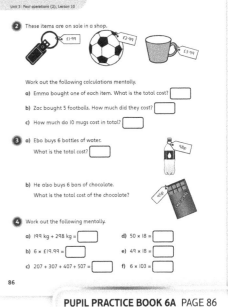

PUPIL PRACTICE BOOK 6A PAGE 85

PUPIL PRACTICE BOOK 6A PAGE 86

Reflect

WAYS OF WORKING Independent thinking

IN FOCUS This question assesses whether children can reliably explain how to use a mental method when it is the most appropriate and efficient choice. It will give evidence that children's fluency and reasoning with mental calculations will enable them to use the mathematics they know flexibly and confidently.

ASSESSMENT CHECKPOINT Children should be able to explain the kinds of clues to look for when using a mental method with a number near a multiple of ten.

ANSWERS Answers for the **Reflect** part of the lesson can be found in the *Power Maths* online subscription.

After the lesson

- Were children less confident at any particular operation?
- How will you support this in future lessons?
- Could this lesson have been made more practical?

PUPIL PRACTICE BOOK 6A PAGE 87

Mental calculations ❷

Learning focus

In this lesson, children will learn efficient mental methods for solving calculations with larger numbers, up to 1,000,000s.

Before you teach ⏸

- How confident were children at visualising problems mentally in the previous lesson?
- Does this skill need support before or during this lesson?

NATIONAL CURRICULUM LINKS

Year 6 Number – addition, subtraction, multiplication and division

Perform mental calculations, including with mixed operations and large numbers.

ASSESSING MASTERY

Children can use efficient mental methods to confidently and fluently solve calculations with larger numbers, up to 1,000,000s. They can use these mental methods to help them solve number problems and puzzles, and can describe the methods they used, explaining how they were helpful.

COMMON MISCONCEPTIONS

As numbers get larger, children are more likely to confuse the place value of a digit. This may result in incorrectly reading or writing numbers. Ask:

- *Can you identify the place value of each digit in this number?*
- *Having done that, can you read it again?*
- *Does what you have read match what you wanted to write? Explain.*

STRENGTHENING UNDERSTANDING

Before the lesson, it would be beneficial to practise multiplying and dividing larger numbers by 10, 100 and 1,000 through activities such as 'Follow me' cards or bingo games. This will ensure that all children are prepared for the mental methods that will be introduced in this lesson.

GOING DEEPER

Children could be encouraged to design their own word problems using larger numbers. Ask: *Can you create a word problem for your partner to solve? Now can you create a problem that has more than one step and at least two different operations?*

KEY LANGUAGE

In lesson: reduce, column subtraction, mental method, difference, increase, add, subtract, column method, exchange, reasoning, inverse operation, calculation, more than, less than, double, halve, take away

STRUCTURES AND REPRESENTATIONS

Column subtractions, place value grids, bar models, number lines, tables

RESOURCES

Optional: 'Follow me' cards, bingo game, number lines, real house sale advertisements

 In the eTextbook of this lesson, you will find interactive links to a selection of teaching tools.

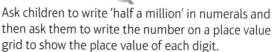

Quick recap 🔁

Ask children to write 'half a million' in numerals and then ask them to write the number on a place value grid to show the place value of each digit.

Discover

WAYS OF WORKING Pair work

ASK

- Question ① a): *What type of calculation is needed to work out a reduction?*
- Question ① a): *What is the easiest way of solving each subtraction?*
- Question ① a): *How did the numbers change when you solved the subtractions? Explain.*
- Question ① b): *Can you change the numbers in any way to make the calculation easier?*
- Question ① b): *Which house has the price that is easiest to calculate with? Explain.*

IN FOCUS Question ① a) encourages children to begin considering the most efficient method of solving the given problem. While discussing the question, it would be interesting to consider whether any of the methods children learnt in the last lesson will help them. Can they identify where these methods will be useful and where they will not?

PRACTICAL TIPS This part of the lesson could be easily geared towards the interests of your class, to ensure children are fully engaged. For example, the sale items could be changed to sports cars, jewellery, footballers, breeds of horse and so on. You could use real advertisements from a local newspaper.

ANSWERS

Question ① a): This requires two subtractions. Written or mental methods can be used. A mental method works well with these numbers.

Question ① b): House A is £800,000 more expensive than House C.

Share

WAYS OF WORKING Whole class teacher led

ASK

- Question ① a): *Which method is more efficient in this example: written or mental? Explain.*
- Question ① b): *How does looking at 950,000 as 950 thousands make calculating easier?*
- Question ① b): *Can you solve 760,000 – 240,000 using this method? Explain.*

IN FOCUS At this point in the lesson, make sure children recognise how their understanding of place value can be very powerful when solving calculations mentally. In question ① b), discuss the similarities between 950 – 150 and 950,000 – 150,000; what is similar and what is different about the two calculations? Some children may notice they are dividing each of the numbers by 1,000 to make them easier to calculate with mentally.

DEEPEN The discussion about question ① b) can be continued to consider how the mental method used in this question can be applied to other numbers.

Mental calculations ②

Discover

① a) The estate agent reduces the prices of House B and House D by £10,000. What methods can you use to work out the new costs?

b) What is the difference in price between the most expensive house and the least expensive house?

120

PUPIL TEXTBOOK 6A PAGE 120

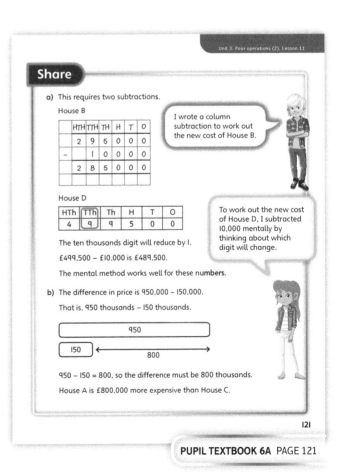

PUPIL TEXTBOOK 6A PAGE 121

Think together

ASK

- Question **1** a): *Is Dexter's idea a good one? Explain.*
- Question **1** b): *What mental method will you use to solve this question?*
- Question **1** b): *How is the mental method you are using more efficient than a written one?*
- Question **2**: *How can you use easier number facts to help you with these problems?*

IN FOCUS Use Dexter's comment in question **1** a) to reiterate the importance of looking at numbers in the 1,000s as *x* number of 1,000s. Again, discuss how this can make tackling numbers like those in the **Discover** picture easier. While solving question **1** b), discuss with children how visualising the number line can help them solve similar problems. Can they combine this with the method of dividing by 1,000 to solve the problem as efficiently as possible?

STRENGTHEN When solving question **2**, it may help children who are struggling to ask: *How could you write each number so it is easier to calculate with? Can you write the calculation that is being described? How will you solve it mentally?*

DEEPEN Children could be challenged to come up with more than one mental method to solve the calculations in question **3**. Once they have done so, encourage them to think critically about the methods they have come up with. Ask: *Which method is more efficient? Explain.*

ASSESSMENT CHECKPOINT At this point in the lesson, children should be able to mentally calculate addition and subtraction problems. They should be able to explain the mental method they used to do so. Question **3** assesses children's ability to link their learning from this lesson and the last, and their understanding of place value, to efficiently and accurately solve each calculation.

ANSWERS

Question **1** a): £75,000

Question **1** b): £50,000

Question **2** a): two hundred and fifty-six thousand (256,000)

Question **2** b): 1,450,000

Question **2** c): fifty thousand (50,000)

Question **2** d): You need to add 501,000 to 499,000 to make a million.

Question **3**: Look for children using their knowledge and understanding from the **Discover** and **Share** sections to help create mental methods for these calculations:

1,000 − 10 = 990
10,000 − 10 = 9,990
100,000 − 100 = 99,900
10,000,000 − 10,000 = 9,990,000

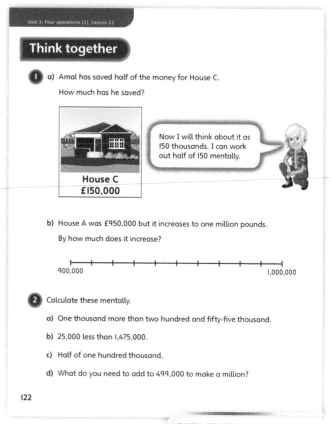

PUPIL TEXTBOOK 6A PAGE 122

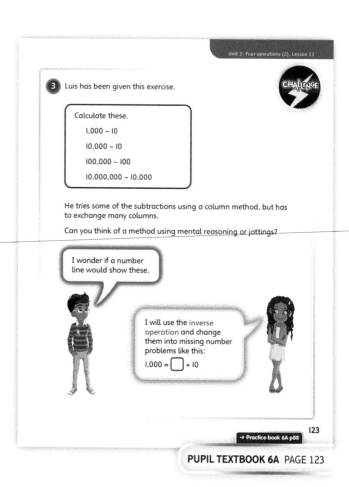

PUPIL TEXTBOOK 6A PAGE 123

Practice

WAYS OF WORKING Independent thinking

IN FOCUS While children are solving each of the questions in this part of the lesson, it will be interesting and valuable to stop them at regular intervals and discuss how they approached solving each question. Focus on where children have chosen different methods. Ask: *Why did you choose to solve it in that way? Do you think there was a more efficient method? Explain. Was your method as efficient as your partner's? Explain. How did you use the mental strategies you have learnt today to help you?*

STRENGTHEN For question ④, it may help to provide children with a blank number line with divisions drawn on. Ask: *How could you use this to represent the problems? Can you show me a representation of your mental method using this number line?*

DEEPEN Children could be given multiplication or division problems involving large numbers as seen throughout the lesson. Ask: *How can you use the methods you have learnt today to help you solve these calculations mentally?*

ASSESSMENT CHECKPOINT At this point in the lesson, children should be able to confidently solve calculations using larger numbers. They should be able to explain the method they used and how it was efficient at solving the problem they were working on. Question ⑦ assesses children's fluency and flexibility with the mental methods covered in this lesson and the last, by asking them to mentally calculate backwards through a problem.

ANSWERS Answers for the **Practice** part of the lesson can be found in the *Power Maths* online subscription.

Reflect

WAYS OF WORKING Independent thinking

IN FOCUS This question offers the opportunity to assess children's understanding of the mental methods covered in the lesson. If they are able to create a question that can be solved mentally and one that cannot, then they are showing good understanding of the necessary properties of a calculation that can be solved mentally. Once children have written their calculations, get them to swap with a partner. Ask: *Did your partner identify the calculation that you thought could not be solved mentally? Can they prove to you that they can solve the other one mentally?*

ASSESSMENT CHECKPOINT Look for children's understanding of mental methods through their ability to make a calculation that cannot be solved with them.

ANSWERS Answers for the **Reflect** part of the lesson can be found in the *Power Maths* online subscription.

After the lesson

- How will you make sure children continue to use and develop mental methods?
- How confident are you that all children were able to fluently solve the problems mentally?

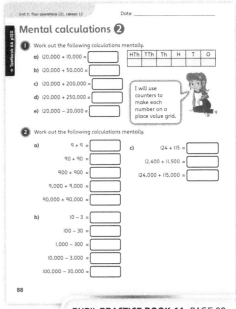

PUPIL PRACTICE BOOK 6A PAGE 88

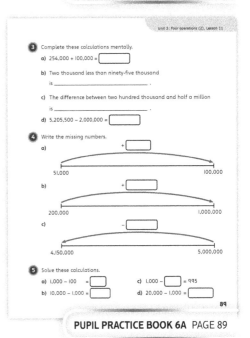

PUPIL PRACTICE BOOK 6A PAGE 89

Page 90 content:

6 Work out the following calculations mentally.
a) 8 × 3 =
8,000 × 3 =
b) 19 × 4 =
190 × 4 =
c) £199 × 3 =
£199,000 × 3 =

7 What number is Mrs Dean thinking of?
I add 100 and then double the result. After that, I subtract a quarter of a million. My final number is one more than 599,999.

Reflect

Write one calculation, using large numbers, that can be solved mentally.
Write one calculation that is difficult to solve mentally. Explain why this one is difficult to solve mentally.

PUPIL PRACTICE BOOK 6A PAGE 90

Reason from known facts

Learning focus

In this lesson, children will draw upon their learning throughout the unit to read, understand and solve mathematical puzzles and problems. They will use number facts they know to help them solve more complicated problems.

Before you teach

- How have the children responded to problem-solving activities in the past?
- How will this influence your teaching approach in this lesson?

NATIONAL CURRICULUM LINKS

Year 6 Number – addition, subtraction, multiplication and division

Use their knowledge of the order of operations to carry out calculations involving the four operations.

Solve problems involving addition, subtraction, multiplication and division.

ASSESSING MASTERY

Children can fluently and confidently draw upon their knowledge of number facts to understand and solve mathematical puzzles and problems. They can share their reasoning, confidently expressing their knowledge and understanding of numbers, and use this to help them solve more complicated problems.

COMMON MISCONCEPTIONS

Children may rely too heavily on trial and error, rather than thinking about the number facts they can see in the question and those they know. Ask:

- *Are you solving this problem in the most efficient way?*
- *Are there any clues in the question you could use?*
- *Are there any number facts you already know that would help you to solve this?*

STRENGTHENING UNDERSTANDING

Before teaching this lesson, it may help children to practise calculation strings. For example, 6 × 3 = 18, 6 × 30 = 180, 6 × 300 = 1,800 and so on. Ask: *What other calculations can you find using 6 × 3?*

GOING DEEPER

Challenge children to find three or more different methods that they could use to solve 7,500 ÷ 25. They should use diagrams to show how each of the different methods works.

KEY LANGUAGE

In lesson: fact, related facts, mind map, equation

Other language to be used by the teacher: compare, long division, inverse, addition, subtraction, multiplication

STRUCTURES AND REPRESENTATIONS

Column methods of multiplication and division

RESOURCES

Optional: multiplication grids, mini whiteboards

 In the eTextbook of this lesson, you will find interactive links to a selection of teaching tools.

Quick recap

Write on the board the calculation 17 × 5 = 85.

Ask: *How can you use this number fact to work out 17 × 6? What about 17 × 4?*

Discover

Reason from known facts

WAYS OF WORKING Pair work

ASK

- Question ❶ a): *What does the equals sign mean in this number sentence?*
- Question ❶ b): *Is the multiplication correct? How can you check?*

IN FOCUS Children are presented with two calculations which they can use to explore the relationship between numbers in equations and the use of the inverse operations. They consider how known facts and comparison can be used to derive other information.

PRACTICAL TIPS Work with children to complete each of the calculations. Encourage children to use diagrams to show each step of the process.

ANSWERS

Question ❶ a): 187

Question ❶ b): 25

Discover

❶ a) What number is missing from the addition calculation?

b) What is $925 \div 37$?

124

PUPIL TEXTBOOK 6A PAGE 124

Share

WAYS OF WORKING Whole class teacher led

ASK

- Question ❶ a): *What is the same and what is different about the top bar and the bottom bar?*
- Question ❶ b): *Can you show any other examples like this?*

IN FOCUS In question ❶ a), children explore addition equations that are given as a comparison statement with a missing number problem. They are shown how comparison or adjustment could be used to find the missing number. In question ❶ b), children are shown a long multiplication and consider how the inverse operation would be shown as a long division.

PUPIL TEXTBOOK 6A PAGE 125

Think together

Whole class teacher led (I do, We do, You do)

ASK

- Question **1** a): *What diagram could you draw to show your thinking?*
- Question **2**: *Can you think of a family of facts to go with Kate's calculation?*
- Question **3**: *Do you need to use any inverse operations?*

IN FOCUS In question **1**, children compare and adjust related addition equations in order to find missing numbers. In question **2**, they explore the use of inverse operations to derive division facts from a given multiplication. Question **3** requires children to consider different methods that they could use to solve missing number problems.

STRENGTHEN Build children's confidence by giving them the opportunity to first work on similar problems but using smaller numbers.

DEEPEN Challenge children to explore how they might solve 49 × 51 if they start by using the fact that 5 × 5 = 25. Can they explain how the facts are related and why this is helpful?

ASSESSMENT CHECKPOINT Assess whether children can justify their reasoning when selecting each method that they use to solve calculations with related facts.

ANSWERS

Question **1** a): 177

Question **1** b): 190

Question **1** c): 287

Question **1** d): $186\frac{1}{2}$

Question **2**: 174 × 8 = 1,392; 175 × 7 = 1,225,
1,218 ÷ 7 = 174; 1,218 ÷ 174 = 7

Question **3** a): 2,240 × 16 = 35,840
225 × 1,600 = 360,000

Question **3** b): 224 × 8 = **1,792**
112 × **32** = 3,584
224 × 16 = **222** × 16 + 32

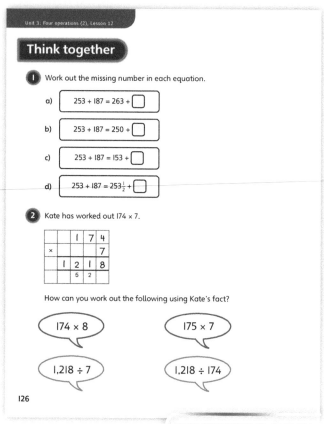

PUPIL TEXTBOOK 6A PAGE 126

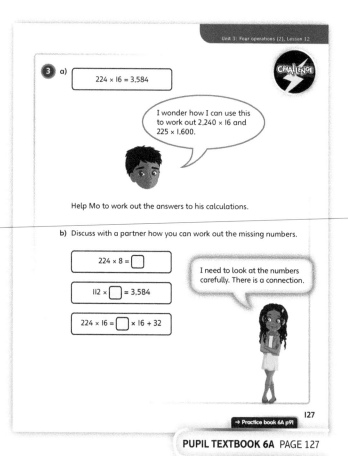

PUPIL TEXTBOOK 6A PAGE 127

Practice

WAYS OF WORKING Independent thinking

IN FOCUS In question **1**, children work through a series of addition equations to find missing numbers. In question **2**, they explore possible errors when solving a missing number equation. In question **3**, children use related facts to solve multiplications and divisions and, in question **4**, they use given facts to work out related facts.

STRENGTHEN Ask children to write families of related facts for a given times-table fact, for example 5 × 7 = 35.

DEEPEN If children finish the mind map in question **6**, ask: *Can you create your own mind map? What calculation will you put in the middle of your mind map? How many related facts can you find?*

THINK DIFFERENTLY Question **5** challenges assumptions about multiplying by 100. Children are likely to assume that, as one of the numbers in the calculation has an extra 100, then the answer should be multiplied by 100. The reality is, however, that the calculation will result in 100 more lots of 6, not 100 more lots of 288.

ASSESSMENT CHECKPOINT Children should be fluently using number facts they know to solve mathematical puzzles. They should be able to confidently explain how they use known number facts to solve each problem.

ANSWERS Answers for the **Practice** part of the lesson can be found in the *Power Maths* online subscription.

Reflect

WAYS OF WORKING Pair work

IN FOCUS Successfully writing their own number facts will demonstrate that children are able to make links between related calculations. After children have written three other facts on their own, allow them time to share with a partner. Have they found the same facts? Ask: *Can you explain the link between the given calculation and your partner's number facts? Why have they chosen those facts?*

ASSESSMENT CHECKPOINT Children should be able to identify how the three facts link to the original calculation given in the question.

ANSWERS Answers for the **Reflect** part of the lesson can be found in the *Power Maths* online subscription.

After the lesson ⏸

- Where could the problems in this lesson have been given real-life contexts to appeal more to the children?
- How resilient were the children when they were problem solving?
- If problem solving was a challenge, how will you support and develop this skill in the future?

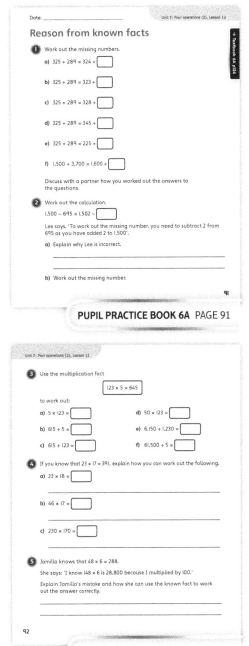

PUPIL PRACTICE BOOK 6A PAGE 91

PUPIL PRACTICE BOOK 6A PAGE 92

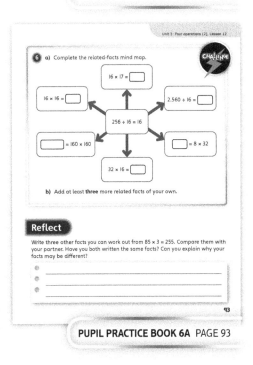

PUPIL PRACTICE BOOK 6A PAGE 93

End of unit check

> **Don't forget the unit assessment grid in your _Power Maths_ online subscription.**

WAYS OF WORKING Group work teacher led

IN FOCUS

- Question **1** assesses children's understanding of the order of operations and the effect of using brackets.
- Question **2** assesses children's mental methods and understanding of place value.
- Question **3** assesses whether children can use known facts to multiply a 4-digit number by a 2-digit number.
- Question **4** assesses if children are able to identify an error in divisions that use either factors or the long division method.
- Question **5** assesses if children can apply a suitable division method to a word problem and show an understanding of how remainders can be used in a contextual division problem.
- Question **6** is a SATS-style question that assesses children's ability to use a known number fact to solve a more challenging calculation.

ANSWERS AND COMMENTARY

Children will be able to efficiently and accurately solve the multiplication of a 4-digit number by a 1- or 2-digit number using a columnar method. Additionally, they will be able to use multiple methods to solve division calculations, including short and long division, explaining how their understanding of multiples and factors can help them. They will demonstrate mastery of all the written methods through using them fluently to solve real-life word problems. Children will be able to fluently adhere to the correct order of operations, including use of brackets. Finally, they will be able to solve problems using efficient mental methods and explain how a known fact can help to solve a related calculation.

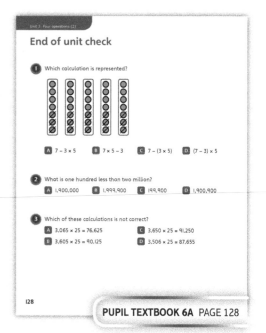

PUPIL TEXTBOOK 6A PAGE 128

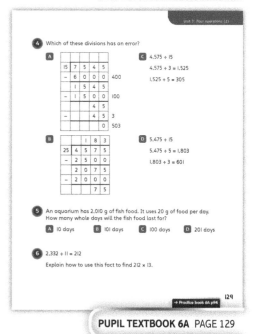

PUPIL TEXTBOOK 6A PAGE 129

Q	A	WRONG ANSWERS AND MISCONCEPTIONS	STRENGTHENING UNDERSTANDING
1	D	A suggests that the child knows the operations required but is not confident using brackets. B or C indicates the child has not mastered the order in which to write or calculate operations.	When solving calculations, like those in question **3**, ask: • _What do we know about the numbers?_ • _If we are looking at multiples of 25, what can we say will always be the case about those numbers?_
2	B	A, C or D indicates a place value error.	
3	D	Incorrect answers suggest children are not using known number facts to help them.	When children are trying to use the factor method to solve a division calculation, ask: • _What factors of the divisor can you list?_ • _Will the dividend easily divide into any of these factors?_
4	D	A, B or C suggests insufficient understanding of long division.	
5	C	A may indicate children have mistaken the remainder for the solution. B suggests children were unsure how the remainder fitted the context of the question. D suggests an insufficient understanding of long division.	• _What will you need to divide the resulting number by? Explain why._ • _Why is it important to factorise and not partition the divisor?_
6		Children should be able to find the related fact 212 × 11 = 2,332 and add two more lots of 212. 2,332 + (2 × 212) = 2,756.	

My journal

WAYS OF WORKING Independent thinking

ANSWERS AND COMMENTARY Children should recognise that they will need to work backwards, finding numbers that divide by 25 to leave a remainder of 10 by adding 10 to a multiple of 25. Following this thought process, they will be able to complete the given calculations. If children do not know where to begin, ask:

- *What do you know about every number that, when divided by 25, leaves a remainder of 10?*
- *How can you use this to begin investigating what numbers can be created?*
- *Could this be solved by trial and improvement? Is that the most efficient way to solve it? Why?*

Power check

WAYS OF WORKING Independent thinking

ASK

- *How confident are you that you could solve a division calculation efficiently and accurately?*
- *What checking methods would you use to check a multiplication calculation?*
- *How confident are you about the order of operations? Do you think you could explain it to someone else?*
- *Did you know what brackets are used for before starting this unit? How confident are you using them now?*

Power puzzle

WAYS OF WORKING Pair work

IN FOCUS Use this **Power puzzle** to assess whether children are confident multiplying a 3-digit number by a 2-digit number, using a written method. Ask:

- *What method did you use to solve the multiplication?*
- *Has your partner used a different method? What is the same and what is different about your methods?*

ANSWERS AND COMMENTARY Answers will vary, depending on the numbers the children have chosen.

Children should be able to use a written method such as the grid method or column method to solve their multiplications. In order to score the greater total, they may strategically place the numbers rolled on the dice within their multiplications. For example: larger numbers should be placed in the tens or hundreds boxes and smaller numbers should be placed in the ones boxes. In this way, children also strengthen their understanding of place value.

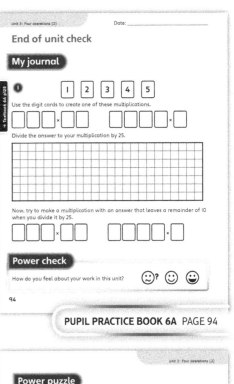

PUPIL PRACTICE BOOK 6A PAGE 94

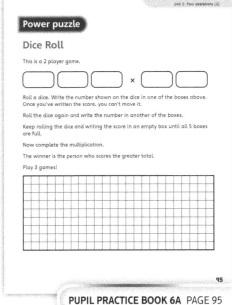

PUPIL PRACTICE BOOK 6A PAGE 95

After the unit ⏸

- How will you weave the learning from this unit into children's future problem-solving activities? For example, using the order of operations and brackets to more easily generalise about number patterns.
- Did the **End of unit** check show any misconception that the class still has? How will you support and develop this area of learning?

Strengthen and **Deepen** activities for this unit can be found in the *Power Maths* online subscription.

Unit 4
Fractions ①

Mastery Expert tip! 'To develop confidence in adding and subtracting fractions and mixed numbers, I encourage children to use fractions in different areas of mathematics. For example, when finding perimeters of shapes, use fractional lengths.'

Don't forget to watch the Unit 4 video!

WHY THIS UNIT IS IMPORTANT

This unit is important because it develops children's understanding of fractions, moving on to comparing, adding and subtracting unrelated fractions using common denominators. It encourages problem solving of fraction problems while exploring efficient methods.

WHERE THIS UNIT FITS

→ Unit 3 – Four operations (2)

→ **Unit 4 – Fractions (1)**

→ Unit 5 – Fractions (2)

In this unit, children extend their understanding of fractions and mixed numbers by adding and subtracting unrelated fractions by finding common denominators. Children continue to develop their reasoning and problem-solving skills while exploring efficient methods.

Before they start this unit, it is expected that children:
- can find factors and multiples of numbers using multiplication facts
- can find equivalent fractions and convert between improper fractions and mixed numbers
- can compare and order fractions and add and subtract fractions which have the same denominator.

ASSESSING MASTERY

Children who have mastered this unit will be able to add and subtract fractions and mixed numbers confidently using several formal written methods. They will be able to solve multi-step problems and explain which method is most efficient.

COMMON MISCONCEPTIONS	STRENGTHENING UNDERSTANDING	GOING DEEPER
Children may simplify to find an equivalent fraction but not fully simplify the fraction.	Encourage children to use fraction strips and to write out factors of the numerator and denominator to identify common factors.	Give children some fractions and ask them to identify which ones have been simplified fully. Encourage them to explain why.
Children may compare the numbers in fractions rather than comparing the overall fractions.	Encourage children to initially compare fractions with the same denominator, so they understand that they can compare the numerators **only if** the denominators are the same.	Encourage children to identify the most efficient way of comparing fractions. This may not always be to find a common denominator, such as $\frac{2}{3}$ and $\frac{16}{17}$ ($\frac{16}{17}$ is closer to 1).
Children may simply add or subtract the numerators and denominators when adding or subtracting fractions.	Use fraction strips to show that the denominators must be the same before adding or subtracting the numerators. Start with fractions that have the same denominator to clarify why they do not add or subtract the denominators.	Encourage children to add and subtract fractions and mixed numbers using the most efficient method, for example $\frac{1}{2} + \frac{1}{4}$ (could be done mentally) and $5\frac{1}{6} - 4\frac{7}{8}$ (think about adding $\frac{1}{8}$ and $\frac{1}{6}$).

UNIT STARTER PAGES

Introduce this unit using teacher-led discussion. Allow children time to discuss questions in pairs or small groups and share ideas as a whole class. Children should be encouraged to use representations to visualise the fractions.

STRUCTURES AND REPRESENTATIONS

Number lines: The number line will allow children to see which numbers a fraction sits between and the fractional divisions. It can help children to visualise fractional increases or decreases.

$$0 \qquad \frac{3}{5} \qquad 1$$

Fraction strips: These models allow children to see how to split up fractions so they can be added or subtracted using common denominators.

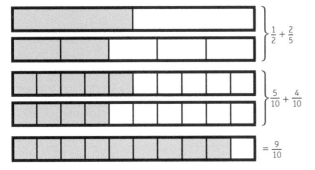

$$\frac{1}{2} + \frac{2}{5}$$

$$\frac{5}{10} + \frac{4}{10}$$

$$= \frac{9}{10}$$

KEY LANGUAGE

There is some key language that children will need to know as part of the learning in this unit.

→ whole, part
→ numerator, denominator, common denominator, lowest common denominator
→ equivalent
→ simplify, simplest form
→ factor, common factor, highest common factor, lowest common multiple (LCM)
→ compare
→ order
→ less than, greater than
→ proper fraction, improper fraction
→ mixed number
→ convert

PUPIL TEXTBOOK 6A PAGE 130

PUPIL TEXTBOOK 6A PAGE 131

Equivalent fractions and simplifying

Learning focus

In this lesson, children will apply their knowledge of factors to use common factors to simplify fractions. They will also extend their understanding of simplifying fractions to simplify mixed numbers and improper fractions.

NATIONAL CURRICULUM LINKS

Year 6 Number – fractions

Use common factors to simplify fractions; use common multiples to express fractions in the same denomination.

ASSESSING MASTERY

Children can confidently simplify fractions by dividing the numerator and denominator by a common factor. They can show fluency in using known multiplication facts, finding factors and demonstrating an understanding of the terms simplify, numerator, denominator and equivalent. They can also simplify a fraction (including mixed numbers and improper fractions), either by finding and using the highest common factor of the numerator and the denominator, or by using repeated division.

COMMON MISCONCEPTIONS

A common misconception among children is that they believe that, to simplify a fraction, you just divide the numerator and denominator by 2; for example, to simplify $\frac{9}{12}$, children may think the answer is $\frac{4 \cdot 5}{6}$. Conversely, they may realise that the numerator is not divisible by 2 and may incorrectly conclude that the fraction cannot be simplified because of this. To address this misconception, it is important to expose children to problems where the numerator or denominator is not divisible by 2, but the fraction can be simplified. Ask:

- *Are the numerator and denominator odd or even?*
- *If 2 is not a common factor, are there any odd common factors?*

Another common mistake that children make when simplifying an improper fraction is to convert it to a mixed number without simplifying, or to mix up the numerator and denominator. For example, some children may simplify $\frac{8}{4}$ to $\frac{1}{2}$ instead of $\frac{2}{1}$ or 2. Showing this on a fraction strip will help children to understand why the fraction simplifies to 2 and not $\frac{1}{2}$.

STRENGTHENING UNDERSTANDING

Children who are struggling with simplifying fractions fully should be encouraged to represent the fractions on a fraction strip. It may be beneficial to practise this with fractions that have only one possible way of being simplified. Encourage children to identify the factors of the numerator and denominator and to consider the common factors before moving on to fractions that can be simplified in several steps.

GOING DEEPER

Deepen learning by giving children a fraction or mixed number, for example $\frac{7}{8}$ or $2\frac{1}{3}$, and asking them to find at least three fractions that simplify to the original fraction.

KEY LANGUAGE

In lesson: numerator, denominator, improper fraction, factor, **simplest form**, common factor, **simplify**, equivalent, fully

Other language to be used by the teacher: highest common factor, HCF, improper fraction, mixed number

STRUCTURES AND REPRESENTATIONS

Fraction strips

 In the eTextbook of this lesson, you will find interactive links to a selection of teaching tools.

Quick recap

Ask children to draw three or more different diagrams to show the following fractions:

$$\frac{1}{4} \qquad \frac{1}{3} \qquad \frac{2}{3}$$

Compare and contrast the different diagrams together as a class.

Discover

Equivalent fractions and simplifying

Discover

WAYS OF WORKING Pair work

ASK

- Question ❶ a): *How many people are there? How many are children? How can you write this as a fraction?*
- Question ❶ b): *What do you notice about the numerator and the denominator? Is the numerator in any times-tables? Is the denominator in any times-tables?*

IN FOCUS In question ❶ a), children will generate a fraction from a real-life context presented as a word problem. In question ❶ b), they will start to recognise when fractions can be simplified because the numerator and the denominator share a common factor.

PRACTICAL TIPS Ask children to make a 3 × 6 array of counters to represent the total number of people and then to show what fraction of this total is the number of children.

ANSWERS

Question ❶ a): $\frac{12}{18}$ of the people are children.

Question ❶ b): Various equivalent fractions are possible, including $\frac{6}{9}$, $\frac{4}{6}$ or $\frac{2}{3}$. ($\frac{2}{3}$ is the simplest form of $\frac{12}{18}$.)

❶ a) There are 18 people in total.
   What fraction are children?

   $\boxed{}$
   $\overline{18}$

b) Find equivalent fractions for the fraction of children.

132

PUPIL TEXTBOOK 6A PAGE 132

Share

WAYS OF WORKING Whole class teacher led

ASK

- Question ❶ a): *How many people are there in total? How many of them are children?*
- Question ❶ b): *Can you complete this stem sentence?*

   '…. out of … is equivalent to $\frac{12}{18}$.'

- Question ❶ b): *Which equivalent fractions can you see? Which of them can you find by multiplying the numerator and denominator of $\frac{12}{18}$? Which equivalent fractions do you get by dividing the numerator and denominator of $\frac{12}{18}$?*

IN FOCUS Question ❶ a) supports children in understanding fractions as one number 'out of' a larger total number. In question ❶ b), children will need to recognise how to multiply in order to find equivalent fractions and will then be introduced to finding equivalent fractions by dividing or simplifying.

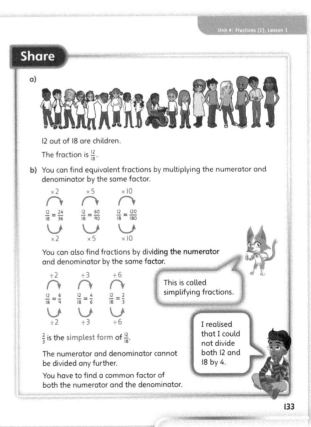

Share

a)

12 out of 18 are children.
The fraction is $\frac{12}{18}$.

b) You can find equivalent fractions by multiplying the numerator and denominator by the same factor.

$$\frac{12}{18} \overset{\times 2}{\underset{\times 2}{=}} \frac{24}{36} \qquad \frac{12}{18} \overset{\times 5}{\underset{\times 5}{=}} \frac{60}{90} \qquad \frac{12}{18} \overset{\times 10}{\underset{\times 10}{=}} \frac{120}{180}$$

You can also find fractions by dividing the numerator and denominator by the same factor.

$$\frac{12}{18} \overset{\div 2}{\underset{\div 2}{=}} \frac{6}{9} \qquad \frac{12}{18} \overset{\div 3}{\underset{\div 3}{=}} \frac{4}{6} \qquad \frac{12}{18} \overset{\div 6}{\underset{\div 6}{=}} \frac{2}{3}$$

This is called simplifying fractions.

$\frac{2}{3}$ is the simplest form of $\frac{12}{18}$.

The numerator and denominator cannot be divided any further.

You have to find a common factor of both the numerator and the denominator.

I realised that I could not divide both 12 and 18 by 4.

133

PUPIL TEXTBOOK 6A PAGE 133

Think together

WAYS OF WORKING Whole class teacher led (I do, We do, You do)

ASK

- Question **1** a): *What is the common factor of 4 and 6? How does the fraction strip help?*
- Question **2**: *What are the common factors of the numerator and denominator in this question? What multiplication facts do you know to help you find the common factors?*

IN FOCUS Question **1** supports children to develop their understanding of simplifying fractions by dividing both the numerator and the denominator by a common factor, while using fraction strips as scaffolding. In question **1** b), it may be beneficial to discuss how 5 is half of 10, so $\frac{5}{10}$ can be simplified to $\frac{1}{2}$.

Question **2** gives children an opportunity to simplify fractions without using fraction strips as support. Encourage children to discuss what they are dividing the numerator and denominator by, and to use their known multiplication facts and division rules to find common factors.

STRENGTHEN If children are struggling to simplify fractions, encourage them to draw their own fraction strip to help work out the answer. Some children might struggle to recall multiplication facts. In this instance, encourage them to write these down when thinking about common factors and to use their known multiplication facts to link with other facts (for example, doubling to connect the 2, 4 and 8 times-tables).

DEEPEN Extend question **3** by asking children to write similar equivalent fractions with mistakes for a partner to explain and correct.

ASSESSMENT CHECKPOINT Questions **1** and **2** will assess children's ability to simplify fractions using a common factor; question **2** withdraws the pictorial support of fraction strips. Look for children using multiplication facts to find factors.

ANSWERS

Question **1** a): $\frac{4}{6} = \frac{2}{3}$

Question **1** b): $\div 5$ $\frac{5}{10} = \frac{1}{2}$

Question **1** c): $\div 3$ $\frac{9}{15} = \frac{3}{5}$

Question **2** a): $\frac{3}{4}$

Question **2** b): $\frac{1}{10}$

Question **2** c): 1

Question **3**: Kate can simplify again by dividing by 3:

$$\frac{12}{30} = \frac{6}{15} = \frac{2}{5}$$

Max has tried to simplify the mixed number instead of just the fraction:

$2\frac{3}{9}$ simplifies to $2\frac{1}{3}$.

Reena thinks that only proper fractions can be simplified, but this is not the case:

$$\frac{24}{8} = \frac{3}{1} = 3$$

Ebo made a mistake in the multiplication facts for the 7 times-table:

$56 \div 7 = 8$, not 9

$$\frac{7}{56} = \frac{1}{8}$$

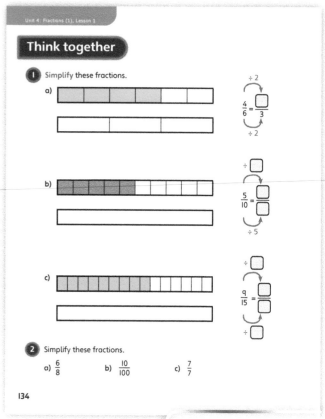

Think together

1 Simplify these fractions.

a)

b)

c)

2 Simplify these fractions.

a) $\frac{6}{8}$ b) $\frac{10}{100}$ c) $\frac{7}{7}$

134

PUPIL TEXTBOOK 6A PAGE 134

3 Each pupil has made a mistake. Discuss with a partner their ideas and how to simplify each fraction fully.

$\frac{12}{30} = \frac{6}{15}$

Kate: This cannot simplify further, because the numerator divides by 2 but the denominator doesn't. $\frac{6}{15}$ is the simplest form of $\frac{12}{30}$.

$2\frac{3}{9}$

Max: There is no common factor of 2, 3 and 9, so $2\frac{3}{9}$ cannot be simplified.

$\frac{24}{8}$

Reena: This cannot simplify because it is an improper fraction.

$\frac{7}{56} = \frac{1}{9}$

Ebo: I used my 7 times-tables to simplify this fraction.

135

→ Practice book 6A p96

PUPIL TEXTBOOK 6A PAGE 135

Practice

WAYS OF WORKING Independent thinking

IN FOCUS Questions **1** and **2** consolidate children's understanding of simplifying fractions by dividing the numerator and denominator by a common factor, using limited pictorial representation as support. Question **4** deepens children's conceptual understanding by asking them to identify the fractions that are not equivalent (non-examples). In question **6**, they simplify mixed numbers and improper fractions. Question **8** requires children to identify whether or not given fractions are equivalent to a whole number. They should notice that one of the fractions is not an improper fraction and so will not have a whole number answer when simplified.

STRENGTHEN If children are struggling with simplifying fractions in questions **1** and **2**, encourage them to use fraction strips and to explain their steps. Ask: *What number is a common factor of both the numerator and denominator?* Strengthen learning by encouraging children to simplify in multiple steps if they are struggling to find the highest common factor. Ask: *What number do you know goes into the numerator and denominator? How do you know? Can you simplify further?*

DEEPEN To build on question **2**, give children more fractions that can be simplified using different methods. For example, $\frac{24}{30}$ could be simplified by dividing the numerator and denominator by 2 to get $\frac{12}{15}$, then dividing by 3 to get $\frac{4}{5}$; alternatively, it could be simplified by dividing the numerator and denominator by the highest common factor of 6 to get $\frac{4}{5}$. Ask: *What method did you use? Could you use a different method?*

ASSESSMENT CHECKPOINT Questions **1** and **2** assess children's ability to simplify fractions using a common factor with some pictorial support. Question **3** assesses children's ability to find and identify equivalent fractions. Question **5** will assess children's ability to fully simplify a proper fraction, an improper fraction and a mixed number in abstract form. Look for children confidently using the lowest common multiple (not yet formally introduced) to simplify.

ANSWERS Answers for the **Practice** part of the lesson can be found in the *Power Maths* online subscription.

Reflect

WAYS OF WORKING Independent thinking

IN FOCUS The **Reflect** part of the lesson shows children's understanding by asking them to explain how they would simplify a fraction.

ASSESSMENT CHECKPOINT Assess whether children can explain clearly how to simplify a fraction. Do they use terminology correctly?

ANSWERS Answers for the **Reflect** part of the lesson can be found in the *Power Maths* online subscription.

After the lesson ⏸

- Are children confident in using the terms 'simplify', 'numerator', 'denominator' and 'equivalent'?
- Can children use known multiplication facts to find common factors and use multiples to find equivalent fractions?
- Can children fully simplify a proper fraction, an improper fraction and a mixed number?
- Do children understand that in a mixed number the whole number does not need to be simplified?

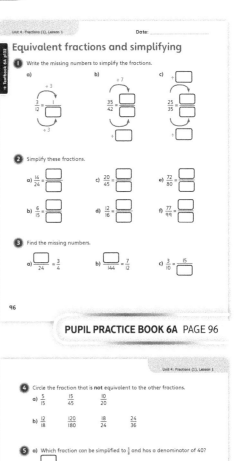

PUPIL PRACTICE BOOK 6A PAGE 96

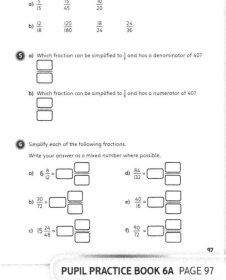

PUPIL PRACTICE BOOK 6A PAGE 97

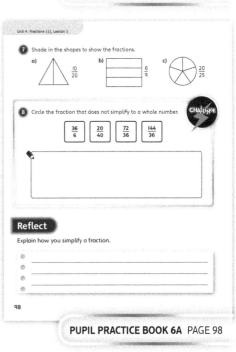

PUPIL PRACTICE BOOK 6A PAGE 98

Equivalent fractions on a number line

Learning focus

In this lesson, children use their understanding of fractions to count up and down on a number line, place missing fractions on a number line and find missing numbers in a fractional sequence.

NATIONAL CURRICULUM LINKS

Year 6 Number – fractions

Compare and order fractions, including fractions > 1.

ASSESSING MASTERY

Children can reliably count up and down fractional increases or decreases on a number line, find missing fractions in a sequence and place missing fractions on a number line.

COMMON MISCONCEPTIONS

Some children may become confused when counting beyond 1. For example, when counting in steps of $\frac{2}{3}$, they may say: $\frac{2}{3}$, $1\frac{2}{3}$, $2\frac{2}{3}$, $3\frac{2}{3}$ etc., instead of bridging the whole number correctly. Ask:
- *What fractional divisions is the number line divided into?*

STRENGTHENING UNDERSTANDING

Children who are struggling to place fractions on a number line can be encouraged to strengthen their understanding by counting on or back in fractions where the numerator is 1 (unit fractions) on number lines where all the numbers are labelled, before moving on to counting on or back in non-unit fractions using number lines where some of the numbers are missing. Encourage children to fill in all the unknown numbers on a number line before working out the missing numbers in a pattern.

GOING DEEPER

Children can be encouraged to deepen their understanding of using number lines for fractions, by simplifying fractions before placing them on the number line. For example, children could be asked to draw a number line from 2 to 4 that is divided into thirds, and then place on it numbers such as $2\frac{3}{9}$, $3\frac{8}{12}$, $2\frac{11}{33}$.

Encourage children to use their knowledge of equivalent fractions when placing fractions on a number line. For example, ask children to draw a number line from 5 to 7, divided into twelfths, and then place numbers on it such as $5\frac{2}{3}$, $6\frac{3}{4}$, $5\frac{1}{2}$ and $6\frac{5}{6}$.

KEY LANGUAGE

In lesson: number line, divide, fraction, greatest, more, less, gaps

Other language to be used by the teacher: division, pattern, simplify, equivalent

STRUCTURES AND REPRESENTATIONS

Number lines with fractional divisions

 In the eTextbook of this lesson, you will find interactive links to a selection of teaching tools.

Discover

Equivalent fractions on a number line

WAYS OF WORKING Pair work

ASK

- Question ① a): *How much water would Max collect after one hour? After two hours? After three hours?*
- Question ① a): *How long will it be until Max's jug is full?*

IN FOCUS This scenario gives children the opportunity to count up in steps of a given unit fraction, including mixed numbers and improper fractions.

PRACTICAL TIPS Ask children to shade in or cover up sections of simple fraction diagrams as you count up together in steps of one chosen unit fraction.

ANSWERS

Question ① a): Children count in eighths, then should also try counting using simplified fractions.

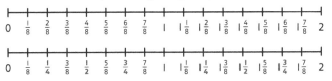

Question ① b): Children count on in sixths.

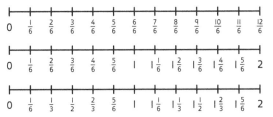

Discover

① a) Max collects $\frac{1}{8}$ litre every hour.
   Count in $\frac{1}{8}$s on a number line.

   b) Max's dad collects $\frac{1}{6}$ litre every hour.
   Count in $\frac{1}{6}$s on a number line.

136

PUPIL TEXTBOOK 6A PAGE 136

Share

WAYS OF WORKING Whole class teacher led

ASK

- Question ① a): *Can you count up and keep the denominator as 8 each time?*
- Question ① a): *Can you count up but simplify the fraction each time? Which did you find trickier? Which did you enjoy more?*
- Question ① a): *Can you use both methods to do both counts? Which did you find trickier, counting in eighths or counting in sixths?*
- Question ① b): *What would happen if you counted up in quarters?*

IN FOCUS In question ① a), children use the number line model to count up in eighths. The count is first shown as mixed numbers with eighths and then as mixed numbers with the fractions simplified where possible. In question ① b), children use the number line model to count up in sixths. This time the count is first shown as improper fractions with sixths, then as mixed numbers with sixths and then as mixed numbers with the fractions simplified where possible.

Share

a) You can count in $\frac{1}{8}$s.

You can also use the simplest form of each fraction.

b) Here are different ways to count in $\frac{1}{6}$s.

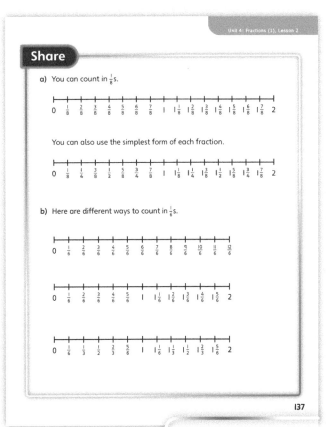

137

PUPIL TEXTBOOK 6A PAGE 137

Think together

Whole class teacher led (I do, We do, You do)

ASK

- Question **1**: *What fractions can you see? What simplified fractions can you see?*
- Question **2**: *Why do you think twelfths have so many different simplified fractions?*
- Question **3**: *What do you notice about the fractions that cannot be simplified?*

IN FOCUS In question **1**, children are simplifying tenths, with support from the number line model. In question **2**, children draw their own number line to show how to simplify twelfths. In question **3**, they explore which fractions can have simplified forms. They should find that fractions where the denominator is a prime number can never be simplified.

STRENGTHEN Rehearse counting together in steps of fractions that are not being simplified. Then gradually add simplified fractions (for example, $\frac{1}{2}$) into the count.

DEEPEN Challenge children to explore the pattern of simplified fractions when counting up in twentieths.

ASSESSMENT CHECKPOINT Question **2** will allow you to assess whether children can represent equivalent fractions on a number line and work flexibly with simplified fractions.

ANSWERS

Question **1** a):

Question **1** b): $\frac{1}{10}, \frac{1}{5}, \frac{3}{10}, \frac{2}{5}, \frac{1}{2}, \frac{3}{5}, \frac{7}{10}, \frac{4}{5}, \frac{9}{10}, 1$

Question **2**: The fractions should be changed to:

$\frac{1}{12}, \frac{1}{6}, \frac{1}{4}, \frac{1}{3}, \frac{5}{12}, \frac{1}{2}, \frac{7}{12}, \frac{2}{3}, \frac{3}{4}, \frac{5}{6}, \frac{11}{12}, 1$

$1\frac{1}{12}, 1\frac{1}{6}, 1\frac{1}{4}, 1\frac{1}{3}, 1\frac{5}{12}, 1\frac{1}{2}, 1\frac{7}{12}, 1\frac{2}{3}, 1\frac{3}{4}, 1\frac{5}{6}, 1\frac{11}{12}, 2$

Question **3**: Answers will vary, but children should express that:
- Fractions with a prime number denominator do not simplify.
- Fractions can be simplified if their denominator has more than two factors (is a composite number).
- Denominators with a lot of factors have the most interesting simplifications.

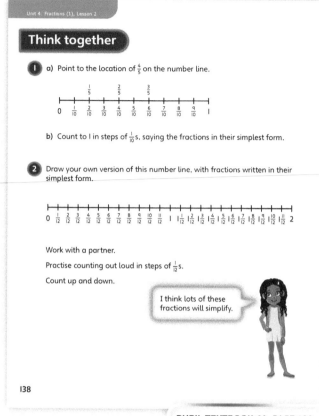

Think together

1 a) Point to the location of $\frac{4}{5}$ on the number line.

b) Count to 1 in steps of $\frac{1}{10}$s, saying the fractions in their simplest form.

2 Draw your own version of this number line, with fractions written in their simplest form.

Work with a partner.

Practise counting out loud in steps of $\frac{1}{12}$s.

Count up and down.

> I think lots of these fractions will simplify.

138

PUPIL TEXTBOOK 6A PAGE 138

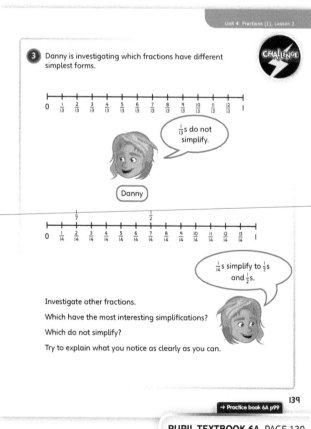

3 Danny is investigating which fractions have different simplest forms.

> $\frac{1}{13}$s do not simplify.

Danny

> $\frac{1}{14}$s simplify to $\frac{1}{7}$s and $\frac{1}{2}$s.

Investigate other fractions.

Which have the most interesting simplifications?

Which do not simplify?

Try to explain what you notice as clearly as you can.

139

→ Practice book 6A p99

PUPIL TEXTBOOK 6A PAGE 139

Practice

WAYS OF WORKING Independent thinking

IN FOCUS In question ①, children use the number line to model simple counting in fraction steps, including fractions that can be simplified. Question ② requires children to recognise the location of given fractions on a marked number line, including those that have been simplified. In question ③, children count in fraction steps, simplifying where possible, and in question ④ they simplify mixed number fractions.

STRENGTHEN Ask children to draw number lines for quarters and then to label the unsimplified fractions below the line and the simplified fractions above the line.

DEEPEN In question ⑤, children explore which simplified fractions they can make when counting in $\frac{1}{36}$ths. Deepen this by asking them to identify which times-tables contain the number 36 and to explain how this helps them to simplify the fractions in their count.

ASSESSMENT CHECKPOINT Use question ③ to assess whether children can simplify to show equivalent fractions on given number lines.

ANSWERS Answers for the **Practice** part of the lesson can be found in the *Power Maths* online subscription.

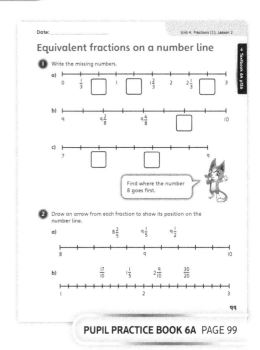

PUPIL PRACTICE BOOK 6A PAGE 99

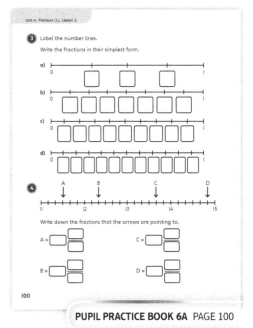

PUPIL PRACTICE BOOK 6A PAGE 100

Reflect

WAYS OF WORKING Pair work

IN FOCUS Encourage children to write all the unknown numbers on the number line and to explain how they know their answers are correct. Children may need support with identifying where an eighth lies on a number line that is divided into quarters.

ASSESSMENT CHECKPOINT Look for children fluently explaining where a fraction lies on a number line, reliably converting between mixed numbers and improper fractions as needed.

ANSWERS Answers for the **Reflect** part of the lesson can be found in the *Power Maths* online subscription.

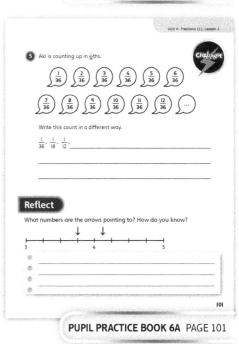

PUPIL PRACTICE BOOK 6A PAGE 101

After the lesson ⏸

- Can children fully simplify a proper fraction, an improper fraction and a mixed number?
- Can children confidently use repeated division to fully simplify a fraction?
- Do children understand that in a mixed number the whole number does not need to be simplified?

Compare and order fractions

Learning focus

In this lesson, children will use their understanding of fractions to develop their ability to compare and order fractions by making the denominators the same and comparing the numerators.

Before you teach ⏸

- Can children compare and order fractions that have the same denominator?
- Can children find the lowest common multiple (LCM) of two or more numbers?
- Can children find equivalent fractions?

NATIONAL CURRICULUM LINKS

Year 6 Number – fractions

Compare and order fractions, including fractions > 1.

Use common factors to simplify fractions; use common multiples to express fractions in the same denomination.

ASSESSING MASTERY

Children can confidently compare and order more than two fractions by finding the LCM and comparing the numerators.

COMMON MISCONCEPTIONS

Children may focus on the numbers in the fractions and compare these, rather than looking at the overall fractions. For example, when comparing $\frac{2}{3}$ and $\frac{3}{5}$ children may think $\frac{2}{3}$ is bigger because 3 and 5 are bigger than 2 and 3. Ask:

- *How can you compare the fractions, not just the digits?*
- *Do you need to find the LCM of the denominators?*
- *How can you use equivalent fractions to compare the fractions?*
- *Would a fraction strip help?*

Encourage children to realise that in order to compare and order fractions, they need to consider the value of the fractions rather than the individual digits, using the LCM of the denominators and equivalent fractions to make the comparison. Support children with representations on a fraction strip to secure understanding and avoid this misconception.

STRENGTHENING UNDERSTANDING

Children who are struggling should be encouraged to consolidate learning by comparing and ordering fractions that have the same denominator. Encourage them to show these fractions on a fraction strip and use the numerators to compare the fractions. Move on to compare fractions where one of the denominators is a multiple of the other, such as $\frac{1}{2}$ and $\frac{3}{4}$ or $\frac{2}{3}$ and $\frac{5}{12}$, so they can easily find the common denominator.

GOING DEEPER

Children could be challenged to find missing fractions: give children some fractions and ask them to find the missing fraction so that the fractions are in order, for example $\frac{1}{8}, \frac{1}{4}, \frac{1}{2}, \square, \frac{7}{8}$.

KEY LANGUAGE

In lesson: lowest common denominator, lowest common multiple (LCM), greater than, less than, order, denominator, numerator, equivalent

Other language to be used by the teacher: ascending, inequality symbol, comparison

STRUCTURES AND REPRESENTATIONS

Fraction strips

 In the eTextbook of this lesson, you will find interactive links to a selection of teaching tools.

Quick recap

Count together as a class from 0 to 2 in steps of sixths. Ensure children include simplified fractions in their count.

Discover

Unit 4: Fractions (1), Lesson 3

Compare and order fractions

Discover

Group A   Group B

1. a) Which group has a greater fraction of people wearing glasses?

   b) Another group, Group C, has $\frac{2}{3}$ of people wearing glasses.

      Which group now has the greatest fraction of people wearing glasses?

140

PUPIL TEXTBOOK 6A PAGE 140

WAYS OF WORKING Pair work

ASK

- Question 1 a): *How many people wear glasses in each group? Out of how many? How can you write this as a fraction?*
- Question 1 b): *Could you use a model to help you compare the fractions?*
- Question 1 b): *Which fraction do you need to compare $\frac{2}{3}$ with, if you are looking to find the greatest fraction of people who wear glasses?*

IN FOCUS Question 1 a) and b) will introduce children to comparing fractions. In question 1 a), one denominator is a multiple of the other (related fractions) whereas in question 1 b), the denominators are unrelated; hence, children are required to use the LCM to find equivalent fractions to compare.

PRACTICAL TIPS This situation could be recreated in a class setting using children from the class. Children could be asked to line up in groups, with children wearing glasses on one side and children not wearing glasses on the other side, and then to recreate a fraction strip. However, it may be best to do this after children have tried the problem themselves, to reinforce the concepts introduced.

ANSWERS

Question 1 a): $\frac{3}{4}$ is bigger than $\frac{5}{8}$, so Group A has a greater fraction of people who wear glasses.

Question 1 b): $\frac{3}{4} > \frac{2}{3}$, so Group A has the greatest fraction of people who wear glasses.

Share

WAYS OF WORKING Whole class teacher led

ASK

- Question 1 a): *What do you need to do to the denominators in order to compare the fractions?*
- Question 1 b): *Which fractions do you need to compare now? Why don't you need to compare all three?*
- Question 1 b): *How can you compare these fractions now? What is the lowest common multiple of 4 and 3?*

IN FOCUS Ensure that children realise they need to make the denominators the same in order to compare the fractions. Question 1 a) introduces children to the comparison of related fractions. Question 1 b) then gives children a third, unrelated, fraction and asks them to compare it with the first two fractions, using the LCM to find equivalent fractions. Explore with children why they need to make the denominators the same and check that they know how to do this.

In question 1 b), guide the children to decide which fractions they need to compare and ensure they understand why it is not necessary to compare all three. Emphasise the need to find the LCM of 4 and 3, and remind children to use the inequality symbol.

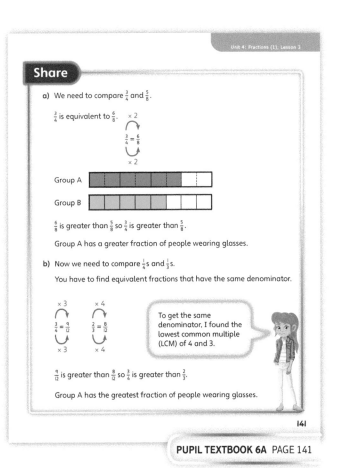

Share

a) We need to compare $\frac{3}{4}$ and $\frac{5}{8}$.

$\frac{3}{4}$ is equivalent to $\frac{6}{8}$.   × 2

$$\frac{3}{4} = \frac{6}{8}$$

× 2

Group A

Group B

$\frac{6}{8}$ is greater than $\frac{5}{8}$ so $\frac{3}{4}$ is greater than $\frac{5}{8}$.

Group A has a greater fraction of people wearing glasses.

b) Now we need to compare $\frac{1}{4}$s and $\frac{1}{3}$s.

You have to find equivalent fractions that have the same denominator.

× 3   × 4

$$\frac{3}{4} = \frac{9}{12}$$ $$\frac{2}{3} = \frac{8}{12}$$

× 3   × 4

To get the same denominator, I found the lowest common multiple (LCM) of 4 and 3.

$\frac{9}{12}$ is greater than $\frac{8}{12}$ so $\frac{3}{4}$ is greater than $\frac{2}{3}$.

Group A has the greatest fraction of people wearing glasses.

141

PUPIL TEXTBOOK 6A PAGE 141

Think together

Whole class teacher led (I do, We do, You do)

ASK

- Question **1**: *What equivalent fractions could you use? Do you always have to change both fractions?*
- Question **2**: *What do you notice about the numerators and denominators?*
- Question **3** b): *Does comparing the whole numbers help? Can you make the denominators on some of the fractions the same to help you compare them? How can you make the fraction on the final card greater than $2\frac{2}{6}$?*

IN FOCUS In question **1**, children compare fractions by choosing related equivalent fractions. Question **2** requires them to choose an appropriate method to compare given pairs of fractions. In question **3**, children compare a selection of mixed numbers and improper fractions.

STRENGTHEN Start by asking children to compare unit fractions to develop their 'fraction sense'. Then move on to comparing fractions that have the same denominator.

DEEPEN Question **3** b) can be explored further by giving children another mixed number or an improper fraction and asking them to decide which of the fraction cards are greater or smaller than each other.

ASSESSMENT CHECKPOINT Question **3** b) will assess children's ability to compare more than three mixed numbers or improper fractions in abstract form with no pictorial support. Children should fluently convert between mixed numbers and improper fractions and find the lowest common denominator to compare fractions, while confidently using inequality symbols. Look for children being able to explain their steps and their reasoning.

ANSWERS

Question **1** a): $\frac{2}{3} = \frac{4}{6}$ so $\frac{5}{6} > \frac{2}{3}$

Question **1** b): $\frac{1}{2} = \frac{4}{8}$ so $\frac{1}{2} > \frac{3}{8}$

Question **1** c): $\frac{3}{5} = \frac{9}{15}$; $\frac{2}{3} = \frac{10}{15}$ so $\frac{3}{5} < \frac{2}{3}$

Question **1** d): $\frac{5}{6} = \frac{20}{24}$; $\frac{7}{8} = \frac{21}{24}$ so $\frac{5}{6} < \frac{7}{8}$

Question **1** e): $\frac{5}{11} > \frac{5}{12}$ because elevenths are greater than twelfths.

Question **1** f): $\frac{7}{9} = \frac{49}{63}$; $\frac{6}{7} = \frac{54}{63}$ so $\frac{7}{9} < \frac{6}{7}$

Question **2**: $\frac{1}{2}, \frac{3}{5}, \frac{7}{10}$

Question **3** a): $\frac{5}{12}, \frac{1}{2}, \frac{3}{4}, \frac{5}{6}$

Question **3** b): $4\frac{1}{8}; \frac{21}{4}; 2\frac{2}{3}; 2\frac{*}{20}$ are all greater than $2\frac{3}{5}$, given that $* = 13, 14, 15, 16, 17, 18$ or 19.

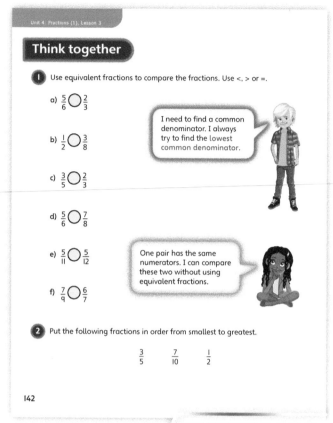

Think together

1 Use equivalent fractions to compare the fractions. Use <, > or =.

a) $\frac{5}{6} \bigcirc \frac{2}{3}$

b) $\frac{1}{2} \bigcirc \frac{3}{8}$

c) $\frac{3}{5} \bigcirc \frac{2}{3}$

d) $\frac{5}{6} \bigcirc \frac{7}{8}$

e) $\frac{5}{11} \bigcirc \frac{5}{12}$

f) $\frac{7}{9} \bigcirc \frac{6}{7}$

> I need to find a common denominator. I always try to find the lowest common denominator.

> One pair has the same numerators. I can compare these two without using equivalent fractions.

2 Put the following fractions in order from smallest to greatest.

$$\frac{3}{5} \qquad \frac{7}{10} \qquad \frac{1}{2}$$

142

PUPIL TEXTBOOK 6A PAGE 142

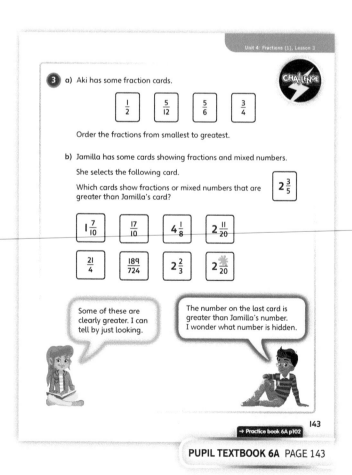

3 a) Aki has some fraction cards.

$\boxed{\frac{1}{2}}$ $\boxed{\frac{5}{12}}$ $\boxed{\frac{5}{6}}$ $\boxed{\frac{3}{4}}$

CHALLENGE

Order the fractions from smallest to greatest.

b) Jamilla has some cards showing fractions and mixed numbers.

She selects the following card.

Which cards show fractions or mixed numbers that are greater than Jamilla's card?

$\boxed{2\frac{3}{5}}$

$\boxed{1\frac{7}{10}}$ $\boxed{\frac{17}{10}}$ $\boxed{4\frac{1}{8}}$ $\boxed{2\frac{11}{20}}$

$\boxed{\frac{21}{4}}$ $\boxed{\frac{189}{724}}$ $\boxed{2\frac{2}{3}}$ $\boxed{2\frac{*}{20}}$

> Some of these are clearly greater. I can tell by just looking.

> The number on the last card is greater than Jamilla's number. I wonder what number is hidden.

→ Practice book 6A p102   143

PUPIL TEXTBOOK 6A PAGE 143

Practice

WAYS OF WORKING Independent thinking

IN FOCUS Question ① aims to consolidate children's understanding of comparing fractions by finding the LCM. The question progresses from providing support and structure for question ① a) – where one denominator is a multiple of the other – to requiring children to follow and write out the method independently by question ① d). Question ③ uses a visual representation of fractions and asks children to identify fractions and then put them in order. This variation ensures they can apply the skill of comparing fractions in a new context. Encourage children to find the LCM and not to rely on the diagrams to order the fractions. This is further developed in question ④, which moves to the abstract with no pictorial representations given.

STRENGTHEN When comparing fractions, encourage children to represent the fractions on a fraction strip. If necessary, use the structure from question ① for the other questions. Ask: *How can you represent the fractions on a fraction strip? How many parts does each model need to be divided into? What part of the fraction does this relate to?*

DEEPEN In question ⑥, children locate fractions and improper fractions on a number line by selecting numerators and denominators. Challenge them to explain each answer, including recognising fractions greater than 1, where the numerator must be greater than the denominator.

ASSESSMENT CHECKPOINT Question ① provides an opportunity to assess children's ability to compare two fractions by finding the LCM.

Question ④ develops this further and will assess children's ability to compare and order more than two fractions by finding the LCM. Look for children confidently using known multiplication facts to find multiples and demonstrating understanding that the LCM is not always the denominators multiplied together.

ANSWERS Answers for the **Practice** part of the lesson can be found in the *Power Maths* online subscription.

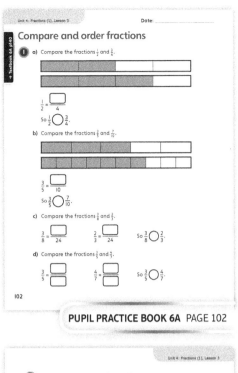

PUPIL PRACTICE BOOK 6A PAGE 102

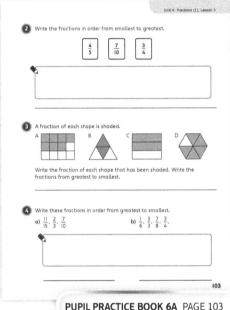

PUPIL PRACTICE BOOK 6A PAGE 103

Reflect

WAYS OF WORKING Independent thinking

IN FOCUS The **Reflect** part of the activity provides an opportunity to check children's understanding of comparing fractions where the numerators are the same. Children should be able to explain that if the numerators are the same, the smaller the denominator, the bigger the fraction.

ASSESSMENT CHECKPOINT Look for children confidently explaining how to compare fractions that have the same numerators.

ANSWERS Answers for the **Reflect** part of the lesson can be found in the *Power Maths* online subscription.

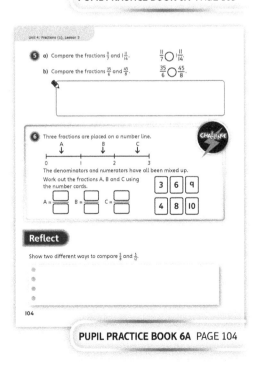

PUPIL PRACTICE BOOK 6A PAGE 104

After the lesson ⏸

- Can children compare fractions that have different denominators?
- Can children find the LCM?
- Do children understand that to order fractions they need to find the LCM and compare the numerators?

Add and subtract simple fractions

Learning focus

In this lesson, children revisit how to add and subtract fractions with the same denominator, or with easily related denominators. This is a chance to revisit some common misconceptions.

Before you teach

- Can children identify the numerator and denominator of fractions?
- Can children explain what numerator and denominator mean when dividing a whole into equal parts?
- Can children demonstrate how to locate fractions on a number line, based on the numerator and the denominator?

NATIONAL CURRICULUM LINKS

Year 6 Number – fractions

Add and subtract fractions with different denominators and mixed numbers, using the concept of equivalent fractions.

ASSESSING MASTERY

Children can add and subtract fractions where both fractions have the same denominator, or where one fraction has a denominator that is a multiple of the denominator of the other fraction.

COMMON MISCONCEPTIONS

Children may add both the numerator and the denominator, for example, saying that $\frac{1}{6} + \frac{3}{6} = \frac{4}{12}$. Ask:

- *What is 1 apple plus 3 apples?*
- *What is 1 elephant plus 3 elephants?*
- *What is 1 ten plus 3 tens?*
- *What is 1 sixth plus 3 sixths?*

STRENGTHENING UNDERSTANDING

Ask children to draw and shade simple fraction strips or fraction wheels to illustrate what happens when adding fractions with the same denominator.

GOING DEEPER

Challenge children to explore addition and subtraction of fractions where the answer is a fraction that has been simplified, for example $\frac{2}{5} + \frac{2}{10} = \frac{1}{2}$.

KEY LANGUAGE

In lesson: numerator, denominator

Other language to be used by the teacher: simplify, simplest form

STRUCTURES AND REPRESENTATIONS

Fraction strip, number line

RESOURCES

Optional: fraction strips, fraction wheels

 In the eTextbook of this lesson, you will find interactive links to a selection of teaching tools.

Quick recap ⟳

Ask children to write three or more equivalent fractions for each of these fractions:

$\frac{1}{6}$ $\frac{3}{4}$ $\frac{2}{3}$

Discover

Unit 4: Fractions (1), Lesson 4

Add and subtract simple fractions

WAYS OF WORKING Pair work

Discover

ASK

- Question ① a): *What calculations can you see? What numerators? What denominators?*
- Question ① a): *Can you describe the steps that Bella might have taken? Does this seem correct?*
- Question ① a): *Can you describe the steps that Andy might have taken. Does his answer make sense?*

IN FOCUS The errors that the pupils in this scenario have made will give children the opportunity to explore common misconceptions that can occur when adding and subtracting fractions.

PRACTICAL TIPS Ask children to draw and shade simple fraction diagrams, for example fraction strips or fraction wheels, to model each of the calculations and identify what the correct answers should be and why.

ANSWERS

Question ① a): Andy and Bella have added or subtracted both the numerators and the denominators, instead of just the numerator.

Question ① b): $\frac{5}{8} + \frac{2}{8} = \frac{7}{8}$

$\frac{5}{8} - \frac{2}{8} = \frac{3}{8}$

① a) What mistakes have Andy and Bella made?

b) Complete the calculations correctly.

144

PUPIL TEXTBOOK 6A PAGE 144

Share

WAYS OF WORKING Whole class teacher led

ASK

- Question ① a): *How do you feel when you make a mistake or when someone helps you in a good way?*
- Question ① a): *How could you support Bella and Andy and explain why it is OK to make mistakes?*
- Question ① b): *Why have the numerators not changed? Could you show why, using a diagram?*

IN FOCUS Question ① a) requires children to identify common errors that occur when adding and subtracting fractions, namely adding or subtracting with both the numerators and the denominators, rather than just the numerators. In order to correct the errors in question ① b), children will need to recognise how to accurately add and subtract fractions with the same denominator.

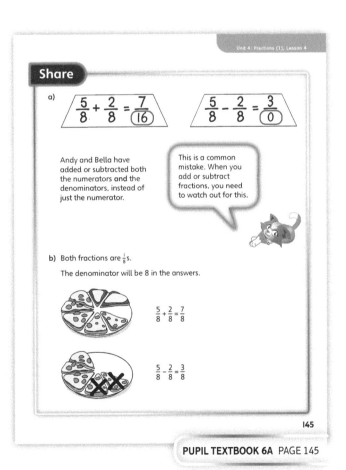

PUPIL TEXTBOOK 6A PAGE 145

Think together

WAYS OF WORKING Whole class teacher led (I do, We do, You do)

ASK

- Question ①: *What diagram could you draw to find if there is a mistake?*
- Question ②: *What errors should you try not to make when you complete these number sentences?*
- Question ②: *How can you use equivalent fractions to help you?*

IN FOCUS Question ① draws attention to the common misconception that a simplified fraction cannot be the correct answer to an addition of two fractions that have the same denominator. Explain that if the fraction in the answer can be simplified, then the denominator of the answer might be different to the denominator of the two fractions you are subtracting. Explore together how $\frac{5}{10}$ is the same as $\frac{1}{2}$ and why they are both correct answers. In question ②, children complete simple addition and subtraction of fractions with the same denominator. Prompt them to simplify their answers where possible. In question ③, children use equivalent fractions to add and subtract fractions when one denominator is a multiple of the other.

STRENGTHEN Ask children to draw a simple fraction model of their choosing for each calculation, shading the relevant parts to support their reasoning as they work through.

DEEPEN Challenge children to explore fraction additions that have a total of exactly one whole, for example $\frac{?}{5} + \frac{?}{10} = 1$, $\frac{?}{6} + \frac{?}{24} = 1$.

ASSESSMENT CHECKPOINT Question ② will give you the opportunity to assess whether children can fluently add and subtract fractions with the same denominator.

ANSWERS

Question ①: Olivia and Luis are both correct.
Luis has simplified his answer.

Question ② a): $\frac{2}{5} + \frac{1}{5} = \frac{3}{5}$

Question ② b): $\frac{4}{9} - \frac{1}{9} = \frac{3}{9}$ or $\frac{1}{3}$

Question ② c): $\frac{7}{15} + \frac{3}{15} = \frac{10}{15}$ or $\frac{2}{3}$

Question ② d): $\frac{95}{100} - \frac{70}{100} = \frac{25}{100}$ or $\frac{1}{4}$

Question ③: $\frac{1}{2} + \frac{1}{4} = \frac{2}{4} + \frac{1}{4} = \frac{3}{4}$

$\frac{9}{10} - \frac{1}{2} = \frac{9}{10} - \frac{5}{10} = \frac{4}{10} = \frac{2}{5}$

$\frac{1}{5} + \frac{7}{20} = \frac{4}{20} + \frac{7}{20} = \frac{11}{20}$

$\frac{1}{4} = \frac{3}{12}$

$\frac{3}{12} + \frac{1}{12} = \frac{4}{12}; \frac{4}{12} = \frac{1}{3}$

$\frac{1}{3} - \frac{1}{4} = \frac{1}{12}$

PUPIL TEXTBOOK 6A PAGE 146

PUPIL TEXTBOOK 6A PAGE 147

182

Practice

WAYS OF WORKING Independent thinking

IN FOCUS In question ❶, children are carrying out basic additions and subtractions of fractions with the same denominator. Question ❷ requires them to recognise a common error that can occur when adding fractions, where a pupil has added both the numerators and the denominators, rather than just adding the numerators. In question ❸, children complete missing number calculations with pairs of fractions that have the same denominator. In question ❹, they find pairs of fractions that sum to a given simplified fraction, and in question ❻, they complete missing number calculations that sum to a given simplified fraction. In question ❺, children add and subtract pairs of fractions that have the same denominator, giving their answers in the simplest form.

STRENGTHEN Ask children to draw and shade in simple fraction diagrams to represent each of the calculations.

DEEPEN Challenge children to explore how many solutions they can find to this calculation: ☐ + ☐ = $\frac{1}{3}$.

ASSESSMENT CHECKPOINT Use question ❷ to assess whether children can recognise and correct a common mistake that can occur when adding fractions.

ANSWERS Answers for the **Practice** part of the lesson can be found in the *Power Maths* online subscription.

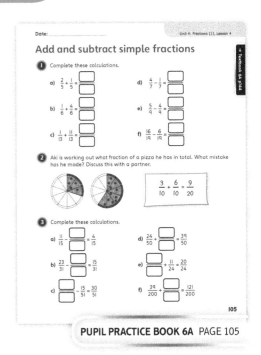

PUPIL PRACTICE BOOK 6A PAGE 105

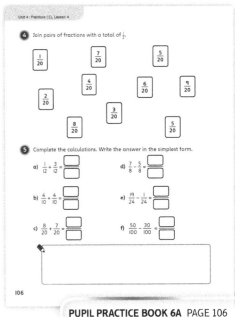

PUPIL PRACTICE BOOK 6A PAGE 106

Reflect

WAYS OF WORKING Independent thinking

IN FOCUS The **Reflect** part of the lesson prompts children to provide an explanation of the steps required to calculate with fractions that have the same denominator.

ASSESSMENT CHECKPOINT Assess whether children can break their thinking down and explain each of the steps that they carry out when adding two fractions that both have the same denominator.

ANSWERS Answers for the **Reflect** part of the lesson can be found in the *Power Maths* online subscription.

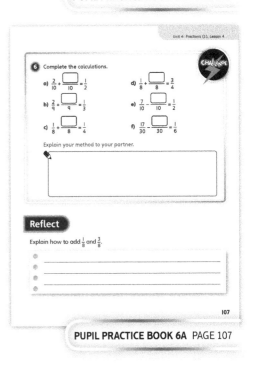

PUPIL PRACTICE BOOK 6A PAGE 107

After the lesson ⏸

- Were children able to recognise and address all of the common misconceptions?
- What opportunities can you find for children to regularly rehearse simple addition and subtraction as part of fluency practice?

183

Add and subtract any two fractions

Learning focus

In this lesson, children link their prior knowledge of finding equivalent fractions with common denominators to adding and subtracting fractions where the answer is between 0 and 1.

Before you teach ⏸

- Can children add/subtract fractions that have the same denominator?
- Can children find the lowest common multiple of two or more numbers?
- Can children find equivalent fractions?

NATIONAL CURRICULUM LINKS

Year 6 Number – fractions

Add and subtract fractions with different denominators and mixed numbers, using the concept of equivalent fractions.

ASSESSING MASTERY

Children can fluently add and subtract fractions by using a common multiple to create equivalent fractions with a common denominator and then adding and subtracting the numerators, explaining the steps in their own words.

COMMON MISCONCEPTIONS

Children may simply add or subtract the numbers in the fractions, for example, children may give an answer of $\frac{2}{7}$ to the calculation $\frac{1}{3} + \frac{1}{4}$. It will be beneficial to use fraction strips to secure understanding and avoid this misconception, and to encourage children to understand that they can only add and subtract fractions that have the same denominator.

Another common mistake is that children may find a common denominator but forget to convert the numerators; for example, children may rewrite the calculation $\frac{3}{4} + \frac{1}{5}$ as $\frac{3}{20} + \frac{1}{20}$ to get an answer of $\frac{4}{20}$. Ask:

- *Can you find the lowest common multiple of the denominators?*
- *What do you do to the denominator and the numerator?*
- *Can you write down the equivalent fractions before carrying out the calculation?*

STRENGTHENING UNDERSTANDING

If children are struggling with adding and subtracting fractions, strengthen understanding by adding and subtracting fractions with the same denominator using fraction strips. Next, encourage children to add and subtract fractions where one of the denominators is a multiple of the other, so they can easily find the common denominator. Explain that the denominators need to be the same when adding and subtracting fractions and encourage children to write out the equivalent fractions before adding or subtracting.

GOING DEEPER

Children can be encouraged to solve number problems finding missing fractions, for example $\square + \frac{2}{5} = \frac{24}{35}$.

KEY LANGUAGE

In lesson: common denominator, equivalent, difference, lowest common denominator

Other language to be used by the teacher: numerator

STRUCTURES AND REPRESENTATIONS

Fraction strips, number lines

In the eTextbook of this lesson, you will find interactive links to a selection of teaching tools.

Quick recap 🔁

Ask children to complete these equivalent fractions:

$\frac{1}{5} = \frac{?}{20}$ $\frac{3}{4} = \frac{?}{20}$ $\frac{3}{5} = \frac{?}{20}$ $\frac{9}{10} = \frac{?}{20}$

Discover

WAYS OF WORKING Pair work

ASK

- Question ① a): *What fraction of the hay does Hattie eat in the morning? What fraction of the hay does she eat in the evening? How can you work out how much Hattie eats in the whole day? What calculation do you need to do?*
- Question ① b): *What is the same and what is different about the fractions eaten by Molly and Hattie? What is the same and what is different about how you can approach question ① a) and question ① b)?*

IN FOCUS Question ① introduces children to adding fractions, linking with their knowledge of finding equivalent fractions to convert two fractions so they have a common denominator before completing the calculation.

PRACTICAL TIPS There are many ways in which fractions can be added in a practical context, for instance, cooking using cups. Encourage children to use fraction strips so they have pictorial representations for support.

ANSWERS

Question ① a): Hattie eats $\frac{5}{6}$ of a bale of hay in a day.

Question ① b): Molly eats $\frac{5}{12}$ of a bale of hay in a day.

Discover

I need to feed Hattie $\frac{2}{3}$ of a bale of hay in the morning and $\frac{1}{6}$ in the evening.

Hattie   Molly

① a) What fraction of a bale of hay does Hattie eat in a day?

b) Molly eats $\frac{1}{4}$ of a bale in the morning and $\frac{1}{6}$ of a bale in the evening. Calculate the fraction of a bale of hay Molly eats in a day.

148

PUPIL TEXTBOOK 6A PAGE 148

Share

WAYS OF WORKING Whole class teacher led

ASK

- Question ① a): *How can you add $\frac{2}{3}$ and $\frac{1}{6}$? How does the fraction strip help?*
- Question ① a): *Can you find a common denominator? Will finding the lowest common multiple help?*
- Question ① a): *What fraction is equivalent to $\frac{2}{3}$ with a denominator of 6? Why is it necessary to add the numerators and keep the denominator as 6?*
- Question ① b): *How can you choose which equivalent fractions to use?*
- Question ① b): *How do you know you need to change both fractions?*

IN FOCUS With this question, it is important that children recognise the need to make the denominators the same using the LCM and then to add the numerators. Encourage children to use the fraction strip so they have a visual representation of why they need to make the denominators the same in order to add the fractions. Explore with children how to make the denominators the same, and highlight the conversion calculations shown.

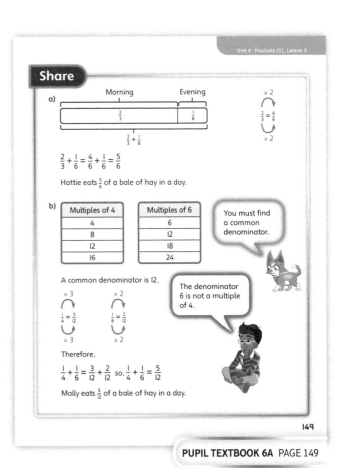

Share

a)

| Morning | Evening |
| $\frac{2}{3}$ | $\frac{1}{6}$ |

$\frac{2}{3} + \frac{1}{6}$

$\times 2$
$\frac{2}{3} = \frac{4}{6}$
$\times 2$

$\frac{2}{3} + \frac{1}{6} = \frac{4}{6} + \frac{1}{6} = \frac{5}{6}$

Hattie eats $\frac{5}{6}$ of a bale of hay in a day.

b)

Multiples of 4	Multiples of 6
4	6
8	12
12	18
16	24

You must find a common denominator.

A common denominator is 12.

$\times 3$
$\frac{1}{4} = \frac{3}{12}$
$\times 3$

$\times 2$
$\frac{1}{6} = \frac{2}{12}$
$\times 2$

The denominator 6 is not a multiple of 4.

Therefore,

$\frac{1}{4} + \frac{1}{6} = \frac{3}{12} + \frac{2}{12}$ so, $\frac{1}{4} + \frac{1}{6} = \frac{5}{12}$

Molly eats $\frac{5}{12}$ of a bale of hay in a day.

149

PUPIL TEXTBOOK 6A PAGE 149

Think together

Whole class teacher led (I do, We do, You do)

ASK

- Question **1**: *Can you find the lowest common multiple of the denominators?*
- Question **2**: *How will you choose the denominators?*
- Question **3**: *Does this method always work? Is it always efficient? What simple calculations could you try to test it out?*

IN FOCUS In question **1**, children find simple equivalent fractions in order to add two fractions that have different denominators. In question **2**, they use equivalence to find common denominators and then carry out subtractions. Question **3** requires children to explore a generalised method for finding fractions that have common denominators.

STRENGTHEN To strengthen understanding, encourage children to represent the calculations with fraction strips. Clearly break down the calculations into steps, by asking: *What do you need to do first? What do you need to make the same before you can add or subtract?*

DEEPEN Challenge children to explore efficient methods for adding fractions. They could take it in turns to roll a dice to generate denominators for the following fractions:

$$\frac{1}{?} + \frac{1}{?}$$

They could then choose and justify an efficient method to complete their addition.

ASSESSMENT CHECKPOINT Question **1** assesses children's ability to add fractions using the LCM to find a common denominator and adding the numerators with prompting.

ANSWERS

Question **1** a): $\frac{2}{5} = \frac{6}{15}$ $\frac{1}{3} = \frac{5}{15}$ $\frac{6}{15} + \frac{5}{15} = \frac{11}{15}$

Question **1** b): $\frac{1}{6} = \frac{4}{24}$ $\frac{3}{8} = \frac{9}{24}$ $\frac{4}{24} + \frac{9}{24} = \frac{13}{24}$

Question **1** c): $\frac{1}{10} = \frac{2}{20}$ $\frac{3}{4} = \frac{15}{20}$ $\frac{2}{20} + \frac{15}{20} = \frac{17}{20}$

Question **2** a): $\frac{1}{2} - \frac{1}{3} = \frac{3}{6} - \frac{2}{6} = \frac{1}{6}$

Question **2** b): $\frac{2}{3} - \frac{1}{4} = \frac{8}{12} - \frac{3}{12} = \frac{5}{12}$

Question **2** c): $\frac{12}{15} - \frac{3}{10} = \frac{24}{30} - \frac{9}{30} = \frac{15}{30} = \frac{1}{2}$

Question **3**: The method of multiplying the denominators and cross multiplying will always work, but using the lowest common multiple of the denominators avoids having to multiply such large numbers and having to simplify the answers so much.

For example:

$$\frac{7}{8} - \frac{5}{12} = \frac{84}{96} - \frac{40}{96} = \frac{44}{96} = \frac{11}{24}$$
$$\frac{7}{8} - \frac{5}{12} = \frac{21}{24} - \frac{10}{24} = \frac{11}{24}$$

PUPIL TEXTBOOK 6A PAGE 150

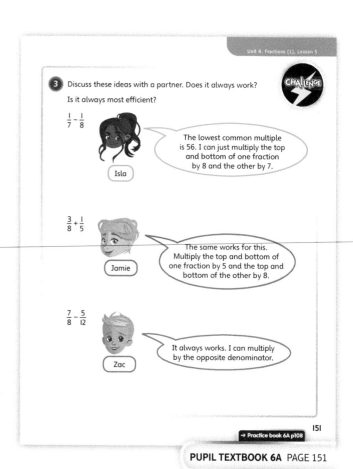

PUPIL TEXTBOOK 6A PAGE 151

Practice

WAYS OF WORKING Independent thinking

IN FOCUS Questions ① and ② develop children's understanding of adding and subtracting fractions using the LCM and equivalent fractions.

Questions ③ and ④ introduce children to adding and subtracting fraction problems in abstract form and then in words, with no scaffolding. Encourage children to show their method for these questions.

STRENGTHEN If children are struggling with questions ③ and ④, or with adding and subtracting fractions in general, encourage them to represent the fractions on a fraction strip and to work through the steps systematically, writing down the multiples to find the LCM.

DEEPEN Question ④ can be explored with more worded questions, for example: 'I think of a fraction and add $\frac{3}{4}$. My answer is $\frac{11}{12}$. What fraction was I thinking of?'. Children could also work in pairs, setting each other questions.

ASSESSMENT CHECKPOINT Questions ① and ② assess children's ability to add and subtract fractions using the LCM to find equivalent fractions. It will be useful to observe whether children rely on fraction strips to make the calculations, or if they are struggling to find the LCM and use equivalent fractions; if so, they may need more support.

ANSWERS Answers for the **Practice** part of the lesson can be found in the *Power Maths* online subscription.

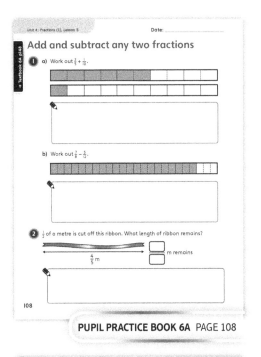

PUPIL PRACTICE BOOK 6A PAGE 108

PUPIL PRACTICE BOOK 6A PAGE 109

Reflect

WAYS OF WORKING Pair work

IN FOCUS The **Reflect** activity prompts children to provide an explanation of the steps add or subtract fractions with different denominators.

ASSESSMENT CHECKPOINT Children should be able to explain how to change denominators to a common denominator. Assess whether children can break their thinking down and explain each of the steps clearly.

ANSWERS Answers for the **Reflect** part of the lesson can be found in the *Power Maths* online subscription.

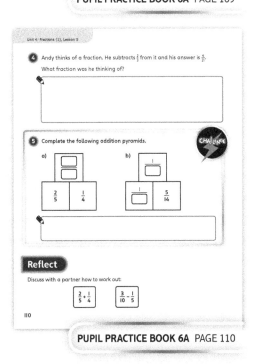

After the lesson ⏸

- Can children convert two fractions to equivalent fractions with the same denominator?
- Can children add and subtract fractions that have different denominators, without support such as fraction strips?
- Can children explain the steps for adding and subtracting fractions in their own words?

PUPIL PRACTICE BOOK 6A PAGE 110

Add mixed numbers

Learning focus

In this lesson, children extend their knowledge of adding mixed numbers and fractions by using two methods to add mixed fractions where the fractional answer is greater than 1.

Before you teach ▮▮

- Can children add fractions with different denominators?
- Can children convert between mixed numbers and improper fractions?
- Can children add mixed numbers where the fractional parts total less than 1?

NATIONAL CURRICULUM LINKS

Year 6 Number – fractions

Add and subtract fractions with different denominators and mixed numbers, using the concept of equivalent fractions.

ASSESSING MASTERY

Children can fluently add and subtract mixed numbers, either by adding or subtracting the wholes and fractional parts or by converting to improper fractions and adding these, while clearly explaining the steps for both methods.

COMMON MISCONCEPTIONS

Children may add the wholes and add the fractions separately but then fail to convert the fractional answer to a mixed number; for example, when adding $1\frac{2}{3}$ and $2\frac{3}{4}$ children may correctly do $1 + 2 = 3$ and $\frac{2}{3} + \frac{3}{4} = \frac{17}{12}$ but then incorrectly write the final answer as $3\frac{17}{12}$. Use fraction strips to provide pictorial support to secure this understanding, while encouraging children to realise that $\frac{17}{12}$ is an improper fraction requiring conversion to a mixed number. Ask:

- *What type of fraction is your answer?*
- *Do you need to convert this further?*

STRENGTHENING UNDERSTANDING

When adding mixed numbers, strengthen understanding by showing children the numbers on a fraction strip so they can see why they can add the wholes and add the parts and why they can convert to improper fractions and add these.

GOING DEEPER

Deepen learning using diagrams of quadrilaterals and triangles that have mixed numbers as their lengths, asking children to find the perimeter of each shape.

Encourage children to solve missing number problems that involve working backwards, such as $\boxed{} - 3\frac{2}{5} = 6\frac{3}{4}$.

KEY LANGUAGE

In lesson: wholes, parts, mixed number, improper fraction, convert, add

STRUCTURES AND REPRESENTATIONS

Fraction strips

 In the eTextbook of this lesson, you will find interactive links to a selection of teaching tools.

Quick recap

Count together as a class up to 5 in steps of one quarter.

Ensure children include simplified fractions and mixed numbers in their count.

Discover

Unit 4: Fractions (1), Lesson 6

Add mixed numbers

WAYS OF WORKING Pair work

ASK

- Question **1** a): *How can you work out how many carrots the farmer has harvested so far?*
- Question **1** a): *What calculation do you need to do?*
- Question **1** b): *How can you work out how many carrots the farmer harvests next?*
- Question **1** b): *What calculation do you need to do?*

IN FOCUS Question **1** develops the concept of adding or subtracting fractions by introducing calculations where the fractional answer is greater than 1. An extra conversion step is required to complete the calculation.

PRACTICAL TIPS Adding mixed numbers can be introduced in a practical setting in the classroom, for instance by adding fractions of the distance between local places of interest. Encourage children to use pictorial supports such as fraction strips to give a visual representation.

ANSWERS

Question **1** a): The farmer has harvested $4\frac{1}{4}$ tonnes of carrots so far.

Question **1** b): The farmer has harvested $5\frac{1}{20}$ tonnes, so he has harvested enough.

Discover

There are $2\frac{3}{4}$ tonnes of carrots on one trailer and $1\frac{1}{2}$ tonnes on the other.

1 a) What is the total mass of carrots the farmer has harvested so far?

b) A supermarket orders 5 tonnes of carrots.
The farmer harvests another $\frac{4}{5}$ tonnes of carrots from a different field.
Has the farmer harvested enough carrots to fulfil the order?

152

PUPIL TEXTBOOK 6A PAGE 152

Share

WAYS OF WORKING Whole class teacher led

ASK

- Question **1** a): *How can you add the parts? What is the common denominator? What do you notice about the answer? Is this a mixed number or an improper fraction?*
- Question **1** a): *Which method is more efficient?*

IN FOCUS Question **1** a) introduces two methods for adding mixed numbers where the total of the fractional part is greater than 1. Explore both methods using the fraction strips, emphasising that in Dexter's method, $\frac{5}{4}$ is an improper fraction and needs to be converted to a mixed number before the final step of combining the different parts. With Flo's method, encourage children to work through the calculations converting the mixed numbers to improper fractions. Discuss adding the fractions by finding a common denominator, ensuring that the children understand the type of fraction this results in. Check that they clearly recognise and can explain the difference between mixed numbers and improper fractions.

Question **1** b) requires children to add a mixed number and a fraction together before comparing their answer with a whole number. Encourage children to work through the calculations of adding the fractions and converting to a mixed number before adding on 4. Explore with children if they could have used the other method, i.e. converting $4\frac{1}{4}$ to an improper fraction before adding $\frac{4}{5}$. Discuss which method is more efficient, ensuring children recognise that the answer would be the same for both methods.

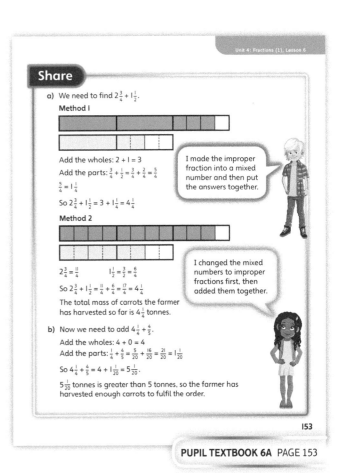

Share

a) We need to find $2\frac{3}{4} + 1\frac{1}{2}$.

Method 1

Add the wholes: $2 + 1 = 3$

Add the parts: $\frac{3}{4} + \frac{1}{2} = \frac{3}{4} + \frac{2}{4} = \frac{5}{4}$

$\frac{5}{4} = 1\frac{1}{4}$

So $2\frac{3}{4} + 1\frac{1}{2} = 3 + 1\frac{1}{4} = 4\frac{1}{4}$

I made the improper fraction into a mixed number and then put the answers together.

Method 2

$2\frac{3}{4} = \frac{11}{4}$ $1\frac{1}{2} = \frac{3}{2} = \frac{6}{4}$

So $2\frac{3}{4} + 1\frac{1}{2} = \frac{11}{4} + \frac{6}{4} = \frac{17}{4} = 4\frac{1}{4}$

The total mass of carrots the farmer has harvested so far is $4\frac{1}{4}$ tonnes.

I changed the mixed numbers to improper fractions first, then added them together.

b) Now we need to add $4\frac{1}{4} + \frac{4}{5}$.

Add the wholes: $4 + 0 = 4$

Add the parts: $\frac{1}{4} + \frac{4}{5} = \frac{5}{20} + \frac{16}{20} = \frac{21}{20} = 1\frac{1}{20}$

So $4\frac{1}{4} + \frac{4}{5} = 4 + 1\frac{1}{20} = 5\frac{1}{20}$.

$5\frac{1}{20}$ tonnes is greater than 5 tonnes, so the farmer has harvested enough carrots to fulfil the order.

153

PUPIL TEXTBOOK 6A PAGE 153

Think together

WAYS OF WORKING Whole class teacher led (I do, We do, You do)

ASK

- Question **1** a): *How can you add the wholes and the parts? How can you find the lowest common multiple of the denominators?*
- Question **1** b): *How can you convert the mixed numbers to improper fractions?*
- Question **2**: *How can you add the wholes and the parts? What do you need to do to the parts so you can add them? Would it be easier to convert the mixed numbers to improper fractions first?*
- Question **3**: *Which numbers might you add to make $11\frac{13}{24}$? Could you start by trying to add some of the numbers?*

IN FOCUS Question **2** allows children to practise applying the techniques they learnt in question **1**. They can either add the wholes and the parts separately, or change to improper fractions first.

Question **3** involves problem solving, requiring children to select the correct numbers to make $11\frac{13}{24}$. The numbers have been carefully selected so that the common denominator could be 24 for all fractions and some of the whole numbers add to 11. Look out for a common mistake where children may think the answer is $7\frac{3}{4}$ and $4\frac{2}{3}$ because 7 + 4 = 11, i.e. they have not realised that the total of the fractions will cross the whole and make the answer >12.

STRENGTHEN Encourage children to represent the numbers on fraction strips and to work systematically through the calculations following the structure used in question **1**.

DEEPEN Question **3** can be explored further by giving children another total and asking them which two fractions added together make this total. Alternatively, ask children to add three of the cards together using their preferred method.

ASSESSMENT CHECKPOINT Question **1** assesses children's ability to use both methods for adding mixed numbers where the fractional part makes a total greater than 1.

Question **3** provides an opportunity to assess children's ability to solve word problems involving adding mixed numbers. Look for children confidently translating the problem into number sentences and solving the calculations while clearly explaining the method they have used and why; children who can do this will have mastered the topic.

ANSWERS

Question **1** a): $1\frac{2}{3} + 2\frac{1}{2} = 3 + 1\frac{1}{6} = 4\frac{1}{6}$

Question **1** b): $\frac{5}{3} + \frac{5}{2} = \frac{10}{6} + \frac{15}{6} = \frac{25}{6} = 4\frac{1}{6}$

Question **2** a): $2\frac{1}{3} + 1\frac{2}{9} = 3 + \frac{3}{9} + \frac{2}{9} = 3\frac{5}{9}$

Question **2** b): $3\frac{2}{5} + 1\frac{9}{10} = 4 + \frac{4}{10} + \frac{9}{10} = 4 + \frac{13}{10} = 4 + 1\frac{3}{10} = 5\frac{3}{10}$

Question **2** c): $7\frac{2}{3} + 2\frac{4}{5} = 9 + \frac{10}{15} + \frac{12}{15} = 9 + \frac{22}{15} = 9 + 1\frac{7}{15} = 10\frac{7}{15}$

Question **3** a): Isla chose $4\frac{2}{3}$ and $6\frac{7}{8}$.

Question **3** b): $7\frac{3}{4} + 27\frac{17}{24} = 35\frac{11}{24}$; use the method of adding the wholes and adding the parts, as converting the mixed numbers to improper fractions is inefficient because of the large denominator involved.

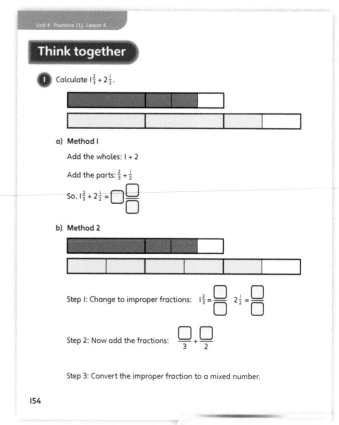

Think together

1 Calculate $1\frac{2}{3} + 2\frac{1}{2}$.

a) **Method 1**

Add the wholes: 1 + 2

Add the parts: $\frac{2}{3} + \frac{1}{2}$

So, $1\frac{2}{3} + 2\frac{1}{2} = \square\frac{\square}{\square}$

b) **Method 2**

Step 1: Change to improper fractions: $1\frac{2}{3} = \frac{\square}{\square}$ $2\frac{1}{2} = \frac{\square}{\square}$

Step 2: Now add the fractions: $\frac{\square}{3} + \frac{\square}{2}$

Step 3: Convert the improper fraction to a mixed number.

154

PUPIL TEXTBOOK 6A PAGE 154

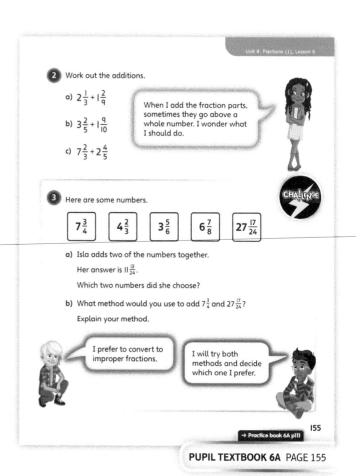

2 Work out the additions.

a) $2\frac{1}{3} + 1\frac{2}{9}$

When I add the fraction parts, sometimes they go above a whole number. I wonder what I should do.

b) $3\frac{2}{5} + 1\frac{9}{10}$

c) $7\frac{2}{3} + 2\frac{4}{5}$

3 Here are some numbers.

| $7\frac{3}{4}$ | $4\frac{2}{3}$ | $3\frac{5}{6}$ | $6\frac{7}{8}$ | $27\frac{17}{24}$ |

CHALLENGE

a) Isla adds two of the numbers together.

Her answer is $11\frac{13}{24}$.

Which two numbers did she choose?

b) What method would you use to add $7\frac{3}{4}$ and $27\frac{17}{24}$?

Explain your method.

I prefer to convert to improper fractions.

I will try both methods and decide which one I prefer.

155

→ Practice book 6A p111

PUPIL TEXTBOOK 6A PAGE 155

Practice

WAYS OF WORKING Independent thinking

IN FOCUS Question ① aims to consolidate children's understanding of adding mixed numbers with pictorial representation as support. Encourage children to use their preferred method and to write down their calculations.

Question ② requires children to add mixed numbers with no pictorial support. Encourage children to look carefully at the numbers and to consider if they can simplify any fractions.

Question ⑥ incorporates problem solving, with children required to carry out more than one addition and then decide how many packs of fencing are needed. Encourage children to find the distance around the edge of the garden before working out how many packs of fencing are required. Explore with children whether they need to round up or down so that there is enough fencing.

STRENGTHEN Encourage children to use pictorial supports such as fraction strips to support them with the calculations.

ASSESSMENT CHECKPOINT Questions ① and ② assess children's ability to add mixed numbers where the fractional part makes a total > 1. Question ① is supported with prompting and pictorial representations, while question ② has no support. Look for children breaking down question ② into the steps shown in question ① and systematically adding the parts, using common denominators.

Questions ④, ⑤ and ⑥ assess children's ability to solve word problems which involve adding mixed numbers where the fraction part crosses the whole. Look for children confidently translating the problem into number sentences and solving the calculations, while clearly explaining the methods they have used and why.

ANSWERS Answers for the **Practice** part of the lesson can be found in the *Power Maths* online subscription.

Reflect

WAYS OF WORKING Pair work

IN FOCUS This **Reflect** activity will evaluate children's understanding of adding two mixed numbers. Encourage children to use their preferred method and to explain why they have chosen that method, as well as calculating the answer.

ASSESSMENT CHECKPOINT This gives an opportunity to assess children's ability to add two mixed numbers where the fractional part is more than 1. Look for children who can confidently explain how to add two mixed numbers, showing understanding of improper fractions and mixed numbers and explaining why they have chosen the method they have used.

ANSWERS Answers for the **Reflect** part of the lesson can be found in the *Power Maths* online subscription.

After the lesson ⏸

- Can children add mixed numbers using both methods?
- Can they convert between mixed numbers and improper fractions?
- Can they explain which method is more efficient for different calculations?

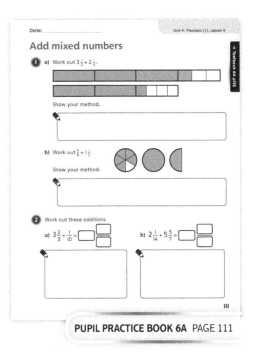

PUPIL PRACTICE BOOK 6A PAGE 111

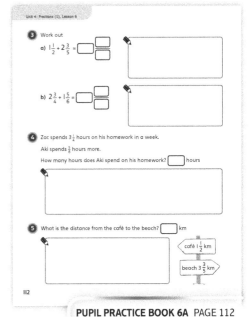

PUPIL PRACTICE BOOK 6A PAGE 112

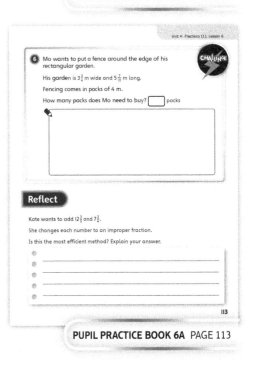

PUPIL PRACTICE BOOK 6A PAGE 113

Subtract mixed numbers

Learning focus

In this lesson, children extend their understanding of subtracting mixed numbers and fractions to calculations where the fractional answer crosses the whole and they cannot simply subtract the wholes and subtract the parts.

Before you teach

- Can children subtract fractions with different denominators?
- Can children convert between mixed numbers and improper fractions?
- Can children break down mixed numbers into alternative combinations of wholes and parts, for example, $3\frac{1}{2} = 2 + 1\frac{1}{2} = 2 + \frac{3}{2}$?

NATIONAL CURRICULUM LINKS

Year 6 Number – fractions

Add and subtract fractions with different denominators and mixed numbers, using the concept of equivalent fractions.

ASSESSING MASTERY

Children can reliably subtract mixed numbers by rewriting the calculation and subtracting the wholes and subtracting the fractions, or by converting to improper fractions. Children can confidently use a number line to subtract mixed numbers.

COMMON MISCONCEPTIONS

Children may subtract the wholes and then the fractions separately but incorrectly change the order of the fractions so that the smaller fraction is always subtracted from the larger fraction. For example, when calculating $5\frac{1}{3} - 2\frac{3}{4}$, children find the difference between 5 and 2 (= 3) and then find the difference between $\frac{1}{3}$ and $\frac{3}{4}$ ($= \frac{5}{12}$), incorrectly writing their answer as 3. Encourage children to realise that the order of the numbers does matter when subtracting and that if the second fraction is greater than the first one, they cannot just subtract the smaller from the larger. Use fraction strips to aid understanding and address this misconception. Ask:

- *Does the order in which you subtract matter?*
- *Can you change the order when subtracting?*
- *Can you use a fraction strip to help you visualise the calculation?*

STRENGTHENING UNDERSTANDING

When rewriting a mixed number to make it easier to subtract, strengthen understanding using fraction strips so that children can see why the change is needed. It may also be beneficial to encourage children to use a number line to count on to carry out a subtraction.

GOING DEEPER

Deepen learning with missing number problems involving subtraction, for example $?\frac{2}{4} + 2\frac{4}{5} = 6\frac{11}{20}$. Give children some verbal 'think of a number' problems involving subtracting mixed numbers, for example: *I think of a number and add $5\frac{2}{3}$ to it. My answer is $8\frac{13}{24}$. What was my number?*

KEY LANGUAGE

In lesson: wholes, parts, mixed number, improper fraction, convert, subtract, lowest common multiple, common denominator

Other language to be used by the teacher: add on

STRUCTURES AND REPRESENTATIONS

Fraction strips, number lines, part-whole models

 In the eTextbook of this lesson, you will find interactive links to a selection of teaching tools.

Quick recap

Count together as a class down from 5 in steps of one sixth.

Ensure children include simplified fractions and mixed numbers in their count.

Discover

Unit 4: Fractions (1), Lesson 7

Subtract mixed numbers

WAYS OF WORKING Pair work

ASK

- Question ① a): *Why does Max think you can't do the calculation? What can you do to work out the subtraction?*
- Question ① b): *How can you convert the mixed numbers to improper fractions? What calculation is needed?*

IN FOCUS Question ① introduces children to subtraction of mixed numbers where the first fractional part is smaller than the second fractional part. In these cases, the fraction must be changed – either to an alternative mixed number or to an improper fraction – to complete the subtraction. It is important for children to understand why this change is needed and to recognise that they cannot simply change the order of the fractions.

PRACTICAL TIPS This concept could be introduced in a practical cooking session in the classroom – for example, making fruit smoothies using cups as measures. It will be important with this concept to use visual representations such as fraction strips and number lines to strengthen understanding that the order of fractions matters when subtracting.

ANSWERS

Question ① a): Max is not correct. He can do the subtraction. $1\frac{5}{6}$ more cups of cherries are needed.

Question ① b): $\frac{10}{3} - \frac{3}{2} = \frac{20}{6} - \frac{9}{6} = \frac{11}{6} = 1\frac{5}{6}$

Discover

① a) Is Max correct?

b) Show how Isla's method will give you the answer to $3\frac{1}{3} - 1\frac{1}{2}$.

156

PUPIL TEXTBOOK 6A PAGE 156

Share

WAYS OF WORKING Whole class teacher led

ASK

- Question ① a): *How could you rewrite $3\frac{1}{3}$?*
- Question ① a): *What do you need to subtract first? What next?*
- Question ① b): *How can you convert the mixed numbers to improper fractions? How can you now subtract? What sort of fraction is the answer?*

IN FOCUS This question develops children's understanding of how to subtract mixed numbers and fractions that cross the whole. Question ① a) introduces the concept that when subtracting mixed numbers where the first fractional part is smaller than the second fractional part, it is necessary to change the fractions to carry out the subtraction. Discuss with children the first method of subtracting the wholes and subtracting the parts they learnt for addition in Lesson 6. For subtraction, it will be important to highlight the fraction strips, emphasising that $\frac{1}{2}$ is greater than $\frac{1}{3}$, which is why Max thinks the subtraction is impossible. Explore how to rewrite $3\frac{1}{3}$ to make the fraction part bigger, so that the subtraction is easier.

Question ① b) focuses on the alternative method of converting mixed numbers to improper fractions and subtracting the fractions. Again, highlight the fraction strips showing the numbers as improper fractions and discuss the calculations needed to convert the mixed numbers to improper fractions.

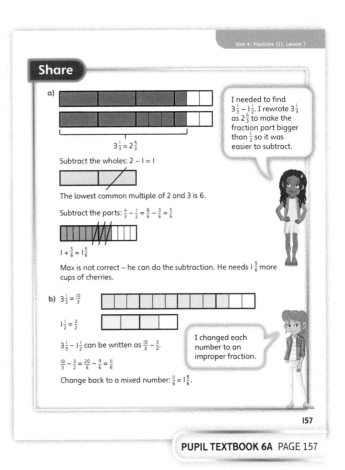

PUPIL TEXTBOOK 6A PAGE 157

Think together

Whole class teacher led (I do We do, You do)

ASK

- Question **1**: *How can you find the lowest common multiple of each denominator? How can you convert the mixed numbers to improper fractions?*
- Question **2**: *What calculation do you need to do to find $\frac{2}{3}$ less? What method will you use?*
- Question **3**: *What do you add to $2\frac{5}{6}$ to get to 3? What do you add to 3 to get to 5? What do you add to 5 to get to $5\frac{3}{10}$? What do you need to do with all your answers?*

IN FOCUS Question **2** is a word problem where children identify the required calculation with no structure or support. Encourage them to write the fraction in numbers and to show their method. Some children may choose to convert the hours into minutes and calculate the answer this way. This could be a good way for children to check their answer but encourage them to use fractions to find the answer.

Question **3** introduces a third method: using a number line to subtract fractions. Encourage children to explain how this method works, then ask them to use all three methods to work out the answers to part b). Encourage them to explain which method is most efficient.

STRENGTHEN To strengthen learning, encourage children to write down their calculations, breaking the methods into steps and using pictorial support.

DEEPEN Question **2** can be explored further by giving children another time and asking them to find the difference; for example, *Alex completes the puzzle in $1\frac{5}{6}$ hours. How much quicker was Alex than Andy? How much quicker was Alex than Jamilla?*

ASSESSMENT CHECKPOINT Question **2** provides an opportunity to assess children's ability to subtract mixed numbers where the fractional parts cross the whole. Look for children using either method, confidently working through the steps.

Question **3** provides an opportunity to assess children's ability to subtract mixed numbers. Look for children who can clearly explain the steps and who show fluency with all methods and an understanding of why it is necessary to convert the numbers before completing the calculations.

ANSWERS

Question **1** a): Subtract the wholes: $4 - 2 = 2$.

Subtract the parts: $\frac{2}{3} - \frac{1}{6} = \frac{4}{6} - \frac{1}{6} = \frac{3}{6} = \frac{1}{2}$

$2 + \frac{1}{2} = 2\frac{1}{2}$

Question **1** b): $4\frac{1}{3} = \frac{13}{3}$; $2\frac{3}{4} = \frac{11}{4}$

$4\frac{1}{3} - 2\frac{3}{4} = \frac{13}{3} - \frac{11}{4} = \frac{52}{12} - \frac{33}{12} = \frac{19}{12} = 1\frac{7}{12}$

Question **1** c): Subtract the wholes: $5 - 1 = 4$.

Subtract the parts: $\frac{3}{10} - \frac{1}{4} = \frac{6}{20} - \frac{5}{20} = \frac{1}{20}$

$4 + \frac{1}{20} = 4\frac{1}{20}$

Question **2**: It takes Andy $1\frac{7}{12}$ hours to complete the puzzle.

Question **3** a): $5\frac{3}{10} - 2\frac{5}{6} = 2\frac{7}{15}$

Question **3** b): $3\frac{1}{2} - 1\frac{7}{10} = 1\frac{4}{5}$; $26\frac{1}{2} - 18\frac{4}{5} = 7\frac{7}{10}$

Think together

1 a) Work out $4\frac{2}{3} - 2\frac{1}{6}$.

b) Work out $4\frac{1}{3} - 2\frac{3}{4}$.

c) Work out $5\frac{3}{10} - 1\frac{1}{4}$.

I am going to try and subtract the wholes and then the fraction parts.

I think it is easier to change each number to an improper fraction first.

158

PUPIL TEXTBOOK 6A PAGE 158

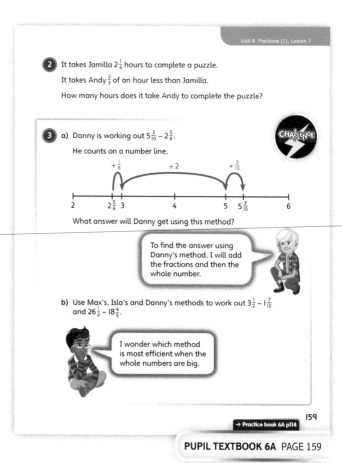

2 It takes Jamilla $2\frac{1}{4}$ hours to complete a puzzle.

It takes Andy $\frac{2}{3}$ of an hour less than Jamilla.

How many hours does it take Andy to complete the puzzle?

3 a) Danny is working out $5\frac{3}{10} - 2\frac{5}{6}$.

He counts on a number line.

What answer will Danny get using this method?

To find the answer using Danny's method, I will add the fractions and then the whole number.

b) Use Max's, Isla's and Danny's methods to work out $3\frac{1}{2} - 1\frac{7}{10}$ and $26\frac{1}{2} - 18\frac{4}{5}$.

I wonder which method is most efficient when the whole numbers are big.

159

→ Practice book 6A p114

PUPIL TEXTBOOK 6A PAGE 159

Practice

IN FOCUS Question ❶ aims to consolidate children's understanding of subtracting mixed numbers and fractions. Encourage children to use their preferred method and to write down their calculations.

Question ❷ develops subtraction involving mixed numbers and whole numbers. Encourage children to think carefully about how they could work out the calculations, considering which method is most efficient.

Question ❸ uses a part-whole model in a subtraction problem, requiring children to find the missing parts. Again, it will be useful to encourage children to write down the calculation required and to show their method.

Question ❹ introduces a context problem involving heights of giraffes, requiring children to find the height of the baby giraffe. This question adds variation by looking at vertical difference rather than horizontal. Question ❼ develops problem solving in an abstract form, with children required to complete more than one subtraction to work out the value of the symbols.

STRENGTHEN To strengthen learning, encourage children to use fraction strips or number lines for support. Question ❼ could be adapted by giving children the first calculation as $5\frac{3}{4} - 3\frac{5}{6} = $ ☆ so they can see what is required.

DEEPEN Question ❼ can be used to deepen learning and develop abstract and algebraic thinking, by giving children another calculation using ♡ to find the value of another symbol: If $9\frac{3}{8} - $ ⬤ $= $ ♡, what is the value of ⬤?

THINK DIFFERENTLY Question ❻ is a problem-solving question. The fractions are represented as shapes, so children need to work out the numbers before they can complete the subtraction calculation.

ASSESSMENT CHECKPOINT Question ❼ provides an opportunity to assess children's ability to subtract mixed numbers in abstract form, demonstrating algebraic reasoning. Look for children being able to comprehend what steps they need to complete and work backwards while clearly explaining their reasoning. Children should be confident in subtracting mixed numbers and fractions using their preferred method.

ANSWERS Answers for the **Practice** part of the lesson can be found in the *Power Maths* online subscription.

Reflect

IN FOCUS This **Reflect** activity gives children an opportunity to review their learning and describe their preferred method for subtracting mixed numbers, before discussing with a partner why they have chosen that method.

ASSESSMENT CHECKPOINT This activity assesses a child's understanding of subtracting two mixed numbers. Look for children confidently using their preferred method and explaining their reason.

ANSWERS Answers for the **Reflect** part of the lesson can be found in the *Power Maths* online subscription.

After the lesson ⏸

- Can children subtract mixed numbers and fractions that involve subtracting across the whole?
- Can children explain the steps in their preferred method for subtracting mixed numbers and say why they prefer that method?

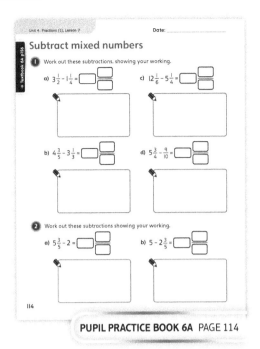

PUPIL PRACTICE BOOK 6A PAGE 114

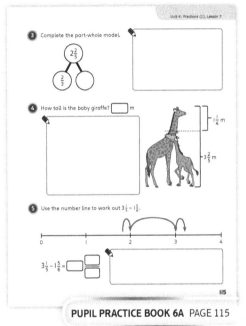

PUPIL PRACTICE BOOK 6A PAGE 115

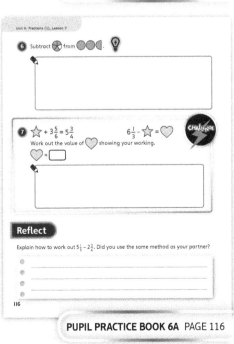

PUPIL PRACTICE BOOK 6A PAGE 116

Multi-step problems

Learning focus

In this lesson, children extend their knowledge of adding and subtracting mixed numbers to solve problems which involve adding and subtracting more than two mixed numbers.

Before you teach

- Can children add and subtract fractions with different denominators?
- Can children add and subtract mixed numbers?
- Can children convert between mixed numbers and improper fractions?

NATIONAL CURRICULUM LINKS

Year 6 Number – fractions

Add and subtract fractions with different denominators and mixed numbers, using the concept of equivalent fractions.

ASSESSING MASTERY

Children can solve problems that incorporate adding and subtracting mixed numbers and fractions, fluently converting between mixed numbers and improper fractions and using equivalent fractions as needed. They can use a variety of methods to solve word and abstract problems, while confidently explaining the steps in their own words.

COMMON MISCONCEPTIONS

Children may struggle to comprehend what calculation a problem is asking them to do; in this case, they will often just add the numbers given. Encourage children to break down the terms and vocabulary used and to use fraction strips or part-whole diagrams to give a pictorial representation of the problem. Ask:
- *What is the problem asking you to do?*
- *What words are used? What do they mean?*
- *Would a representation help?*
- *Can you write down the number sentences?*

STRENGTHENING UNDERSTANDING

Strengthen understanding by encouraging children to translate problems into number sentences and to use number lines and models to support their understanding.

GOING DEEPER

Children can be encouraged to deepen learning with problems based on other areas of mathematics involving addition and subtraction of fractions and mixed numbers. For example, children could work out the perimeter of a field given the length and width, or they could work out the length of a field given the width and the perimeter.

KEY LANGUAGE

In lesson: wholes, parts, equivalent fraction, compare, denominator

Other language to be used by the teacher: numerator, number sentence, perimeter, multi-step

STRUCTURES AND REPRESENTATIONS

Shapes, fraction strips, part-whole models

 In the eTextbook of this lesson, you will find interactive links to a selection of teaching tools.

Quick recap

Ask children to complete this calculation:

$\frac{1}{2} + \frac{1}{3} + \frac{1}{4} + \frac{1}{12} = ?$

Discover

Unit 4: Fractions (1), Lesson 8

Multi-step problems

WAYS OF WORKING Pair work

ASK

- Question ❶ a): *How can they make purple paint? What are you going to need to do with the numbers?*
- Question ❶ b): *How can you work out if there is enough paint?*

IN FOCUS This question introduces children to adding three numbers together, involving mixed numbers and fractions and using two different methods. Children then compare the fractions using equivalent fractions.

PRACTICAL TIPS This lesson uses the context of mixing paint to introduce the addition of three fractions. This could be introduced in a practical session in the classroom – for example, mixing paint to make purple or making mixed fruit juice cups. Encourage children to use pictorial representations as support.

ANSWERS

Question ❶ a): The children will make $4\frac{17}{20}$ litres of purple paint.

Question ❶ b): $4\frac{17}{20} > 4\frac{14}{20}$ so there will be enough purple paint to paint the roof and poles.

Discover

❶ a) The children are going to make some purple paint using the instructions. How many litres of purple paint will they make?

b) $3\frac{1}{2}$ litres of purple paint are needed to fully cover the roof.
$1\frac{1}{5}$ litres of purple paint are needed to paint the wooden poles.
Will there be enough paint to paint both the roof and the poles?

160

PUPIL TEXTBOOK 6A PAGE 160

Share

WAYS OF WORKING Whole class teacher led

ASK

- Question ❶ a): *What calculation do you need to do? Could you start with just two of the numbers?*
- Question ❶ a): *What do you need to convert to add your subtotal to the other fraction? What units do you need in the answer?*
- Question ❶ a): *Is there a different way of working this out?*
- Question ❶ b): *How can you work out how much paint is needed to cover the roof and the poles? What do you need to compare this with? How can you compare two fractions?*
- Question ❶ b): *What do you need to do to the denominators? Can you think of another method to work this out?*

IN FOCUS This question introduces children to adding more than two mixed numbers and fractions that cross the whole, using two different methods. In question ❶ a), it will be useful to build on the previous lesson, by discussing with children the first method of working out the total of red and blue paint – by adding the wholes and adding the parts. Highlight that because $\frac{11}{10}$ is an improper fraction, they need to convert it to a mixed number before adding together the red and blue paint and then adding on the white paint. Next, focus on the alternative method of adding all the numbers together at once; again, highlight the need to convert the improper fraction to a mixed number.

In question ❶ b), encourage children to find the amount of paint needed for the roof and poles together, before comparing with $4\frac{17}{20}$. Discuss any alternative methods that children come up with. For example, children may subtract $3\frac{1}{2}$ and $1\frac{1}{5}$ from $4\frac{17}{20}$ to work out if there is enough paint.

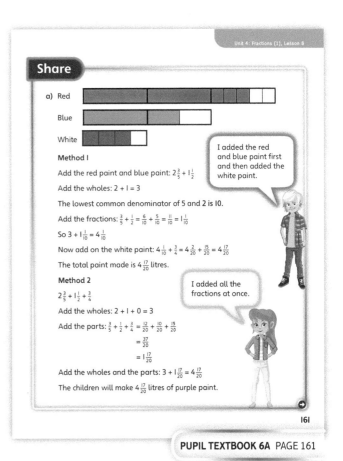

PUPIL TEXTBOOK 6A PAGE 161

Think together

Whole class teacher led (I do, We do, You do)

ASK

- Question ❶: *How could you work out the height of the cake in two different steps? How about in one step?*
- Question ❷: *What calculations do you need to do to find the missing numbers? How can you check your answer?*
- Question ❸: *Is the blue area smaller or bigger? Are you going to need to do an addition or a subtraction?*

IN FOCUS Question ❶ revisits addition of three mixed numbers, using both the methods introduced in the **Share** section. Encourage children to use the correct units of inches in their answer.

Question ❸ involves further problem solving and addresses a common misconception. Children may think that, because the star has been put on top of the square, they must add the numbers together. Encourage children to reason whether the blue area will be bigger or smaller than the rectangle to help them realise that a subtraction calculation is necessary. Encourage children to use the correct units of cm^2 in their answer.

STRENGTHEN In question ❷, encourage children to represent the calculations on a part-whole model to help them identify the calculation needed to solve the problem. Ask: *Can you represent the problem with a part-whole model? What calculation do you need to do to find the answer?*

DEEPEN Question ❶ can be explored further by asking children to find the height differences between the tiers of cakes.

ASSESSMENT CHECKPOINT Question ❶ gives an opportunity to assess children's ability to add three mixed numbers together, using both methods. Look for children confidently converting between mixed numbers and improper fractions and fluently using common denominators, while clearly explaining the steps in the calculations.

Question ❸ will assess children's ability to problem solve by adding or subtracting mixed numbers. Look for children being able to translate the problem into a number sentence and then deduce what they need to do, before competently using whichever method and operation they prefer to find the answer. Children should be able to explain in their own words their steps and reasoning.

ANSWERS

Question ❶: The total height of the cake is $4\frac{19}{24}$ inches.

Question ❷ a): $4\frac{1}{6} - 2\frac{1}{3} = 1\frac{5}{6}$

Question ❷ b): $2\frac{1}{3} - \frac{1}{2} = 1\frac{5}{6}$

Question ❸: The area of the blue background is
$7\frac{1}{6} - 5\frac{4}{9} = 1\frac{13}{18}$ cm^2.

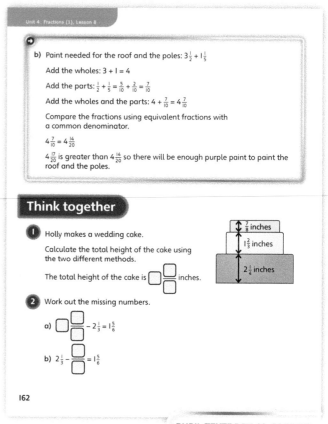

b) Paint needed for the roof and the poles: $3\frac{1}{2} + 1\frac{1}{5}$

Add the wholes: $3 + 1 = 4$

Add the parts: $\frac{1}{2} + \frac{1}{5} = \frac{5}{10} + \frac{2}{10} = \frac{7}{10}$

Add the wholes and the parts: $4 + \frac{7}{10} = 4\frac{7}{10}$

Compare the fractions using equivalent fractions with a common denominator.

$4\frac{7}{10} = 4\frac{14}{20}$

$4\frac{17}{20}$ is greater than $4\frac{14}{20}$ so there will be enough purple paint to paint the roof and the poles.

Think together

❶ Holly makes a wedding cake.

Calculate the total height of the cake using the two different methods.

The total height of the cake is ☐☐ inches.

❷ Work out the missing numbers.

a) ☐☐ $- 2\frac{1}{3} = 1\frac{5}{6}$

b) $2\frac{1}{3} - $ ☐☐ $= 1\frac{5}{6}$

$\frac{7}{8}$ inches
$1\frac{2}{3}$ inches
$2\frac{1}{4}$ inches

162

PUPIL TEXTBOOK 6A PAGE 162

❸ Lee has some sheets of card.

CHALLENGE

Area $= 5\frac{4}{9}$ cm^2 Area $= 7\frac{1}{6}$ cm^2

Lee places the star on the blue card and sticks it down. What is the area of the blue background?

I think you need to add the areas together because you are putting them together.

I am not sure that is correct.

163

→ Practice book 6A p117

PUPIL TEXTBOOK 6A PAGE 163

Practice

WAYS OF WORKING Independent thinking

IN FOCUS Questions **1**, **2** and **3** are multi-step word problems in the context of measure. Question **6** is a multi-step problem that can be solved in more than one way. Ensure children understand what the question is asking and what they need to do to answer it – a common mistake is to only partially answer a problem when it requires multiple steps. Ensure that having found Anna and Georgia's weights they then calculate the difference.

STRENGTHEN In question **6**, strengthen learning by encouraging children to draw three different fraction strips, one to represent the weight of each tortoise. This will help them to visualise the calculations they need to do.

DEEPEN Question **6** can be explored further by giving children a fourth tortoise and asking them to calculate its mass; for example, ask: *Rani weighs $\frac{5}{6}$ lb more than Anna. How much does Rani weigh?*

THINK DIFFERENTLY Question **4** is a missing number problem in a part-whole model. Explore with children the best place to start, and encourage them to write down the number sentences involved and to explain their reasoning.

ASSESSMENT CHECKPOINT Question **5** assesses children's ability to add mixed numbers. Look for children confidently using their preferred method and being able to explain their steps in their own words. Note any children who rely on pictorial representation; they will need support to develop their understanding before they move on to abstract calculations.

Question **6** assesses children's ability to solve problems that involve adding and subtracting mixed numbers and fractions. Children able to translate the problems into number sentences and work systematically to reach a final answer, while explaining their reasoning in their own words, have mastered the topic.

ANSWERS Answers for the **Practice** part of the lesson can be found in the *Power Maths* online subscription.

Reflect

WAYS OF WORKING Pair work

IN FOCUS This **Reflect** activity requires children to work backwards and create a problem with a specific answer. Encourage them to use mixed numbers in their problem and to use both addition and subtraction if possible. Ensure children are confident that their answer is $2\frac{1}{3}$ and ask them to give their problem to a partner to check.

ASSESSMENT CHECKPOINT This activity provides an opportunity to assess children's ability to add and subtract mixed numbers. Look for children fluently working backwards to create a problem in an abstract format, using their understanding of inverse operations while confidently checking their own work and their partner's.

ANSWERS Answers for the **Reflect** part of the lesson can be found in the *Power Maths* online subscription.

After the lesson ⏸

- Can children add and subtract mixed numbers confidently?
- Can they solve problems that involve adding and subtracting mixed numbers and fractions?
- Can they break down multi-step problems into the steps and number sentences they need to solve the problem?

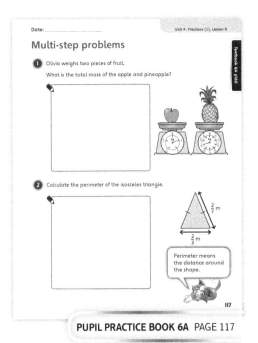

PUPIL PRACTICE BOOK 6A PAGE 117

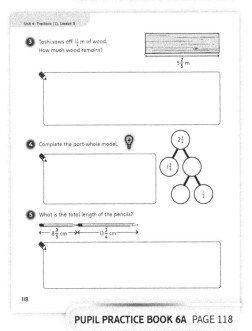

PUPIL PRACTICE BOOK 6A PAGE 118

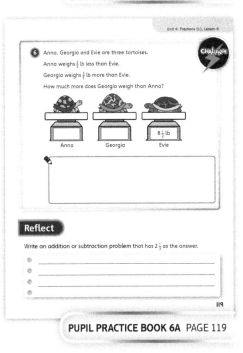

PUPIL PRACTICE BOOK 6A PAGE 119

Problem solving – add and subtract fractions

Learning focus

In this lesson, children understand how to solve more complex problems that involve adding and subtracting mixed numbers and fractions and which have more than one step.

NATIONAL CURRICULUM LINKS

Year 6 Number – fractions

Add and subtract fractions with different denominators and mixed numbers, using the concept of equivalent fractions.

ASSESSING MASTERY

Children can solve problems that incorporate adding and subtracting mixed numbers and fractions using more than one calculation, while fluently converting between mixed numbers and improper fractions. Children can clearly explain their steps and methods in their own words.

COMMON MISCONCEPTIONS

A common mistake is that children may not understand what calculation is required, particularly when the information is represented in a problematic way; they will often rush into the problem, perhaps adding all the numbers given. Encourage children to draw diagrams or use number lines to represent the problem. This will help them to understand which calculation is required and to identify which numbers they are going to use before they carry out any calculations. Ask:

- *Would a representation help?*
- *What is the question asking you to do?*
- *What operations do you need to use?*
- *What information do you need to find?*
- *Which numbers are you going to use?*

STRENGTHENING UNDERSTANDING

Children who are struggling with multi-step problems should focus first on problems that involve either addition or subtraction, but not both. Encourage children to use a fraction strip or a number line to aid their understanding.

GOING DEEPER

Deepen learning with more complex missing number problems that require children to carry out more than one calculation involving the addition and subtraction of fractions and mixed numbers, such as $\frac{5}{8} + \boxed{} = 1 - \frac{1}{4}$ or $2\frac{1}{5} - \frac{2}{3} = \frac{40}{15} - \boxed{}$ or $\boxed{} + \frac{5}{6} = 4\frac{1}{4} - \frac{2}{3}$ or $\frac{7}{8} - \boxed{} = \frac{2}{3} + \boxed{}$.

KEY LANGUAGE

In lesson: fraction, mass, perimeter

Other language to be used by the teacher: multi-step, calculation, operation, mixed number, improper fraction, common denominator, numerator, denominator

STRUCTURES AND REPRESENTATIONS

Fraction strips, number lines

 In the eTextbook of this lesson, you will find interactive links to a selection of teaching tools.

Quick recap 🔄

Ask children to complete this calculation:

$\frac{1}{2} - \frac{1}{3} + \frac{1}{4} - \frac{1}{12} = ?$

Discover

WAYS OF WORKING Pair work

ASK

- Question ① a): *What's the same and what's different about the bowling balls? How can you work out the weight of one yellow bowling ball? Would a fraction strip help?*
- Question ① b): *How could you use your answer from part a) to help with this?*

IN FOCUS Question ① introduces children to solving a multi-step problem involving adding and subtracting with mixed numbers; this requires logical thinking and problem solving. Question ① a) introduces the first step in the problem, while question ① b) requires them to complete the problem using their answer from part a). Encourage children to break down the problem into the steps needed, decoding the language of the question to identify the individual calculations.

PRACTICAL TIPS There are many different everyday contexts for solving multi-step problems involving fractions. This topic uses mass and bowling balls, but other contexts could involve distance, time or capacity. Whatever the context, it will be important to encourage children to use fraction strips to represent problems and break them down into the steps needed to find the answers.

ANSWERS

Question ① a): The weight of one yellow ball is $4\frac{2}{3}$ kg.

Question ① b): The weight of one red striped ball is $1\frac{3}{4}$ kg.

Share

WAYS OF WORKING Whole class teacher led

ASK

- Question ① a): *What do the fraction strips show?*
- Question ① a): *What calculation do you need to do to find the mass of one yellow bowling ball? What do you need to find when subtracting mixed numbers?*
- Question ① b): *How does knowing the mass of the yellow bowling balls help you to find the mass of the red striped ball? How can you check your answer using the fraction strip in part a)?*

IN FOCUS Encourage children to use the fraction strips; discuss how these models have been formed and highlight what is the same and what is different. With question ① a), develop prior knowledge by reminding children that they need to find a common denominator to subtract fractions.

In question ① b), it will be important for children to realise that they can use their answer from part a) to find the solution. Explore with children how they can calculate the mass of two yellow bowling balls by adding, and then find the mass of the red bowling ball by subtracting; rewriting $11\frac{1}{12}$ will make it easier to subtract. Encourage children to consider the alternative method of converting to improper fractions and subtracting.

Problem solving - add and subtract fractions

Discover

① a) Aki's bowling balls have a total mass of $15\frac{3}{4}$ kg.

Bella's bowling balls have a total mass of $11\frac{1}{12}$ kg.

Work out the mass of one yellow ball.

b) Work out the mass of one red striped ball.

164

PUPIL TEXTBOOK 6A PAGE 164

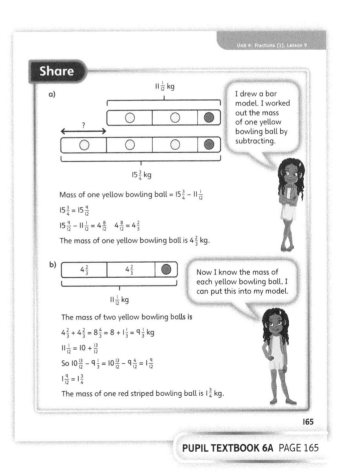

Share

a)

$15\frac{3}{4}$ kg

Mass of one yellow bowling ball $= 15\frac{3}{4} - 11\frac{1}{12}$

$15\frac{3}{4} = 15\frac{9}{12}$

$15\frac{9}{12} - 11\frac{1}{12} = 4\frac{8}{12}$ $4\frac{8}{12} = 4\frac{2}{3}$

The mass of one yellow bowling ball is $4\frac{2}{3}$ kg.

b)

$11\frac{1}{12}$ kg

The mass of two yellow bowling balls is

$4\frac{2}{3} + 4\frac{2}{3} = 8\frac{4}{3} = 8 + 1\frac{1}{3} = 9\frac{1}{3}$ kg

$11\frac{1}{12} = 10 + \frac{13}{12}$

So $10\frac{13}{12} - 9\frac{1}{3} = 10\frac{13}{12} - 9\frac{4}{12} = 1\frac{9}{12}$

$1\frac{9}{12} = 1\frac{3}{4}$

The mass of one red striped bowling ball is $1\frac{3}{4}$ kg.

> I drew a bar model. I worked out the mass of one yellow bowling ball by subtracting.

> Now I know the mass of each yellow bowling ball, I can put this into my model.

165

PUPIL TEXTBOOK 6A PAGE 165

Think together

Unit 4: Fractions (1), Lesson 9

Think together

WAYS OF WORKING Whole class teacher led (I do, We do, You do)

ASK

- Question **1**: *Can you identify what fractions are shaded (or not shaded)? What steps do you need to take to find the answer?*
- Question **3**: *What do you know about the sides of a square? How can you find the perimeter of a square? Of a rectangle? What calculation are you going to need to do to work out the perimeter?*

IN FOCUS Questions **1** and **2** revisit mixed number problems, similar to those in the **Share** section. Question **1** uses the context of shapes, requiring children to find the amount not shaded. Encourage children to explain their method; for example, they could add the amounts together to find what is not shaded, or they could find the amount that is shaded and subtract from one whole. Look for children who erroneously work out the area of the shape that is shaded, rather than the area not shaded. Question **2** requires children to identify the calculation needed to work out the missing numbers. Encourage them to find the value of C first and to write down the number sentence before working anything out.

Question **3** involves further contextual problem solving, finding the perimeter of two rooms, using children's prior knowledge of shapes. Children may need reminding of the definition of perimeter and the properties of squares.

STRENGTHEN Strengthen learning with question **2** by encouraging children to use the number line to count up to help them find the difference.

DEEPEN Question **3** can be explored further by encouraging children to find other ways of calculating the perimeter of the square room, rather than just adding. Children can also be encouraged to find the perimeters of other rooms on the floor plan, using other mixed number questions or even their own mixed numbers.

ASSESSMENT CHECKPOINT Question **3** assesses children's ability to solve multi-step number problems involving adding and subtracting. Look for children confidently translating the problems into number sentences and identifying the steps needed to find the answer, while explaining these steps. Children should be able to convert fluently between mixed numbers and improper fractions and use both methods.

ANSWERS

Question **1**: $\frac{7}{12}$ of the shape is not shaded.

Question **2**: C is $3\frac{2}{3}$ and is $\frac{11}{12}$ greater than A.

Question **3**: The square room has a perimeter of 13 metres. The rectangular room has a perimeter of $12\frac{1}{5}$ metres. The square room has a perimeter that is $\frac{4}{5}$ metres greater.

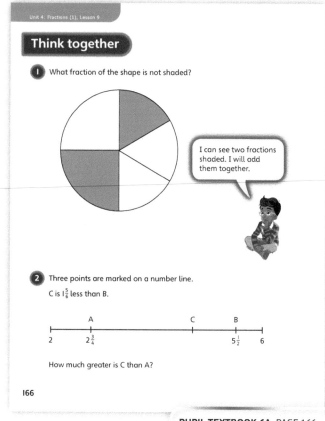

Think together

1 What fraction of the shape is not shaded?

I can see two fractions shaded. I will add them together.

2 Three points are marked on a number line. C is $1\frac{5}{6}$ less than B.

How much greater is C than A?

166

PUPIL TEXTBOOK 6A PAGE 166

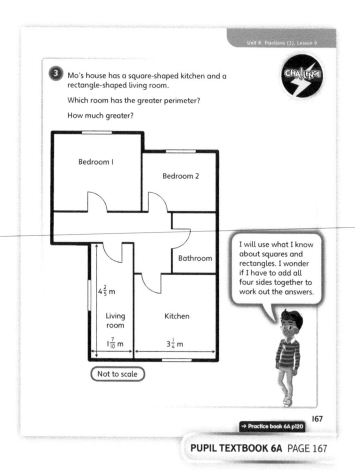

3 Mo's house has a square-shaped kitchen and a rectangle-shaped living room.

Which room has the greater perimeter?

How much greater?

CHALLENGE

Bedroom 1

Bedroom 2

Bathroom

$4\frac{2}{5}$ m

Living room

Kitchen

$1\frac{7}{10}$ m

$3\frac{1}{4}$ m

Not to scale

I will use what I know about squares and rectangles. I wonder if I have to add all four sides together to work out the answers.

167

→ Practice book 6A p120

PUPIL TEXTBOOK 6A PAGE 167

Practice

WAYS OF WORKING Independent thinking

IN FOCUS Question **3** introduces the new unit of a million, and requires children to find a combined total. Look out for children who do not complete the problem and only work out the number of downloads on Saturday.

Question **4** requires children to solve a multi-step problem that incorporates both addition and subtraction. Encourage children to solve the problem one step at a time. Look out for children who do not work out the distance from the top of the drainpipe, but instead just calculate how many metres the spider has climbed up.

Question **5** introduces the new notations of AB and BC to identify distances between points on a number line. Explain that these mean the distance between A and B and the distance between B and C. This problem requires children to carry out several calculations and it will be beneficial to encourage them to solve the problem one stage at a time.

STRENGTHEN Strengthen learning in question **3** by encouraging children to draw two fraction strips to show the number of downloads on Friday and on Saturday; they should label the total of both bars with a question mark.

In question **4**, encourage children to draw a diagram to show what is happening to the spider. This will help them see what calculations they need to do.

DEEPEN Question **5** can be explored further by telling children that B has now been moved to exactly half-way between A and C; ask children to calculate the difference between the old value of B and the new value of B.

ASSESSMENT CHECKPOINT Questions **4** and **5** require children to use multiple steps and calculations to systematically solve a problem. Children should be able to convert fluently between mixed numbers and improper fractions and use whichever method for adding and subtracting is most suitable for the calculation.

ANSWERS Answers for the **Practice** part of the lesson can be found in the *Power Maths* online subscription.

Reflect

WAYS OF WORKING Pair work

IN FOCUS This **Reflect** activity encourages children to identify which questions they found difficult. This will provide an opportunity to discuss the difficulties of problem solving – for example, it is easy to make mistakes or miss a step; reassure them that they should not be put off by this. Encourage children to discuss and explain any stumbling blocks or mistakes they have made and to look for ways to correct them.

ASSESSMENT CHECKPOINT Children are likely to identify many different issues, such as not understanding the question, or not completing the question. Look for children who can clearly identify a mistake or stumbling block and suggest strategies for correcting it.

ANSWERS Answers for the **Reflect** part of the lesson can be found in the *Power Maths* online subscription.

After the lesson ⏸

- Can children confidently translate multi-step problems into the steps and calculations needed?
- Can they solve multi-step problems that incorporate adding and subtracting mixed numbers and fractions?

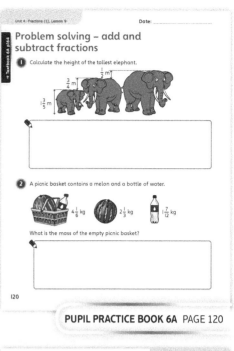

PUPIL PRACTICE BOOK 6A PAGE 120

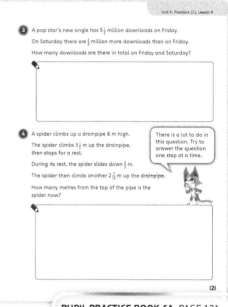

PUPIL PRACTICE BOOK 6A PAGE 121

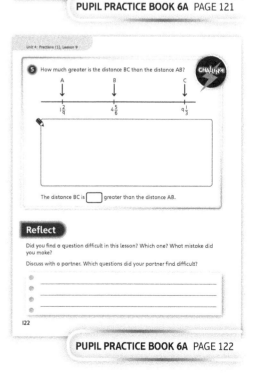

PUPIL PRACTICE BOOK 6A PAGE 122

End of unit check

Don't forget the unit assessment grid in your
Power Maths online subscription.

WAYS OF WORKING Group work adult led and independent working

WAYS OF WORKING Group work adult led and independent working

IN FOCUS

- Questions **1**, **2** and **4** assess children's ability to add or subtract two proper fractions where the answer exceeds 1. These questions test children's understanding of converting an improper fraction to a mixed number.
- Question **5** requires children to find a missing number; children should realise that they only need to focus on the fractions and look for common denominators by making equivalent fractions.
- Question **7** requires children to problem solve by identifying a fraction from a diagram and choosing the correct calculation. Encourage children to use a fraction strip to help them visualise what they need to do.
- Question **8** requires children to identify the calculations needed to subtract two mixed numbers from a given total. They may choose to add the two mixed numbers together before subtracting both from the total, or they may subtract first one and then the other. Whichever method they choose, they will need to find a common denominator, and will need to remember to include the whole numbers in their calculations.

ANSWERS AND COMMENTARY

Children who have mastered this unit will be able to add and subtract fractions and mixed numbers confidently using several formal written methods. They will be able to solve multi-step problems and explain which method is most efficient.

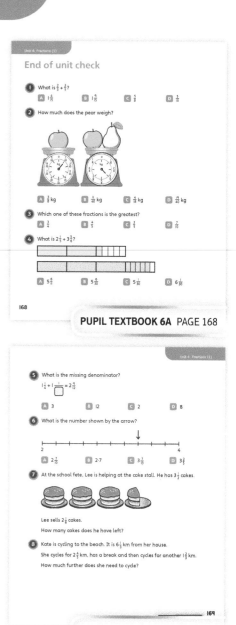

PUPIL TEXTBOOK 6A PAGE 168

PUPIL TEXTBOOK 6A PAGE 169

Q	A	WRONG ANSWERS AND MISCONCEPTIONS	STRENGTHENING UNDERSTANDING
1	A	C suggests adding of fraction parts. D suggests use of a common denominator without converting numerators.	Encourage children to list multiples to find a common denominator.
2	B	A suggests just reading the scale the pear is on. C suggests adding fraction parts. D suggests adding not subtracting.	Encourage children to draw a comparison fraction strip for both groupings.
3	B	D suggests selecting the fraction with the largest numbers.	Reiterate the need to compare numerators when denominators are equal.
4	D	A suggests adding the wholes, then the numerators and the denominators. B suggests omitting to convert the numerators.	Give children mixed number conversions and ask them to identify which are correct.
5	C	A suggests dividing 12 by 4. D suggests subtracting 4 from 12.	Encourage children to work with twelfths.
6	D	A suggests children have misread the small gaps; B suggests they have written their answer as a decimal. C suggests they have identified the arrow as > 3 but still counted 7 tenths.	Encourage children to focus on the part of the line between 2 and 3 and identify how many equal parts the line is divided into.
7	$1\frac{1}{3}$	Children may give an unsimplified answer ($1\frac{2}{6}$). They may add instead of subtracting or may ignore the wholes.	Encourage children to subtract the wholes and fractions separately.
8	$1\frac{13}{15}$	Children may add instead of subtracting, or may not correctly find the common denominator.	Encourage children to draw a bar model to represent the steps of the problem.

My journal

WAYS OF WORKING Independent thinking

ANSWERS AND COMMENTARY

Question **1** assesses children's understanding of the steps needed to add and subtract fractions. Look for children fluently using equivalent fractions and converting their answer to a mixed number, while explaining their method. Observe whether children rely on fraction strips or struggle to find a common denominator.

a) $\frac{2}{3} + \frac{4}{5} = 1\frac{7}{15}$

c) $2\frac{3}{8} - 1\frac{1}{3} = 1\frac{1}{24}$

e) $3\frac{1}{2} + 1\frac{8}{9} = 5\frac{7}{18}$

b) $\frac{7}{10} + \frac{1}{4} = \frac{19}{20}$

d) $3\frac{2}{5} + 4\frac{3}{4} = 8\frac{3}{20}$

f) $\frac{1}{3} + \frac{1}{2} + \frac{1}{4} = 1\frac{1}{12}$

Question **2** assesses children's understanding of both methods for subtracting mixed numbers involving conversion of the fractional parts. Look for children clearly explaining both methods and confidently using common denominators and equivalent fractions.

Jamie's method: $5\frac{1}{4} - 3\frac{2}{5} = 4\frac{5}{4} - 3\frac{2}{5} = 4\frac{25}{20} - 3\frac{8}{20} = 1\frac{17}{20}$.

Danny's method: $5\frac{1}{4} - 3\frac{2}{5} = \frac{21}{4} - \frac{17}{5} = \frac{105}{20} - \frac{68}{20} = \frac{37}{20} = 1\frac{17}{20}$.

Power check

WAYS OF WORKING Independent thinking

ASK

· Can you add and subtract fractions using written methods?
· How confident do you feel about converting an improper fraction to a mixed number? Simplifying a fraction? Finding a common denominator?

Power puzzle

WAYS OF WORKING Independent thinking

IN FOCUS Use this to assess whether children can solve missing number fractional calculations, and whether they show an understanding of the steps needed to add or subtract unrelated fractions using common denominators.

ANSWERS AND COMMENTARY

1 a) The denominators must be 5 and 7, the only factors of 35. The digits needed to make 9 (the wholes) must make 8 or 9 (the fractional parts could be greater than 1); the only options are 3 and 6 or 2 and 6. With 3 and 6, this leaves 2 and 4 for the numerators, giving a fractional answer of $\frac{34}{35}$ (too big) or greater than 1. So the wholes must be 2 and 6 and the fractions must be $\frac{3}{7}$ and $\frac{4}{5}$. Encourage experimentation and answer checking.

1 b) Find the total of the top row to give the total of each row or column (7). Then use this to find the missing numbers, starting with rows or columns where only one value is missing.
Second row: $4\frac{3}{4}, \frac{2}{3}, 1\frac{7}{12}$ Third row: $\frac{19}{20}, 3\frac{5}{6}, 2\frac{13}{60}$

After the unit ⏸

· Can children confidently compare, order and convert fractions, mixed numbers and improper fractions?
· Can they add and subtract fractions and mixed numbers including solving problems?

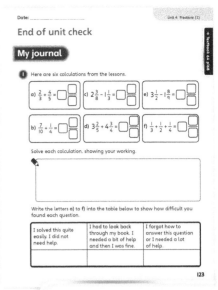

PUPIL PRACTICE BOOK 6A PAGE 123

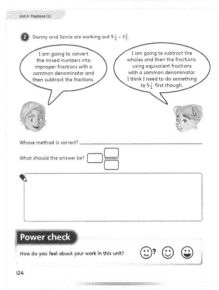

PUPIL PRACTICE BOOK 6A PAGE 124

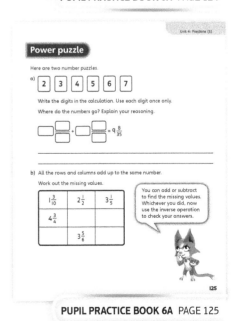

PUPIL PRACTICE BOOK 6A PAGE 125

Strengthen and **Deepen** activities for this unit can be found in the *Power Maths* online subscription.

Unit 5
Fractions ②

Mastery Expert tip! 'I found pictorial methods a great way to help children understand what they were doing and the meaning of the question. Too often we just tell children to follow a rule and it is important for them to understand the why behind the methods that they are using. Eventually we hope children will develop the rule for themselves.'

Don't forget to watch the Unit 5 video!

WHY THIS UNIT IS IMPORTANT

This unit completes children's primary work on fractions, focusing on multiplying fractions and dividing fractions by a whole number and developing their ability to perform any of the four operations with fractions. Pictorial representations are used throughout to ensure children develop a firm understanding rather than just learning rules. Finally, the unit builds on previous work on fractions of amounts, encouraging children to look for the most efficient methods when solving problems.

WHERE THIS UNIT FITS

→ Unit 4: Fractions (1)
→ **Unit 5: Fractions (2)**
→ Unit 6: Measure – imperial and metric measures

This unit builds on children's fraction work in previous units, including multiplying a proper fraction by a whole number and finding a fraction of an amount. It aims to bring together the four operations with fractions and give children confidence in problem solving with fractions.

Before they start this unit, it is expected that children:
- know how to multiply a proper fraction by a whole number and how to find a fraction of an amount
- have seen a fraction strip above a number line to help add and subtract fractions
- are confident with drawing bar models to represent simple problems
- know the rules for the order of operations
- know how to convert between a mixed number and an improper fraction and vice versa.

ASSESSING MASTERY

Children can multiply any fraction by a whole number and by any other fraction, and divide a fraction by a whole number. They can solve simple and multi-step fraction problems, including problems on fractions of an amount where they are given the fraction and need to find the whole.

COMMON MISCONCEPTIONS	STRENGTHENING UNDERSTANDING	GOING DEEPER
When multiplying a fraction by a whole number, children may multiply both the denominator and the numerator, rather than just the numerator.	Children need to understand that the denominator is unaffected. Use bar models so that they understand the reasons behind the method.	Ask children to compile a set of 'Helpful tips' on multiplying and dividing fractions. Encourage them to think about what mistakes others may make.
When presented with questions such as '$\frac{2}{3}$ of a number is 30. What is the number?', children may think that 30 is the whole, not the part, and so find $\frac{2}{3}$ of 30.	Children need to realise that they are finding the whole amount and so need to divide by 2 first and then multiply by 3. Draw a clearly labelled bar model to help them see when they are finding a fraction of an amount and when the whole.	Ask children to explain the most efficient method when solving a problem involving fractions of an amount. For example, when solving 'Which is greater: $\frac{1}{3}$ of 24 or $\frac{1}{4}$ of 24?' children can apply their knowledge of $\frac{1}{3} > \frac{1}{4}$ without working out the values.

Unit 5: Fractions ❷

Use these pages to introduce the unit focus to children. You can use the characters to explore different ways of working.

STRUCTURES AND REPRESENTATIONS

Number line and fraction strip: This model helps children convert between improper fractions and mixed numbers.

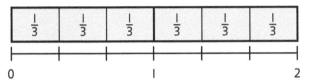

Fraction grid: This model will help children to understand how to multiply fractions, developing the understanding that not only the numerators but also the denominators are multiplied together.

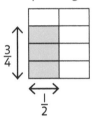

Bar model: This model will help children solve problems involving fractions of an amount, where they are either given the whole or given the fraction of the amount.

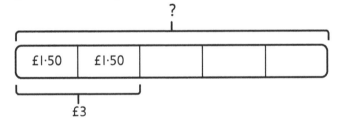

KEY LANGUAGE

There is some key language that children will need to know as part of the learning in this unit:

→ numerator, denominator

→ multiply, divide

→ proper fraction, improper fraction, mixed number, whole number, integer

→ whole, part

→ order of operations

→ convert

→ simplify

PUPIL TEXTBOOK 6A PAGE 170

PUPIL TEXTBOOK 6A PAGE 171

207

Multiply fractions by integers

Learning focus

In this lesson, children will build on their learning from Year 5 to multiply proper and improper fractions and mixed numbers by a whole number.

Before you teach

- Do children know their times-tables?
- Do children remember from year 5 how to multiply a fraction by a whole number?
- Can children represent a fraction using a fraction strip?

NATIONAL CURRICULUM LINKS

Year 5 Number – fractions (including decimals and percentages)

Multiply proper fractions and mixed numbers by whole numbers, supported by materials and diagrams.

ASSESSING MASTERY

Children can multiply any fraction by a whole number, including improper fractions and mixed numbers.

COMMON MISCONCEPTIONS

Children may multiply both the numerator and denominator by the whole number, for example $\frac{2}{3} \times 4 = \frac{8}{12}$. Ask:

- *Can you show me $\frac{2}{3}$ using fraction strips? How can you show $\frac{2}{3} \times 4$ using fraction strips?*
- *The fraction wall shows that $\frac{2}{3}$ and $\frac{8}{12}$ are equivalent. If you multiply a number by 4, will the answer be larger, smaller or the same as the original number? Can $\frac{2}{3} \times 4$ be $\frac{8}{12}$?*

STRENGTHENING UNDERSTANDING

Use fraction strips above number lines and diagrams of fractions to help secure understanding. Discuss with children each step in finding the answer using a pictorial method. Link the representation clearly to the abstract calculation. For example, to show $\frac{2}{3} \times 4$ draw a fraction strip split in thirds, above a number line. Colour in $\frac{2}{3}$ four times. Ask: *What fraction of the strip is shaded?* Use the number line to convert this to a mixed number.

GOING DEEPER

Ask children missing number problems such as $\frac{3}{8} \times \square = \frac{9}{8}$, $\frac{?}{6} \times 3 = \frac{33}{6}$, $\frac{3}{4} \times \square = 3\frac{3}{4}$.

Extend to problems where the answer has been simplified, for example $\frac{5}{6} \times \square = \frac{10}{3}$, $\frac{3}{10} \times \square = 3\frac{3}{5}$.

KEY LANGUAGE

In lesson: whole number, fraction, multiply, mixed number, improper fraction, numerator

Other language to be used by the teacher: denominator, proper fraction

STRUCTURES AND REPRESENTATIONS

Bar model, number line

RESOURCES

Optional: fraction strips, fraction wheels

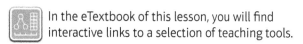 In the eTextbook of this lesson, you will find interactive links to a selection of teaching tools.

Quick recap 🔄

Ask children to count on in unit fractions. Start by counting on from 0 in steps of $\frac{1}{4}$. Then count on from 0 in steps of $\frac{1}{5}$. Finally, count on from 0 in steps of $\frac{1}{7}$. Ask: *What happens when you reach and pass 1?*

Discover

Multiply fractions by integers

Discover

WAYS OF WORKING Pair work

ASK

- Question ① a): *What information do you need? What do you have to do to get the answer?*
- Question ① b): *What is the same about this question? What is different? How will you work out the total time?*

IN FOCUS Question ① a) requires a fraction to be multiplied by a whole number. Some information is given in the question text which needs to be combined with relevant information from the image. Discuss the approaches that children may take to answer this question, for example multiplication or addition. Encourage children to draw diagrams or use fraction shapes to explain their methods. Question ① b) requires a mixed number to be multiplied by a whole number, using all the information given in the image. Again, discuss the different approaches that children may take to answer the question.

PRACTICAL TIPS Have fraction strips or fraction wheels available for children to fit together.

ANSWERS

Question ① a): $\frac{5}{8}$ of a tank of fuel is used in a day.

Question ① b): The total duration of the boat trips is $6\frac{1}{4}$ hours.

① a) The boat uses $\frac{1}{8}$ of a tank of fuel for each trip. How many tanks of fuel are used in a day?

b) What is the total duration of the boat trips in a day?

172

PUPIL TEXTBOOK 6A PAGE 172

Share

WAYS OF WORKING Whole class teacher led

ASK

- Question ① a): *What are the two methods shown? Why do they both give the same answer?*
- Question ① a): *How does the diagram show the answer to the question? Did anyone show it with their equipment in a different way?*
- Question ① b): *Can you draw a diagram to explain method 1? What about method 2? Why do they give the same answer? Which method do you prefer?*

IN FOCUS Question ① a) shows two methods: addition and multiplication. Discuss how the diagram shows both methods, establishing that repeated addition is the same as multiplication. Explain that it makes more sense to give the answer as a mixed number than as an improper fraction.

Question ① b) also shows two methods: multiplying the whole and the fraction separately and then adding them; or converting the mixed number to an improper fraction before multiplying by the number. Both methods require the conversion of an improper fraction into a mixed number. Also use the diagrams to show the total of $\frac{25}{4}$. Explain that it makes sense to change the answer into a mixed number as we would say '6 and a quarter hours' rather than '25 quarter hours'.

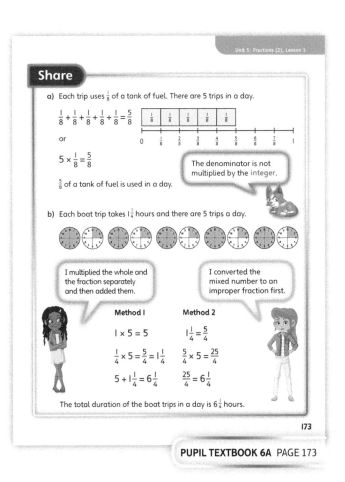

Share

a) Each trip uses $\frac{1}{8}$ of a tank of fuel. There are 5 trips in a day.

$\frac{1}{8} + \frac{1}{8} + \frac{1}{8} + \frac{1}{8} + \frac{1}{8} = \frac{5}{8}$

or

$5 \times \frac{1}{8} = \frac{5}{8}$

The denominator is not multiplied by the integer.

$\frac{5}{8}$ of a tank of fuel is used in a day.

b) Each boat trip takes $1\frac{1}{4}$ hours and there are 5 trips a day.

I multiplied the whole and the fraction separately and then added them.

I converted the mixed number to an improper fraction first.

Method 1

$1 \times 5 = 5$

$\frac{1}{4} \times 5 = \frac{5}{4} = 1\frac{1}{4}$

$5 + 1\frac{1}{4} = 6\frac{1}{4}$

Method 2

$1\frac{1}{4} = \frac{5}{4}$

$\frac{5}{4} \times 5 = \frac{25}{4}$

$\frac{25}{4} = 6\frac{1}{4}$

The total duration of the boat trips in a day is $6\frac{1}{4}$ hours.

173

PUPIL TEXTBOOK 6A PAGE 173

Think together

Whole class teacher led (I do, We do, You do)

ASK

- Question **1**: *How does this question differ from question **1** a) in **Discover**? How many more thirds do you need to add to the fraction strip? How should the answer be presented?*
- Question **2**: *What do you need to multiply? What do you do in the first method? What about the second method? Which method do you prefer? Why?*
- Question **2**: *What do you notice about the numerator of the fraction and the numerator of the final answer?*
- Question **3** a): *What connection can you see between the numerator of the fraction, the whole number and the numerator of the answer?*
- Question **3** b): *How can you use what you discovered in part a) to help you? Can you find more than one answer for each number?*

IN FOCUS Question **2** requires children to use both methods of multiplying a mixed number by a whole number that were described in **Share**. Some children may notice that they multiply the numerator by the whole number when they are working through these methods. Question **3** a) is designed to develop this understanding by asking children to spot patterns. They should see that the numerator of the answer is the same as the numerator of the fraction multiplied by the whole number. They should also see that the denominator does not change. Explore these findings using fraction strips. Children then use this understanding to answer question **3** b).

STRENGTHEN In question **2**, give children fraction wheels to model each method. Discuss how the way that they are manipulating the wheels relates to the steps in the method.

DEEPEN Ask children to explain how they used the patterns they spotted in question **3** a) to find the answers to question **3** b). Challenge them to create their own missing number multiplications for a partner to solve, involving multiplying a fraction and a whole number.

ASSESSMENT CHECKPOINT Use questions **1** and **2** to assess whether children can multiply a fraction and a mixed number by a whole number. They should know different methods for getting to the answer.

ANSWERS

Question **1**: $\frac{1}{8} \times 7 = \frac{7}{8}$. So, $\frac{7}{8}$ of a tank of fuel is used.

Question **2**: Method 1: $1 \times 4 = 4, \frac{2}{5} \times 4 = \frac{8}{5} = 1\frac{3}{5}, 4 + 1\frac{3}{5} = 5\frac{3}{5}$

Method 2: $1\frac{2}{5} = \frac{7}{5}, \frac{7}{5} \times 4 = \frac{28}{5}, \frac{28}{5} = 5\frac{3}{5}$

The boat travels $5\frac{3}{5}$ km.

Question **3** a): $\frac{5}{4}, \frac{9}{4}; \frac{10}{6}, \frac{25}{6}, \frac{35}{6}$

Question **3** b): Numerous answers, for example:

$\frac{1}{8} \times \mathbf{5} = \frac{5}{8}$

$\frac{1}{9} \times \mathbf{10} = \frac{10}{9}$

$\frac{2}{9} \times \mathbf{5} = \frac{10}{9}$

$\frac{1}{5} \times \mathbf{6} = 1\frac{1}{5}$

$\frac{3}{5} \times \mathbf{2} = 1\frac{1}{5}$

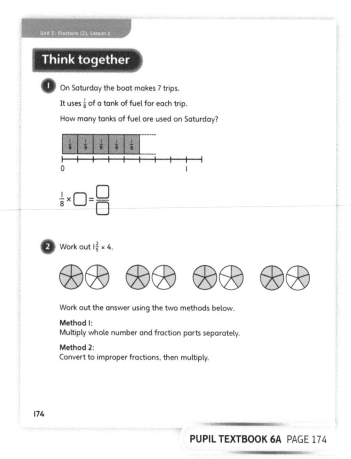

Think together

1 On Saturday the boat makes 7 trips.

It uses $\frac{1}{8}$ of a tank of fuel for each trip.

How many tanks of fuel are used on Saturday?

$\frac{1}{8} \times \square = \frac{\square}{\square}$

2 Work out $1\frac{2}{5} \times 4$.

Work out the answer using the two methods below.

Method 1:
Multiply whole number and fraction parts separately.

Method 2:
Convert to improper fractions, then multiply.

174

PUPIL TEXTBOOK 6A PAGE 174

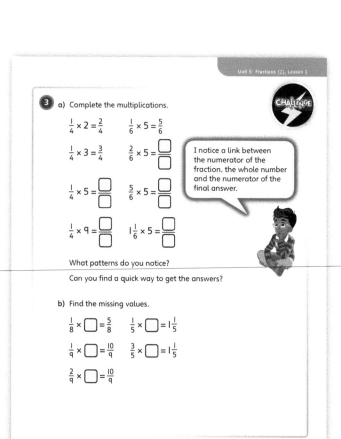

3 a) Complete the multiplications.

$\frac{1}{4} \times 2 = \frac{2}{4}$ $\frac{1}{6} \times 5 = \frac{5}{6}$

$\frac{1}{4} \times 3 = \frac{3}{4}$ $\frac{2}{6} \times 5 = \frac{\square}{\square}$

$\frac{1}{4} \times 5 = \frac{\square}{\square}$ $\frac{5}{6} \times 5 = \frac{\square}{\square}$

$\frac{1}{4} \times 9 = \frac{\square}{\square}$ $1\frac{1}{6} \times 5 = \frac{\square}{\square}$

> I notice a link between the numerator of the fraction, the whole number and the numerator of the final answer.

What patterns do you notice?

Can you find a quick way to get the answers?

b) Find the missing values.

$\frac{1}{8} \times \square = \frac{5}{8}$ $\frac{1}{5} \times \square = 1\frac{1}{5}$

$\frac{1}{9} \times \square = \frac{10}{9}$ $\frac{3}{5} \times \square = 1\frac{1}{5}$

$\frac{2}{9} \times \square = \frac{10}{9}$

175

→ Practice book 6A p126

PUPIL TEXTBOOK 6A PAGE 175

Practice

WAYS OF WORKING Independent thinking

IN FOCUS All the fractions in questions ❶ and ❷ are less than 1. Question ❶ includes images and representations to help children, whereas question ❷ requires them to begin to recognise and use the fact that the numerator of the answer is the same as the numerator multiplied by the whole number.

Question ❸ requires children to use both methods to multiply a mixed number by a whole number. In question ❹, they need to select a method for themselves. Ask: *Which method did you choose? Why?* This could extend into a discussion of times when one method is more appropriate.

Question ❺ has been designed to highlight a common misconception, where children multiply the denominator as well as the numerator by the whole number.

STRENGTHEN Provide children with fraction strips and fraction wheels to model questions ❶ and ❷. Providing a number line with the fraction strips will help them convert between the improper fractions and mixed numbers. Use the fraction strips to establish why the denominator stays the same, but the numerator changes.

DEEPEN Ask children questions that elicit a deeper understanding. For example, ask which method they would use to work out $125\frac{2}{7} \times 8$. Ask: *Can you explain when changing the fraction to an improper fraction would be an efficient method?*

THINK DIFFERENTLY In question ❻, children need to realise that the answer is not $\frac{11}{5}$ or $2\frac{1}{5}$ as the question asks how many bags the owner needs to buy. The owner therefore needs 3 bags.

ASSESSMENT CHECKPOINT Use question ❷ to assess whether children can multiply a proper fraction by a whole number. Use question ❹ to assess whether they can multiply improper fractions and mixed numbers by a whole number. Check whether they are using efficient methods.

ANSWERS Answers for the **Practice** part of the lesson can be found in the *Power Maths* online subscription.

Reflect

WAYS OF WORKING Independent thinking

IN FOCUS Children can use manipulatives, diagrams and words to help explain why $1\frac{2}{3} \times 4$ is equal to $6\frac{2}{3}$. They can use either method of multiplying a mixed number by a whole number or converting the mixed number to an improper fraction before multiplying by the number.

ASSESSMENT CHECKPOINT Children should be able to show that they can multiply a mixed number by a whole number. They must provide some reasoning and not just show the answer. Look for explanations that discuss the rule of multiplying the numerator by the whole number to get the numerator of the final answer.

ANSWERS Answers for the **Reflect** part of the lesson can be found in the *Power Maths* online subscription.

After the lesson ⏸

- Can children use diagrams to explain why a proper fraction multiplied by a whole number gives a particular answer?
- Can they multiply a mixed number by a whole number?

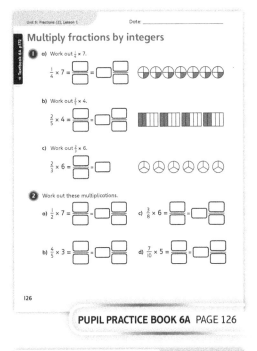

PUPIL PRACTICE BOOK 6A PAGE 126

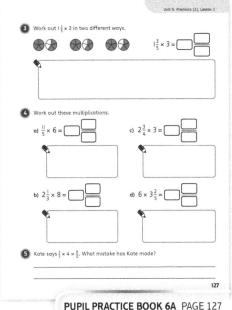

PUPIL PRACTICE BOOK 6A PAGE 127

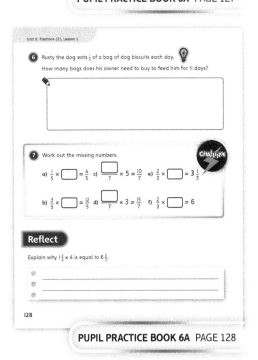

PUPIL PRACTICE BOOK 6A PAGE 128

Multiply fractions by fractions ①

Learning focus

In this lesson, children will learn to multiply a fraction by a fraction. They will use visual aids such as divided squares to support their understanding.

Before you teach ⅠⅠ

- Do children understand what a fraction is?
- Can children draw a diagram to represent a fraction?
- Do children understand the concept of multiplication?

NATIONAL CURRICULUM LINKS

Year 6 Number – fractions (including decimals and percentages)

Multiply simple pairs of proper fractions, writing the answer in its simplest form [for example, $\frac{1}{4} \times \frac{1}{2} = \frac{1}{8}$].

ASSESSING MASTERY

Children can multiply a fraction by a fraction by drawing diagrams, and express their answers in their simplest form. They understand that when a proper fraction is multiplied by a proper fraction the answer will be smaller.

COMMON MISCONCEPTIONS

Children may think that they have to make the denominators the same (as in addition and subtraction). Although this will lead to a correct answer, it is not efficient. Ask:
- *How can you divide a square to show $\frac{2}{3} \times \frac{1}{4}$? What do you need to divide the side and bottom of the square into?*

STRENGTHENING UNDERSTANDING

To strengthen understanding, give children a simple calculation where both numerators are 1 (for example, $\frac{1}{3} \times \frac{1}{2}$). Draw two squares, one divided into three bars and one into two columns. Give children a blank square. In pairs, ask: *How do you need to divide each side of the square to show the calculation?* Establish that the square shows that the answer will be smaller than either of the fractions.

GOING DEEPER

Give children a calculation (for example, $\frac{5}{6} \times \frac{3}{4}$). Ask them to find the answer by drawing two different diagrams (for example, a 4 by 6 square and a 6 by 4 square) and to explain why this shows that the order of the fractions does not matter. Encourage them to simplify their answer when possible.

KEY LANGUAGE

In lesson: whole, proper fraction, numerator, denominator, simplest form

Other language to be used by the teacher: fraction of a fraction

STRUCTURES AND REPRESENTATIONS

Fraction grids

RESOURCES

Optional: square templates, bag of oats and other ingredients

 In the eTextbook of this lesson, you will find interactive links to a selection of teaching tools.

Quick recap 𝒬

Ask children to draw a diagram that shows $\frac{1}{3}$.
Then ask them to split each $\frac{1}{3}$ into two equal parts.
Ask: *What fraction have you split it into now?*

Discover

WAYS OF WORKING Pair work

ASK

- Question ① a): *If Bella and Amal use half of the oats in this bag, will there be any oats left in the bag? Will this be more or less than when they started? How do you know?*
- Question ① b): *What diagram could you draw to help you find out how much butter Bella and Amal need?*
- Question ① b): *In part a), you were asked for a half of a half. Do you think the answer is going to be bigger or smaller than the answer to part a)? How do you know?*

IN FOCUS Question ① a) has been designed to draw out the understanding that a fraction of a fraction is smaller than either of the fractions. Children can see this from the bag of oats because there will only be a quarter of the bag left.

PRACTICAL TIPS Ask children to look at baking ingredients and different recipes. Bring in the half bag of oats and other ingredients to provide a visual aid. Encourage children to draw on diagrams or draw a pictorial representation of the oats.

ANSWERS

Question ① a): Bella and Amal need to use $\frac{1}{4}$ of a bag of oats.

Question ① b): Bella and Amal need to use $\frac{3}{8}$ of a block of butter.

Share

WAYS OF WORKING Whole class teacher led

ASK

- Question ① a): *What does the whole square represent? Why has Dexter divided the square into 4? How many small squares represent the oats in the bag before Bella and Amal start cooking? What does the fraction grid show?*
- Question ① a): *Why does Flo say that $\frac{1}{2}$ of $\frac{1}{2}$ means the same as $\frac{1}{2} \times \frac{1}{2}$?*
- Question ① b): *Is there more than one way to divide the diagram up to show the answer?*

IN FOCUS

Question ① a) makes the link between a 'fraction of a fraction' and a 'fraction multiplied by a fraction'. The fractions in question ① b) have been chosen to be harder to visualise than those in part a), so that children understand the advantage of using a fraction grid. Discuss whether it makes any difference whether the fraction grid is divided into 2 columns and 4 rows or 4 columns and 2 rows.

Multiply fractions by fractions ①

Discover

We have $\frac{1}{2}$ of a bag of oats.

We have $\frac{3}{4}$ of a block of butter.

Amal

Bella

① Bella and Amal are making flapjacks.

a) They have $\frac{1}{2}$ of a bag of oats. They need to use $\frac{1}{2}$ of the oats in the bag.
   What fraction of a whole bag do they need to use?

b) They have $\frac{3}{4}$ of a block of butter.
   They need $\frac{1}{2}$ of this to make the flapjacks.
   What fraction of a whole block do they need to use?

176

PUPIL TEXTBOOK 6A PAGE 176

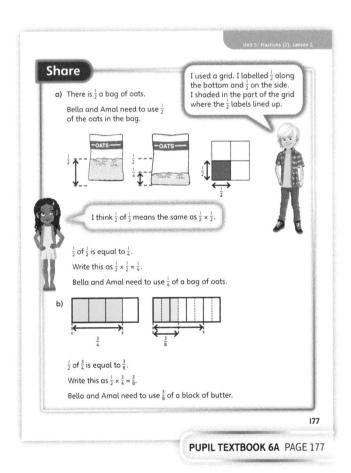

PUPIL TEXTBOOK 6A PAGE 177

Think together

WAYS OF WORKING Whole class teacher led (I do, We do, You do)

ASK

- Question **1**: *What does the whole square represent? Why has it been divided into 9?*
- Question **2**: *How many blocks have you shaded in your grid? How many blocks are there in total? What fraction do the shaded blocks represent? Can you simplify your answer?*
- Question **3** a): *What is the same and what is different about the two fraction grids?*
- Question **3** b): *How else can you write $\frac{2}{5}$ of $\frac{1}{4}$?*

IN FOCUS Question **1** requires children to multiply two unit fractions. Encourage children to draw a fraction grid for themselves, rather than just using the one in the **Textbook**. Question **2** c) uses two non-unit fractions, so children need to shade an array of blocks in a grid. The fractions have been chosen so that the answer can be simplified from $\frac{6}{12}$ to $\frac{1}{2}$. Question **3** also gives children the opportunity to begin to think about multiplying fractions without a context. The two fraction grids both show the same multiplication, to emphasise that multiplication is commutative. All the answers in question **3** can be simplified.

STRENGTHEN To support understanding in question **2**, encourage children to draw grids to model each multiplication. In question **3**, make the link from $\frac{1}{4} \times \frac{2}{5}$ in part a) to $\frac{2}{5}$ of $\frac{1}{4}$ in part b).

DEEPEN Ask children to explore using a circular diagram to build conceptual flexibility. Encourage them to think about which diagram they prefer and which diagram is more efficient for different questions. Challenge them to write a question that will be difficult to represent on a circular diagram.

ASSESSMENT CHECKPOINT Use question **1** to assess whether children can use a diagram to multiply two fractions. Check whether they can explain why the answer is smaller than the starting fractions and whether they recognise the link between 'a fraction of a fraction' and 'a fraction multiplied by a fraction'. Use question **3** to assess whether they are confident working on abstract calculations.

ANSWERS

Question **1**: $\frac{1}{3} \times \frac{1}{3} = \frac{1}{9}$

Bella will use $\frac{1}{9}$ of the bag of sugar.

Question **2** a): $\frac{1}{12}$

Question **2** b): $\frac{3}{12}$

Question **2** c): $\frac{6}{12}$ (or $\frac{1}{2}$)

Question **3** a): Both of the fraction grids are correct, because they both show 2 parts shaded out of 20 equal parts.

Question **3** b): $\frac{1}{2} \times \frac{3}{4} = \frac{3}{8}$

$\frac{2}{5}$ of $\frac{1}{4} = \frac{2}{10}$ (or $\frac{1}{5}$)

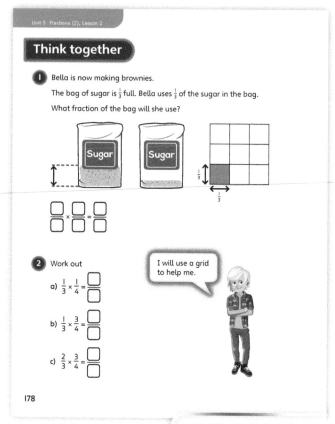

PUPIL TEXTBOOK 6A PAGE 178

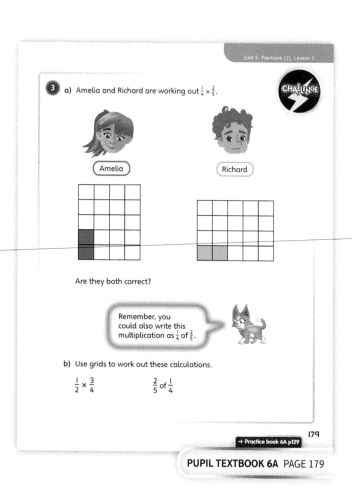

PUPIL TEXTBOOK 6A PAGE 179

Practice

WAYS OF WORKING Independent thinking

IN FOCUS Question ① aims to consolidate children's understanding of calculating a fraction of a fraction and links back to the context of ingredients. Question ② is abstract, gradually reducing the amount of scaffolding provided. It also emphasises the link between '×' and 'of'. The fractions in question ③ have been chosen such that the answers can be simplified. Children may need reminding of the meaning of 'simplest form'. Question ⑤ incorporates a great deal of reasoning to provide an explanation why this is true. Encourage children to explore with various examples, trying to find examples where this is false.

STRENGTHEN Provide children with templates of squares for them to draw on, especially for the challenge question. Encourage them to consolidate their understanding by drawing arrays in all the questions, rather than trying to find short cuts or follow processes they may have discovered.

DEEPEN Encourage children to spot patterns between the fractions in a question and the fraction in the answer. Can they recognise for themselves that the numerators have been multiplied and the denominators have been multiplied? Encourage them to use the diagram to explain this finding.

THINK DIFFERENTLY Question ④ provides the fraction grid and asks what the question *could* be. Encourage children to reflect on whether their answer is the only possibility and, if not, to work out a different answer by rearranging the shaded blocks to show different fractions.

ASSESSMENT CHECKPOINT Use questions ① and ② to assess whether children can use a fraction grid to multiply two fractions. Use questions ② and ③ to assess whether they can draw the fraction grids for themselves without scaffolding. Throughout, children should be able to refer back to a practical context such as a bag of sugar to ensure understanding. Use question ⑤ to assess whether children have a conceptual understanding.

ANSWERS Answers for the **Practice** part of the lesson can be found in the *Power Maths* online subscription.

Reflect

WAYS OF WORKING Independent thinking

IN FOCUS This activity checks that children understand how to multiply two proper fractions. If they cannot draw a suitable fraction grid then their understanding is unlikely to be deep enough to move on.

ASSESSMENT CHECKPOINT Check that children can draw a fraction grid and use it to explain the answer. Can they explain which parts of the fraction have been multiplied and where that can be seen in the fraction grid?

ANSWERS Answers for the **Reflect** part of the lesson can be found in the *Power Maths* online subscription.

After the lesson ⏸

- Can children find a fraction of a fraction by drawing a fraction grid?
- Can children multiply a fraction by a fraction by drawing a fraction grid?
- Can children explain, using a fraction grid, why when they multiply two fractions together the answer will be smaller than either of the starting fractions?

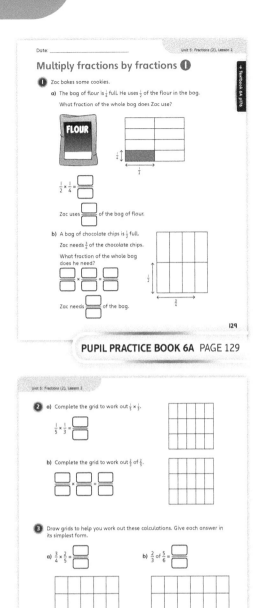

PUPIL PRACTICE BOOK 6A PAGE 129

PUPIL PRACTICE BOOK 6A PAGE 130

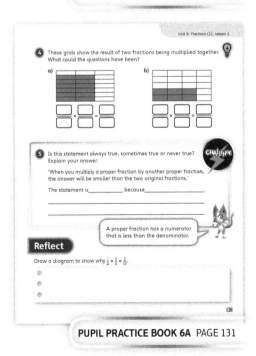

PUPIL PRACTICE BOOK 6A PAGE 131

Multiply fractions by fractions ❷

Learning focus

In this lesson, children will learn to multiply a fraction by a fraction by multiplying the numerators and multiplying the denominators.

Before you teach ⏸

- Can children draw a diagram to represent a fraction?
- Do children understand the concept of multiplication?
- Do children understand equivalent fractions?
- Can children multiply two fractions together using a diagram?

NATIONAL CURRICULUM LINKS

Year 6 Number – fractions (including decimals and percentages)

Multiply simple pairs of proper fractions, writing the answer in its simplest form [for example, $\frac{1}{4} \times \frac{1}{2} = \frac{1}{8}$].

ASSESSING MASTERY

Children can multiply together two or more fractions by multiplying the numerators and multiplying the denominators. They can explain their understanding by drawing a fraction grid.

COMMON MISCONCEPTIONS

Children may think that they have to make the denominators the same (as in addition and subtraction). Although this will lead to a correct answer, it is not efficient. Some children may mix up which process they need to use for multiplying fractions if they do not have a conceptual understanding. As a result, they may try to use the process used for addition, subtraction or division. Ask:

- *How can you divide a square to show $\frac{3}{4} \times \frac{1}{2}$? What will the denominator of the answer be? What will the numerator of the answer be?*

STRENGTHENING UNDERSTANDING

To strengthen understanding, model a calculation such as $\frac{3}{5} \times \frac{3}{4}$ using a square divided into a 5×4 array. Establish that the whole is made up of 5 × 4 parts and link this to multiplying the denominators. Similarly, establish that the number of shaded parts is 3 × 3, which is the two numerators multiplied together.

GOING DEEPER

Challenge children to create a pictorial representation for multiplying together three fractions. They may need to be prompted that the representation should be 3D. Ask: *How does your diagram show that it does not matter in which order the fractions are multiplied together?*

KEY LANGUAGE

In lesson: numerator, denominator, simplest form, simplify

Other language to be used by the teacher: whole, fraction of a fraction, proper fraction, equivalent fractions, area

STRUCTURES AND REPRESENTATIONS

Fraction grids

RESOURCES

Optional: square templates, counters, fraction walls

 In the eTextbook of this lesson, you will find interactive links to a selection of teaching tools.

Quick recap

Ask children to complete these calculations:

$\frac{1}{2} \times \frac{1}{3}$ $\frac{1}{2} \times \frac{1}{4}$ $\frac{1}{2} \times \frac{1}{5}$ $\frac{1}{2} \times \frac{1}{7}$

Then ask them to complete these calculations:

$\frac{1}{5} \times \frac{1}{3}$ $\frac{1}{5} \times \frac{1}{4}$ $\frac{1}{5} \times \frac{1}{5}$ $\frac{1}{5} \times \frac{1}{7}$

Discover

Multiply fractions by fractions ②

Discover

WAYS OF WORKING Pair work

ASK

- Question ❶ a): *What could you draw to help you? What would the diagram show?*
- Question ❶ b): *How could you prove that your answers are correct? What operation have you used?*
- Question ❶ b): *What do you notice about the numerators in the question and the numerator in the answer? What do you notice about the denominators?*

IN FOCUS Question ❶ b) encourages children to look at their answers and the questions, and try to spot links. This gives all children the opportunity to make the link for themselves. By using a fraction grid, children will be able to see what part the denominator represents and what part the numerators represent. By linking to area, children will be able to see why the operation is multiplication. Discussing the two methods will allow you to talk about efficiency and in what circumstances they would want to draw a fraction grid and why it is important to be able to do so.

PRACTICAL TIPS Use templates of squares and counters to illustrate different multiplications. Refer back to the context of baking ingredients from the previous lesson.

❶ a) Use grids to find the answers to the calculations.

b) Look at your answers.
How can you work out the answers without grids?

180

PUPIL TEXTBOOK 6A PAGE 180

ANSWERS

Question ❶ a): $\frac{1}{6}, \frac{3}{20}, \frac{2}{10} (= \frac{1}{5})$

Question ❶ b): The numerators can be multiplied together and the denominators can be multiplied together. The third answer can be simplified.

Share

WAYS OF WORKING Whole class teacher led

ASK

- Question ❶ a): *What does the total number of blocks represent? What do the shaded blocks represent?*
- Question ❶ b): *Is there another way of explaining why the answer is $\frac{1}{5}$? ($\frac{1}{2}$ of $\frac{2}{5}$ is $\frac{1}{5}$) If I said $\frac{1}{2}$ of $\frac{2}{5}$, what diagram could you draw to represent this? How can the answer be $\frac{2}{10}$ and $\frac{1}{5}$?*

IN FOCUS By modelling the multiplication using a square the link is made between the denominator of the answer being the total number of blocks and the numerator being the number of shaded blocks. This develops the understanding that the numerators are multiplied together and similarly the denominators. Links can be made to finding the area of a rectangle. The last calculation has been chosen to show that some answers can be simplified.

Share

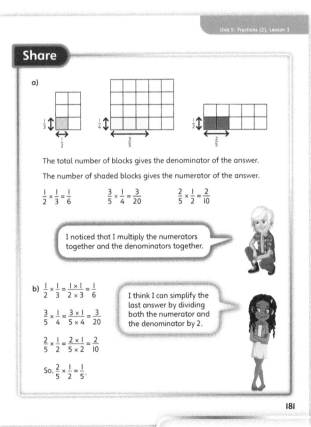

a)

The total number of blocks gives the denominator of the answer.

The number of shaded blocks gives the numerator of the answer.

$\frac{1}{2} \times \frac{1}{3} = \frac{1}{6}$ $\frac{3}{5} \times \frac{1}{4} = \frac{3}{20}$ $\frac{2}{5} \times \frac{1}{2} = \frac{2}{10}$

> I noticed that I multiply the numerators together and the denominators together.

b) $\frac{1}{2} \times \frac{1}{3} = \frac{1 \times 1}{2 \times 3} = \frac{1}{6}$

$\frac{3}{5} \times \frac{1}{4} = \frac{3 \times 1}{5 \times 4} = \frac{3}{20}$

$\frac{2}{5} \times \frac{1}{2} = \frac{2 \times 1}{5 \times 2} = \frac{2}{10}$

So, $\frac{2}{5} \times \frac{1}{2} = \frac{1}{5}$.

> I think I can simplify the last answer by dividing both the numerator and the denominator by 2.

181

PUPIL TEXTBOOK 6A PAGE 181

Think together

WAYS OF WORKING Whole class teacher led (I do, We do, You do)

ASK

- Question **1**: *What is the same and what is different about the two methods?*
- Question **2** c)–f): *How can you simplify your answer?*
- Question **2** b): *How would you find the answer using a diagram? How does this compare with multiplying the numerators and multiplying the denominators?*
- Question **2** f): *How can you use what you know to multiply three fractions? Can you represent this using a fraction grid?*
- Question **3** a): *Do the numerators always need to multiply together to make 5? Why?*

IN FOCUS Question **2** starts to bring in larger denominators and calculations with more than two fractions, showing that a grid is not the most efficient method. Question **3** gives children the opportunity to begin to think about multiplying fractions without a context in a more abstract way: it requires children to use the process of multiplying the numerators and denominators to find possible pairs of fractions.

STRENGTHEN In question **1**, encourage children to reflect on their grid by asking them what is representing the denominator of the answer and what is representing the numerator. Encourage them to think about the operation they are using. Allow children to draw grids for question **2**, because it will help them to understand that it is not an efficient method for $\frac{9}{10} \times \frac{2}{17}$. In question **3**, some children may need to explore equivalent fractions in relation to $\frac{1}{6}$ and $\frac{1}{4}$, for example using a fraction wall.

DEEPEN Ask children what they notice about their answers for question **3** b). They should recognise that the missing fraction will always simplify to $\frac{1}{4}$. Change the known fraction to $\frac{2}{5}$ and ask them to find a possible fraction (for example, $\frac{5}{12}$) and to explain their reasoning.

ASSESSMENT CHECKPOINT Use question **2** to assess whether children can multiply two fractions by multiplying the numerators and the denominators. Use question **3** to assess whether children can work flexibly by recognising equivalent fractions.

ANSWERS

Question **1** a): Answers may vary but should show 8 cells shaded.

Question **1** b): $\frac{2 \times 4}{3 \times 5} = \frac{8}{15}$

Question **2** a): $\frac{3}{8}$ Question **2** d): $\frac{18}{50} = \frac{9}{25}$

Question **2** b): $\frac{5}{42}$ Question **2** e): $\frac{24}{36} = \frac{2}{3}$

Question **2** c): $\frac{15}{24} = \frac{5}{8}$ Question **2** f): $\frac{2}{24} = \frac{1}{12}$

Question **3** a): Possible answers include:

$\frac{1}{1} \times \frac{5}{9}, \frac{2}{3} \times \frac{5}{6}, \frac{4}{6} \times \frac{5}{6}, \frac{2}{3} \times \frac{10}{12}$

Question **3** b): Possible answers include $\frac{1}{4}, \frac{2}{8}, \ldots$ (all fractions equivalent to $\frac{1}{4}$).

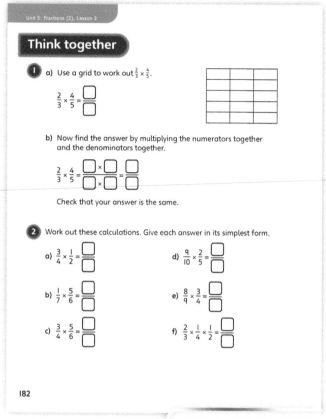

Think together

1 a) Use a grid to work out $\frac{2}{3} \times \frac{4}{5}$.

$\frac{2}{3} \times \frac{4}{5} = \dfrac{\square}{\square}$

b) Now find the answer by multiplying the numerators together and the denominators together.

$\frac{2}{3} \times \frac{4}{5} = \dfrac{\square \times \square}{\square \times \square} = \dfrac{\square}{\square}$

Check that your answer is the same.

2 Work out these calculations. Give each answer in its simplest form.

a) $\frac{3}{4} \times \frac{1}{2} = \dfrac{\square}{\square}$ d) $\frac{9}{10} \times \frac{2}{5} = \dfrac{\square}{\square}$

b) $\frac{1}{7} \times \frac{5}{6} = \dfrac{\square}{\square}$ e) $\frac{8}{9} \times \frac{3}{4} = \dfrac{\square}{\square}$

c) $\frac{3}{4} \times \frac{5}{6} = \dfrac{\square}{\square}$ f) $\frac{2}{3} \times \frac{1}{4} \times \frac{1}{2} = \dfrac{\square}{\square}$

182

PUPIL TEXTBOOK 6A PAGE 182

3 a) Two fractions have been multiplied together.

$\dfrac{\square}{\square} \times \dfrac{\square}{\square} = \dfrac{5}{9}$

What could the fractions be?

How many different answers can you find?

b) Two fractions have been multiplied together.

$\frac{2}{3} \times \dfrac{\square}{\square} = \frac{1}{6}$

What is the other fraction?

Is there more than one answer?

I don't think this works. Both numerators would have to be 1.

I wonder if the answer was simplified.

183

→ Practice book 6A p132

PUPIL TEXTBOOK 6A PAGE 183

Practice

WAYS OF WORKING Independent thinking

IN FOCUS These questions are designed to incorporate variation to ensure children are constantly thinking. Children have the option to refer back to a diagram for support, although question ③ f) will present a problem because the denominators are large. Question ④ relies on children understanding how the numerator and denominator are worked out when multiplying fractions, in order to find missing numbers. Question ⑥ aims to allow all children to reason and problem solve, by requiring more than one possible answer.

STRENGTHEN To strengthen understanding, encourage children to use fraction grids. However, in order for them to be ready to use a more abstract approach, discuss their grids and the link to abstract calculation. In question ⑥, provide support by giving one of the fractions and asking children to find the other fraction.

DEEPEN In question ⑥, ask children to find more than two possible answers and whether it is possible to find them all. This may prompt a discussion about equivalent fractions.

THINK DIFFERENTLY Question ⑤ is designed to address a misconception, where a child has added the numerators instead of multiplying them. Children are then required to identify and explain the correct answer.

ASSESSMENT CHECKPOINT Use questions ② and ③ to assess whether children can multiply two fractions by multiplying the numerators and the denominators. Check that they can explain why they need to multiply the numerators and denominators. Use questions ④ and ⑥ to assess whether children can apply their knowledge of how to multiply fractions and of equivalent fractions to find fractions that multiply to give a specific answer.

ANSWERS Answers for the **Practice** part of the lesson can be found in the *Power Maths* online subscription.

Reflect

WAYS OF WORKING Pair work

IN FOCUS This activity requires children to either use a fraction grid or multiply the numerators together and the denominators together. Comparing their method with a partner may show them an alternative method.

ASSESSMENT CHECKPOINT Check that children can fully explain, in full sentences, what their method is and why it works.

ANSWERS Answers for the **Reflect** part of the lesson can be found in the *Power Maths* online subscription.

After the lesson ⏸

- Can children multiply two fractions by drawing a diagram?
- Can they multiply two fractions without using a diagram?
- Can they explain a more abstract process for multiplying two fractions?
- Can they work fluently with equivalent fractions?

PUPIL PRACTICE BOOK 6A PAGE 132

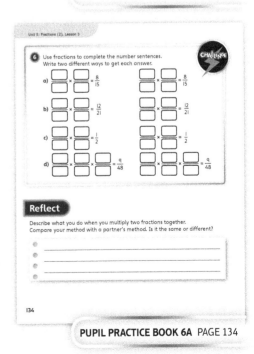

PUPIL PRACTICE BOOK 6A PAGE 133

PUPIL PRACTICE BOOK 6A PAGE 134

Divide a fraction by an integer ❶

Learning focus

In this lesson, children will learn how to divide a non-unit fraction by a whole number when the numerator is a multiple of the whole number. They will build on their work using diagrams in the previous lesson. They will start to identify the pattern between numerators and the number they are dividing by.

NATIONAL CURRICULUM LINKS

Year 6 Number – fractions (including decimals and percentages)
Divide proper fractions by whole numbers [for example, $\frac{1}{3} \div 2 = \frac{1}{6}$].

ASSESSING MASTERY

Children can divide a non-unit fraction by a whole number when the numerator is a multiple of the whole number. First they will use diagrams and then move on to dividing the numerator by the whole number.

COMMON MISCONCEPTIONS

Some children may divide both the numerator and the denominator by the whole number. Ask:
- *What happens if you divide the numerator and denominator of a fraction by the same number? What do we call these fractions? What can you tell me about these fractions?*
- *What happens to a number when you divide it by a whole number – does it get smaller or larger?*

STRENGTHENING UNDERSTANDING

Model $\frac{4}{5} \div 2$ by placing four counters in the segments of a circle divided into five equal parts. Establish that this shows $\frac{4}{5}$. Ask: *How can you divide the number of counters by 2?* Remove half of the counters and ask: *How many counters are left?* Establish that the counters now represent $\frac{2}{5}$. Repeat for other calculations.

GOING DEEPER

Ask children to produce word problems that involve dividing a non-unit fraction by a whole number for a partner to solve. Ensure that they understand that the numerator of the non-unit fraction must be a multiple of the whole number.

KEY LANGUAGE

In lesson: fraction, whole number, numerator, share

Other language to be used by the teacher: non-unit fraction, denominator, part

RESOURCES

Optional: counters, paper circles, paper strips, whiteboard pens, fraction strips, bottle of liquid, cups

 In the eTextbook of this lesson, you will find interactive links to a selection of teaching tools.

Quick recap

Write the following numbers on the board:

10, 12, 18, 24, 25, 26, 30, 31

Ask: *Which of these numbers are multiples of 2? Multiples of 3? Multiples of 6? Multiples of 10? Multiples of 5? Only a multiple of 1 and themselves?*

Discover

Divide a fraction by an integer ❶

WAYS OF WORKING Pair work

ASK

- Question ❶ a): *How many fifths of juice are there? How many are you sharing it between? How are you going to share it? How can you show the sharing?*
- Question ❶ b): *How is this question different from part a)? How much baby food is there to start with? How much will be used?*

IN FOCUS These questions are designed to show whether children understand how to share equally. Their responses will highlight whether they understand that the total they are sharing is $\frac{4}{5}$ and not 1, and that each part they have shared is $\frac{1}{5}$ and not 1. Question ❶ b) draws out this point further by showing a full jar in the picture, but stating that only $\frac{9}{10}$ of the jar is to be used.

PRACTICAL TIPS Give children the opportunity to model question ❶ a) by supplying a bottle partially filled with liquid and two cups to help them investigate and solve the problem. Similarly, for question ❶ b) provide a partially filled jar and three empty bowls.

ANSWERS

Question ❶ a): $\frac{2}{5}$ of the original jug is in each cup.

Question ❶ b): $\frac{3}{10}$ of the jar of baby food should be put into each bowl.

Discover

I have $\frac{4}{5}$ of the jug of juice left.

❶ a) The jug is $\frac{4}{5}$ full of juice.

The juice is divided equally between the 2 empty cups.

What fraction of the original jug is in each of these cups?

b) $\frac{9}{10}$ of the jar of baby food will be enough for 3 equal meals.

What fraction of the jar of baby food should be put into each bowl?

184

PUPIL TEXTBOOK 6A PAGE 184

Share

WAYS OF WORKING Whole class teacher led

ASK

- Question ❶ a): *How does the first diagram represent the jug of juice? Why do the two diagrams for the cups of juice each have two shaded blocks? Can there be more juice in one cup than the other? How can you use the diagrams to work out how much juice goes in each cup? How can you write this as a calculation?*
- Question ❶ b): *How are these diagrams different from those in part a)? Do the diagrams show the same method? Do you find the bars or the wheels easier for dividing a fraction?*

IN FOCUS The diagrams in question ❶ a) emphasise that it is the fraction that is being divided into equal parts, not the whole. Children should see that the parts of the whole can be shared equally without needing to divide each part as they did in the previous lesson. There could be a discussion as to whether the method of dividing each part would work – if so, is it an effective method? Question ❶ b) shows the same method but using wheels instead of bars.

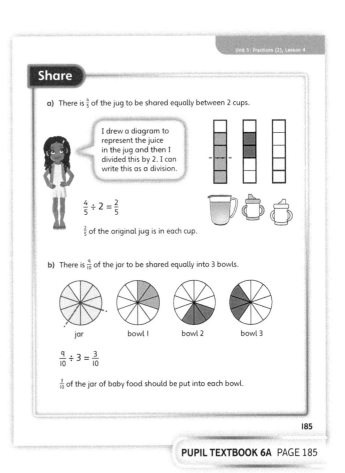

Share

a) There is $\frac{4}{5}$ of the jug to be shared equally between 2 cups.

I drew a diagram to represent the juice in the jug and then I divided this by 2. I can write this as a division.

$$\frac{4}{5} \div 2 = \frac{2}{5}$$

$\frac{2}{5}$ of the original jug is in each cup.

b) There is $\frac{9}{10}$ of the jar to be shared equally into 3 bowls.

jar   bowl 1   bowl 2   bowl 3

$$\frac{9}{10} \div 3 = \frac{3}{10}$$

$\frac{3}{10}$ of the jar of baby food should be put into each bowl.

185

PUPIL TEXTBOOK 6A PAGE 185

Think together

Whole class teacher led (I do, We do, You do)

ASK

- Question **1**: *What is each part worth?*
- Question **1**: *How many parts do you have? What are you dividing/sharing those parts into? Do you need to divide each part by 3 or is there a more efficient way to share?*
- Question **2**: *How many parts has your shape been divided into? How many groups have the parts been shared into? How much are the parts worth?*
- Question **3** a): *Do you agree with Ash's statement? What could the link be?*
- Question **3** b): *How can you work out the fraction for the last question? What operation will you need to use?*

IN FOCUS Question **1** allows children to practise sharing the parts using the most efficient method. Watch to see whether children are identifying how many parts would be in each group straight away or sharing each part one at a time. Encourage children to think about division facts for whole numbers. Question **2** is designed to deepen understanding by requiring children to work out the original fraction, how many groups it was shared into and how many parts are in each group. Question **3** a) requires children to complete four more divisions and then encourages them to spot a pattern between the numerator and the whole number. This enables them to move on to working more abstractly in question **3** b), where they are prompted by Dexter to use pictorial representations to check and prove answers.

STRENGTHEN Provide children with paper circles and strips. Cutting each part up into the number being divided by will help children see that sharing each part gives the same answer as sharing the total number of parts.

DEEPEN Challenge children to investigate whether the pattern between the numerator and whole number works with any numerator, denominator or whole number. Ask them to give examples of when it works and when it does not work. They should be able to explain that it only works when the numerator is a multiple of the whole number.

ASSESSMENT CHECKPOINT Questions **1** and **3** a) assess whether children can divide a non-unit fraction by a whole number. Check whether they can explain the most efficient way to share the fraction. Question **3** assesses whether children can explain the link between the numerator and the whole number and work out divisions using this method.

ANSWERS

Question **1**: $\frac{6}{7} \div 3 = \frac{2}{7}$

Each baby gets $\frac{2}{7}$ of the packet.

Question **2** a): $\frac{9}{11} \div 3 = \frac{3}{11}$

Question **2** b): $\frac{8}{10} \div 2 = \frac{4}{10}$

Question **2** c): $\frac{4}{6} \div 4 = \frac{1}{6}$

Question **3** a): $\frac{3}{5} \div 3 = \frac{1}{5}, \frac{5}{8} \div 5 = \frac{1}{8}, \frac{8}{10} \div 4 = \frac{2}{10} = \frac{1}{5}, \frac{10}{11} \div 5 = \frac{2}{11}$

The numerator divided by the whole number gives the numerator of the answer. The denominator stays the same.

Question **3** b): $\frac{3}{4} \div 3 = \frac{1}{4}, \frac{8}{9} \div 2 = \frac{4}{9}, \frac{12}{25} \div 3 = \frac{4}{25}, \frac{8}{9} \div 4 = \frac{2}{9}$

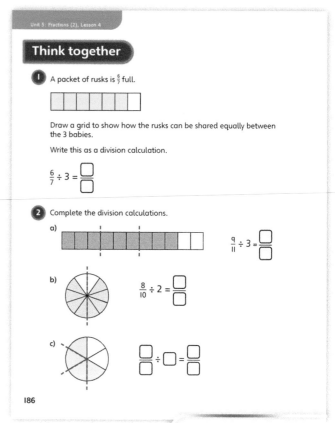

PUPIL TEXTBOOK 6A PAGE 186

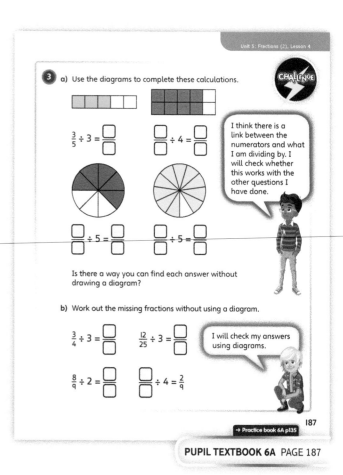

PUPIL TEXTBOOK 6A PAGE 187

Practice

WAYS OF WORKING Independent thinking

IN FOCUS Questions ❶, ❷ and ❸ give children the opportunity to practise the method using a variety of 2D shapes, gradually reducing the amount of scaffolding. Question ❶ is worded to help children understand what is happening at each point and where each number in the calculation comes from. Questions ❹ and ❻ encourage children to work more mentally. In question ❺, children start to work out different combinations of numerators and whole numbers that give a required answer. This could lead into an investigation into how many ways a particular fraction can be divided by a whole number while keeping the denominator the same.

STRENGTHEN Provide children with counters that can be shared into groups. They could use whiteboard pens to write down the worth of each part on the counter to help them remember that the amount in each group is a fraction. When they have shared the counters into groups, they can then add up the number of parts in each group.

DEEPEN When children have completed question ❺, ask them to find other numbers they could use for the numerators and whole number in part d). Ask: *Have you found them all? Explain how you know?* Challenge them to produce similar sets of number sentences, where there are exactly 2, 3 or 4 possible number sentences.

ASSESSMENT CHECKPOINT Use questions ❷ and ❸ to assess whether children can use diagrams to divide non-unit fractions by a whole number. Use questions ❹ and ❻ to assess whether they can divide non-unit fractions by a whole number without the support of a diagram.

ANSWERS Answers for the **Practice** part of the lesson can be found in the *Power Maths* online subscription.

Reflect

WAYS OF WORKING Independent thinking

IN FOCUS This question uses a misconception to provide children with an opportunity to demonstrate they understand the method. Children need to realise that Danny has divided both the numerator and the denominator by 5, rather than just the numerator.

ASSESSMENT CHECKPOINT Check whether children work out the correct answer by drawing a diagram or by dividing the numerator. If they choose to divide the numerator, check that they can explain why this gives the correct answer.

ANSWERS Answers for the **Reflect** part of the lesson can be found in the *Power Maths* online subscription.

After the lesson ⏸

- Can children divide non-unit fractions by whole numbers where the numerator is a multiple of the number? Can they estimate answers and check their answers using diagrams?
- Do children understand the relationship between the numerator and the whole number? Do they recognise that only multiples of the numerator work using the method of dividing the numerator?

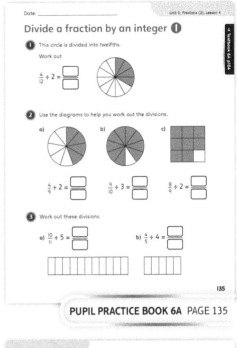

PUPIL PRACTICE BOOK 6A PAGE 135

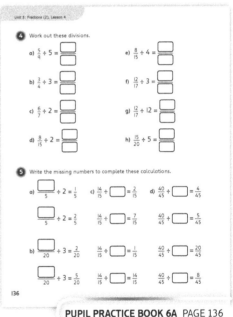

PUPIL PRACTICE BOOK 6A PAGE 136

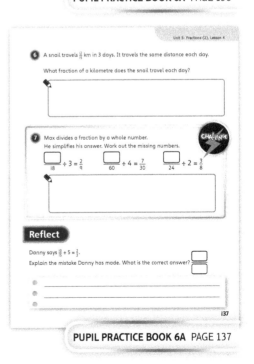

PUPIL PRACTICE BOOK 6A PAGE 137

Divide a fraction by an integer

Learning focus

In this lesson, children will learn to divide unit fractions by a whole number. They will practise recording the original fractions in a diagram and then dividing one of the sections. They will be exposed to a pattern between denominators and the number they are dividing by.

Before you teach ⏸

- Can children divide shapes equally?
- Can they show fractions using shapes?
- Do they understand the importance of equal parts?
- Do they know what happens to a number when it is divided by a whole number?

NATIONAL CURRICULUM LINKS

Year 6 Number – fractions (including decimals and percentages)

Divide proper fractions by whole numbers [for example, $\frac{1}{3} \div 2 = \frac{1}{6}$].

ASSESSING MASTERY

Children can divide a unit fraction by a whole number, first using circular and rectangular diagrams and then multiplying the denominator by the whole number. Children should be able to answer questions in real-life contexts and from given diagrams.

COMMON MISCONCEPTIONS

Some children may divide one section of the shape by the whole number but not the remaining sections so they have unequal parts (for example, divide a third of a circle into 2, but not the other two thirds, giving $\frac{1}{4}$, not $\frac{1}{6}$). Ask:
- *Are all your parts the same size as the shaded part? What is the shaded part as a fraction of the whole?*

Some children may divide the denominator by the whole number instead of multiplying it. Ask:
- *What happens to a number when you divide it by a whole number – does it get smaller or larger?*
- *Does a smaller unit fraction have a larger or smaller denominator?*

STRENGTHENING UNDERSTANDING

Give children paper circles and strips to physically fold and cut. Use a paper strip to model dividing $\frac{1}{4}$ by 3, emphasising the need to create equal parts that are the same size as a third of one of the quarters.

GOING DEEPER

Encourage children to find a variety of ways to create a unit fraction by using the pattern between denominators and the whole number. Ask: *How can you be sure that you have found all the combinations?*

KEY LANGUAGE

In lesson: denominator, whole number, divide, part, whole

Other language to be used by the teacher: numerator, unit fraction, multiply

RESOURCES

Mandatory: fraction strips, paper strips, paper plates

Optional: paper circles, squares and strips

 In the eTextbook of this lesson, you will find interactive links to a selection of teaching tools.

Quick recap

Ask children to complete these divisions:

$\frac{12}{15} \div 2$ $\frac{12}{15} \div 3$ $\frac{12}{15} \div 4$ $\frac{12}{15} \div 6$ $\frac{12}{15} \div 12$

Discover

Divide a fraction by an integer ②

WAYS OF WORKING Pair or small group work

ASK

- Question ① a): *What is each part of the circle worth? How do you know?*
- Question ① b): *What is the same, compared with part a)? What is different?*
- Question ① b): *What is each part of the strip worth? How do you know?*
- Question ① b): *How can you show the section to be glued?*

IN FOCUS Questions ① a) highlights whether children understand that the two parts of one third must be equal. Question ① b) is not more difficult than part a), just set out differently.

PRACTICAL TIPS Take a group of six children and divide them into three equal groups of two children, ensuring that children understand that each group represents $\frac{1}{3}$ of the whole. Divide one of the groups into two single children. Discuss what fraction of the whole each child represents. Establish that the other groups need to be divided in the same way to make equal parts.

ANSWERS

Question ① a): $\frac{1}{6}$ of the penguin's body is white.

Question ① b): $\frac{1}{8}$ of the strip of paper is covered in glue.

Discover

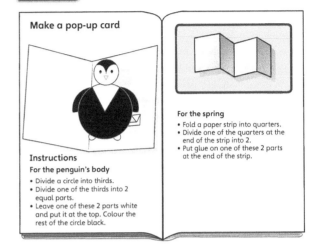

Make a pop-up card

Instructions

For the penguin's body
- Divide a circle into thirds.
- Divide one of the thirds into 2 equal parts.
- Leave one of these 2 parts white and put it at the top. Colour the rest of the circle black.

For the spring
- Fold a paper strip into quarters.
- Divide one of the quarters at the end of the strip into 2.
- Put glue on one of these 2 parts at the end of the strip.

① a) What fraction of the penguin's body is white?

b) What fraction of the strip of paper is covered in glue?

188

PUPIL TEXTBOOK 6A PAGE 188

Share

WAYS OF WORKING Whole class teacher led

ASK

- Question ① a): *The second diagram shows one shaded part and three unshaded – why is the answer not $\frac{1}{4}$? Can you leave only one section with two equal parts? What do you need to do to the other sections? How much is each section worth now? What calculation represents what you have done?*
- Question ① b): *What is each part of the strip worth now? How many parts would be covered in glue?*

IN FOCUS Both parts a) and b) draw attention to the fact dividing into two equal parts is the same as dividing by 2. Emphasis should be placed on the need to divide every part by the same number to give equal parts after the division. Children may spot the pattern that when a fraction is divided by 2 the number of original parts doubles, and so the denominator doubles.

Share

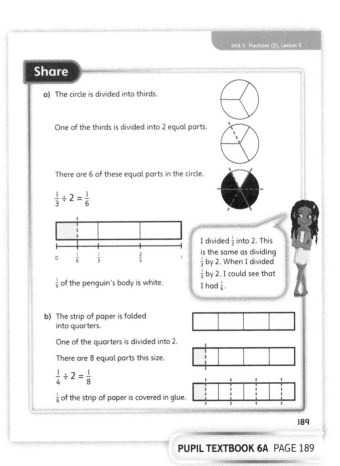

a) The circle is divided into thirds.

One of the thirds is divided into 2 equal parts.

There are 6 of these equal parts in the circle.

$$\frac{1}{3} \div 2 = \frac{1}{6}$$

$\frac{1}{6}$ of the penguin's body is white.

I divided $\frac{1}{3}$ into 2. This is the same as dividing $\frac{1}{3}$ by 2. When I divided $\frac{1}{3}$ by 2, I could see that I had $\frac{1}{6}$.

b) The strip of paper is folded into quarters.

One of the quarters is divided into 2.

There are 8 equal parts this size.

$$\frac{1}{4} \div 2 = \frac{1}{8}$$

$\frac{1}{8}$ of the strip of paper is covered in glue.

189

PUPIL TEXTBOOK 6A PAGE 189

Think together

WAYS OF WORKING Whole class teacher led (I do, We do, You do)

ASK

- Question ❶: *What has the quarter been divided into? What calculation is this representing? What do you need to do to the other quarters? How many parts does this mean the whole is now divided into?*
- Question ❶: *Has the original number of parts been doubled like they were in the **Discover** section? Why not?*
- Question ❷: *How is this similar to question ❶? How is this different from question ❶? Looking at your answer to Question ❶, can you estimate what this answer will be? What has the whole been divided into after you have divided by 3? What is each of these parts worth? Can you spot a pattern when dividing by 3?*
- Question ❸ b): *What patterns can you see? What do you think will happen when you divide a fraction by 5? By 7? Why?*

IN FOCUS Question ❸ draws on all the divisions that children have done in **Share** and in **Think together**. Use questions ❶ and ❷ to encourage them to identify the pattern between the denominators and the number they are dividing by. This then encourages them to use the pattern they have found to move towards a more abstract/mental way of working.

STRENGTHEN Give children paper circles and strips to annotate, fold and/or cut up to help them see the fractions more clearly.

DEEPEN Ask children to investigate whether their 'pattern' works when dividing by any whole number. Ask: *Can you show me with a diagram why the pattern works?*

ASSESSMENT CHECKPOINT Use questions ❶ and ❷ to assess whether children can use diagrams to divide a unit fraction by a whole number. Check whether they can explain what is happening. Use question ❸ to assess whether children can divide a unit fraction by a whole number without using a diagram.

ANSWERS

Question ❶: $\frac{1}{4} \div 3 = \frac{1}{12}$

$\frac{1}{12}$ of the circle is shaded.

Question ❷: $\frac{1}{3} \div 3 = \frac{1}{9}$

$\frac{1}{9}$ of the paper is shaded.

Question ❸ a): All the fractions are unit fractions. All the fractions have been divided by a whole number.
The answers are all smaller than the original fraction.
The first two calculations divide by 2.
The second two calculations divide by 3.

Question ❸ b): The denominator of the answer is the denominator of the original fraction multiplied by the number the fraction is being divided by.
$\frac{1}{6} \div 2 = \frac{1}{12}, \frac{1}{4} \div 4 = \frac{1}{16}, \frac{1}{5} \div 3 = \frac{1}{15}$

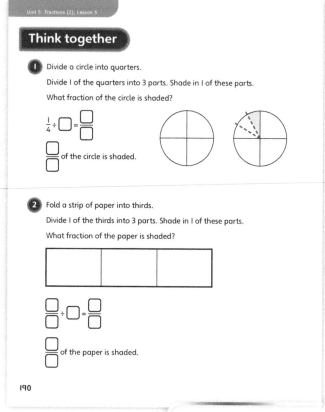

Think together

❶ Divide a circle into quarters.
Divide 1 of the quarters into 3 parts. Shade in 1 of these parts.
What fraction of the circle is shaded?

$\frac{1}{4} \div \square = \frac{\square}{\square}$

$\frac{\square}{\square}$ of the circle is shaded.

❷ Fold a strip of paper into thirds.
Divide 1 of the thirds into 3 parts. Shade in 1 of these parts.
What fraction of the paper is shaded?

$\frac{\square}{\square} \div \square = \frac{\square}{\square}$

$\frac{\square}{\square}$ of the paper is shaded.

190

PUPIL TEXTBOOK 6A PAGE 190

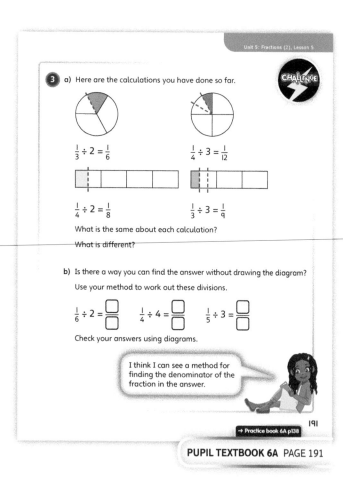

❸ a) Here are the calculations you have done so far.

$\frac{1}{3} \div 2 = \frac{1}{6}$ $\frac{1}{4} \div 3 = \frac{1}{12}$

$\frac{1}{4} \div 2 = \frac{1}{8}$ $\frac{1}{3} \div 3 = \frac{1}{9}$

What is the same about each calculation?
What is different?

b) Is there a way you can find the answer without drawing the diagram?
Use your method to work out these divisions.

$\frac{1}{6} \div 2 = \frac{\square}{\square}$ $\frac{1}{4} \div 4 = \frac{\square}{\square}$ $\frac{1}{5} \div 3 = \frac{\square}{\square}$

Check your answers using diagrams.

I think I can see a method for finding the denominator of the fraction in the answer.

191

→ Practice book 6A p138

PUPIL TEXTBOOK 6A PAGE 191

Practice

WAYS OF WORKING Independent thinking

IN FOCUS Questions ❶, ❷ and ❸ provide more opportunities to practise methods that have been worked on in previous examples. Scaffolding is provided through diagrams that are already divided into fractions, enabling children to focus on the dividing element of the question.

In question ❹, children need to interpret the diagram to work out the division that is shown, requiring them to demonstrate the understanding they have developed through drawing diagrams. In question ❺, the first six parts are standard division of unit fractions by a whole number. The last three parts require children to use the pattern they have identified to find the missing numbers.

STRENGTHEN For question ❺, give children paper circles and strips to annotate, fold and/or cut up. In part g), suggest that they cut the same circle/strip into four and eight parts and compare the size of $\frac{1}{4}$ and $\frac{1}{8}$ to work out the missing number.

DEEPEN Ask those children who spotted the pattern quickly and understand why it works to estimate answers before checking them using the given diagrams in questions ❶ to ❸. In question ❸, ask: *Does the way the shape is divided affect the answer? Does the section you divide affect the answer?* In question ❺, ask: *How can you be sure that you have found all the possible solutions?*

ASSESSMENT CHECKPOINT Use questions ❶, ❷ and ❸ to assess whether children can use diagrams to divide a unit fraction by a whole number. Check that children are aware what section they are dividing and how many parts are in the whole after the division. Use question ❺ to assess whether children can divide a unit fraction by a whole number without using a diagram.

ANSWERS Answers for the **Practice** part of the lesson can be found in the *Power Maths* online subscription.

Reflect

WAYS OF WORKING Independent thinking

IN FOCUS This question explores a common misconception, where children divide the denominator by the whole number instead of multiplying it. Children can practise their reasoning skills to prove why the statement is false.

ASSESSMENT CHECKPOINT Children should be able to answer true or false by working out the answer for themselves, but their reasoning will show their understanding of the concept.

ANSWERS Answers for the **Reflect** part of the lesson can be found in the *Power Maths* online subscription.

After the lesson ⏸

- Can children correctly divide sections?
- Do children understand what the whole has now been divided into?
- Can children see the relationship between the denominators and the whole number?
- Can children use diagrams to solve calculations and/or prove answers?

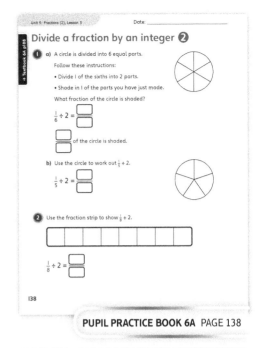

PUPIL PRACTICE BOOK 6A PAGE 138

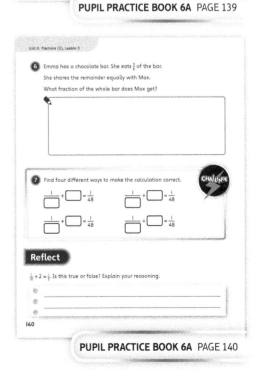

PUPIL PRACTICE BOOK 6A PAGE 139

PUPIL PRACTICE BOOK 6A PAGE 140

Divide a fraction by an integer ③

Learning focus

In this lesson, children will build on the previous two lessons and learn to divide any fraction by a whole number. They will continue to use both diagrams and abstract methods to solve calculations and show their working out.

Before you teach ⏸

- Can children represent fractions using diagrams?
- Are they confident dividing unit fractions by whole numbers and non-unit fractions by multiples of the numerator?
- Can they record working out using diagrams?

NATIONAL CURRICULUM LINKS

Year 6 Number – fractions (including decimals and percentages)

Divide proper fractions by whole numbers [for example, $\frac{1}{3} \div 2 = \frac{1}{6}$].

ASSESSING MASTERY

Children can divide any fraction by a whole number. They understand what is happening when they are sharing a fraction and can use diagrams to explain their thinking.

COMMON MISCONCEPTIONS

Some children may try to divide the denominator by the whole number. Ask:
- *What happens to a fraction when you divide it by a whole number – does it get smaller or larger? If the denominator of a fraction gets smaller, does the fraction get larger or smaller?*

STRENGTHENING UNDERSTANDING

To strengthen children's understanding, provide them with paper circles and strips to annotate, fold, cut up and physically share.

GOING DEEPER

Ask children to produce a poster that explains how to divide any fraction by a whole number, drawing on this lesson and the previous two lessons.

KEY LANGUAGE

In lesson: fraction, whole number, numerator, equally, share

Other language to be used by the teacher: denominator, multiple

RESOURCES

Optional: concrete materials to represent length of bamboo (for example, sticks, wool, ribbon, card, cubes), pictures of pandas, paper circles, paper squares, paper strips, fraction strips, counters, whiteboard pens, sorting circles

 In the eTextbook of this lesson, you will find interactive links to a selection of teaching tools.

Quick recap 🔍

Ask children to complete these divisions:

$\frac{1}{6} \div 3$	$\frac{2}{6} \div 3$	$\frac{3}{6} \div 3$
$\frac{4}{6} \div 3$	$\frac{5}{6} \div 3$	$\frac{6}{6} \div 3$

Discover

WAYS OF WORKING Pair work

ASK

- Question ① a): *What fraction do you have to begin with? Can you draw a diagram to show this fraction?*
- Question ① a): *What are you dividing the fraction into? Can the parts of the fraction be shared equally? Is the numerator a multiple of the whole number? Can you show the division on your diagram?*
- Question ① a): *Which parts of the fraction need splitting up?*
- Question ① b): *How does this new information change the calculation in part a)? How does the diagram need changing? What is the size of each part now?*

IN FOCUS These questions will show whether children recognise that the parts of the fraction cannot be shared equally as in the previous lesson. Children should then realise that each part of the fraction needs to be divided in a similar way to dividing unit fractions in Lesson 4, and that this changes the number of parts. When children are sharing they need to remember that everything they share is a part/fraction, not a whole number.

PRACTICAL TIPS Give children something concrete to represent the bamboo: sticks, wool, ribbon, card, cubes etc. They could also be given pictures of pandas, sorting circles or similar to clearly share the fraction between.

ANSWERS

Question ① a): Each panda will get $\frac{2}{9}$ m of the bamboo shoot.

Question ① b): Each panda will get $\frac{1}{6}$ m of the bamboo shoot.

Share

WAYS OF WORKING Whole class teacher led

ASK

- Question ① a): *What is represented by the diagram? Can the parts of the fraction be shared equally? Explain.*
- Question ① a): *How many parts do you need to divide each part into? What is each part worth now you have divided it further? What is $\frac{2}{3}$ equivalent to? Can the new parts be shared equally? How many parts are in each group? How much does this equal?*
- Question ① b): *Does the diagram need to change? How many parts are in each group now? What is each part worth? What is the total in each group?*
- Question ① b): *What did Ash do that was different? Why did his method work?*

IN FOCUS Question ① a) shows how each part in the diagram can be divided by the whole number and these smaller parts shared equally between the pandas. This links back to Lesson 4, as children need to understand that the size of each smaller part is the unit fraction divided by the whole number, and to Lesson 5, as they divide the numerator of the equivalent fraction by the whole number. In question ① b), Ash shows an alternative way of dividing the diagram when the numerator of the fraction and the whole number share a factor.

Divide a fraction by an integer ❸

Discover

① a) The bamboo shoots are $\frac{2}{3}$ m long.
   If the pandas share one bamboo shoot equally, how much will each panda get?

   b) Another panda comes along to share the bamboo shoot.
   How much will each panda get now?

192

PUPIL TEXTBOOK 6A PAGE 192

Share

a) The bamboo shoot is $\frac{2}{3}$ m long.

3 pandas share the shoot equally.

> I drew a bar model. I shaded in $\frac{2}{3}$. I don't think I can share this equally between 3 pandas, can I?

> Yes, you can. I divided each $\frac{1}{3}$ into 3 so I could give each panda a part of the bamboo. This is the same as $\frac{6}{9}$ m divided by 3.

$\frac{2}{3} \div 3 = \frac{6}{9} \div 3$

$\frac{6}{9} \div 3 = \frac{2}{9}$

Each panda will get $\frac{2}{9}$ m of the bamboo shoot.

193

PUPIL TEXTBOOK 6A PAGE 193

Think together

WAYS OF WORKING Whole class teacher led (I do, We do, You do)

ASK

- Question ❶: *What fraction do you have? How many groups are you sharing into? Can the parts of this fraction be shared equally? Why not? How many are you going to have to divide each part into? How much is each part worth now? How many parts are in each group? What is the total amount in each group?*
- Question ❷: *How can you annotate the bar model to find the answer? What is each part worth? How many groups are you sharing into? What is the total fraction in each group? Are the totals equal?*
- Question ❸ a): *How are the methods different? Can you predict whether the answers will be the same or different? Why do you think that? How will you represent Max's method using a diagram? How will you represent Ambika's method using a diagram? What do you notice about the answers?*

IN FOCUS In question ❶, the fraction and groups to share into are given so children can concentrate on each step within the method. Question ❷ requires children to do a little more for themselves using the information given. Once they have identified the number of groups to share into, children will again be able to practise the method and work through each step. Question ❸ introduces a variation of the method, giving children the chance to investigate a different way to find the answer. Their answers to question ❸ a) will indicate their understanding of equivalent/simplifying fractions.

STRENGTHEN To support children's understanding, provide them with an appropriate fraction strip for each question. Shading and cutting up the fraction strip gives children a clear visual representation of the parts getting smaller and the denominator changing. Physically sharing the parts into piles or sorting circles enables children to check for equal amounts and to see clearly how many parts are in each pile.

DEEPEN In question ❸, challenge children to find other calculations that both Max and Ambika's methods work for. Children may look for and spot patterns linking the numerator, the denominator and the whole number and investigate any patterns or links they find.

ASSESSMENT CHECKPOINT Use question ❷ to assess whether children can divide a fraction by a whole number. Check that they can identify and represent the fraction they are starting with and how many groups they are dividing it into.

ANSWERS

Question ❶: $\frac{5}{6} \div 3 = \frac{15}{18} \div 3 = \frac{5}{18}$

Each panda will get $\frac{5}{18}$ m of the bamboo shoot.

Question ❷: $\frac{3}{30}$ or $\frac{1}{10}$

Question ❸ a): Yes, the children get the same answer. Ambika's answer ($\frac{1}{9}$) is Max's answer ($\frac{2}{18}$) simplified.

Question ❸ b): Max's method: $\frac{3}{24}$, Ambika's method: $\frac{1}{8}$

b) There are now 4 pandas.

$\frac{2}{3} \div 4 = \frac{8}{12} \div 4$

$= \frac{2}{12}$

$= \frac{1}{6}$

Each panda will get $\frac{1}{6}$ m of the bamboo shoot.

I divided each part into 4. I did this by dividing each part into 2.

Think together

❶ A bamboo shoot is $\frac{5}{6}$ m long.

Share this between 3 pandas.

$\frac{5}{6} \div 3 = \frac{\square}{\square} \div 3 = \frac{\square}{\square}$

194

PUPIL TEXTBOOK 6A PAGE 194

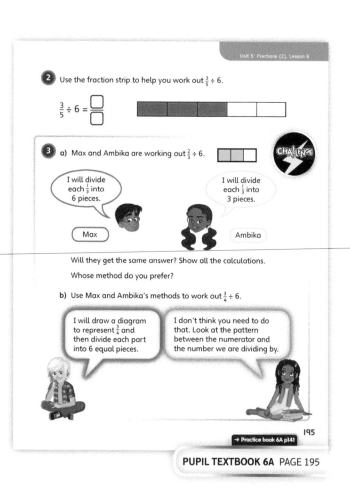

❷ Use the fraction strip to help you work out $\frac{3}{5} \div 6$.

$\frac{3}{5} \div 6 = \frac{\square}{\square}$

❸ a) Max and Ambika are working out $\frac{2}{3} \div 6$.

I will divide each $\frac{1}{3}$ into 6 pieces.

Max

CHALLENGE

I will divide each $\frac{1}{3}$ into 3 pieces.

Ambika

Will they get the same answer? Show all the calculations.

Whose method do you prefer?

b) Use Max and Ambika's methods to work out $\frac{3}{4} \div 6$.

I will draw a diagram to represent $\frac{3}{4}$ and then divide each part into 6 equal pieces.

I don't think you need to do that. Look at the pattern between the numerator and the number we are dividing by.

195

→ Practice book 6A p141

PUPIL TEXTBOOK 6A PAGE 195

Practice

WAYS OF WORKING Independent thinking

IN FOCUS Questions ❶ and ❷ provide diagrams to support children's reasoning. Question ❸ is more abstract, requiring children to understand how to use equivalent fractions. Encourage them to draw diagrams if they lack confidence. The divisions in question ❹ have been chosen to allow more confident children to spot patterns when the numerator and whole number share a common factor.

STRENGTHEN Model a division using fraction strips and counters. For example, when working out $\frac{4}{5} \div 3$, place 3 counters on 4 of the 5 parts on a fifths fraction strip. Use a whiteboard pen to write the fraction represented by each counter ($\frac{1}{5}$) on the counters. Then share the counters into 3 equal piles and establish that each pile represents $\frac{4}{15}$.

DEEPEN In question ❹, ask children whether they can solve these divisions using a diagram in a different way. If necessary, refer them back to Max and Ambika's methods in **Think together** question ❸. Ask: *Can you see a pattern between the numerators, denominators and whole numbers?* Give them a division such as $\frac{4}{7} \div 8$ and ask them to predict the denominator of the answer.

ASSESSMENT CHECKPOINT Use questions ❶ and ❷ to assess whether children can use a diagram to divide a fraction by a whole number. Use questions ❸ and ❹ to assess whether children can use equivalent fractions to divide a fraction by a whole number without using a diagram. Check whether they can explain how the method works.

ANSWERS Answers for the **Practice** part of the lesson can be found in the *Power Maths* online subscription.

Reflect

WAYS OF WORKING Independent thinking

IN FOCUS This question highlights whether children understand how to use the information given by the fractions/numbers in a division calculation. Some children may choose to draw a diagram to help them, while others will go straight to equivalent fractions.

ASSESSMENT CHECKPOINT Assess whether children can confidently divide a fraction by a whole number. Check whether they give the answer in its simplest form. Look for children who realise that they can use the relationship between 2 and 4 to find the answer.

ANSWERS Answers for the **Reflect** part of the lesson can be found in the *Power Maths* online subscription.

After the lesson ⏸

- Can children divide fractions by whole numbers, explaining and modelling using diagrams how each method works?
- Are children aware what each fraction/number represents within a calculation?
- Do they know when parts of a fraction can be shared equally and when they need dividing further?

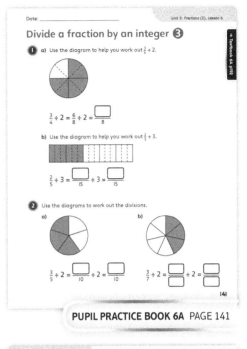

PUPIL PRACTICE BOOK 6A PAGE 141

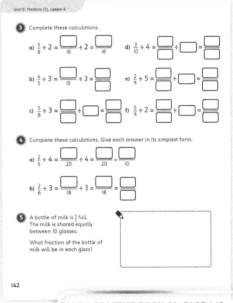

PUPIL PRACTICE BOOK 6A PAGE 142

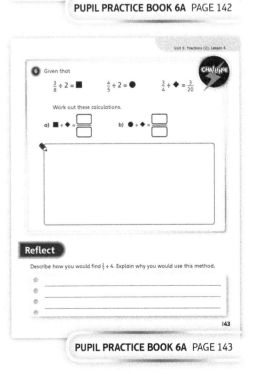

PUPIL PRACTICE BOOK 6A PAGE 143

Mixed questions with fractions

Learning focus

In this lesson, children will solve fraction problems involving addition, subtraction, multiplication and division. They will use the order of operations and visual aids such as bar models to support their understanding.

Before you teach ⏸

- Can children use all four operations with fractions?
- Can children find the areas of rectangles, squares and triangles?
- Can children find the perimeters of polygons?
- Do children understand and follow the order of operations with whole numbers?

Year 6 Number – fractions (including decimals and percentages)

Add and subtract fractions with different denominators and mixed numbers, using the concept of equivalent fractions.

Multiply simple pairs of proper fractions, writing the answer in its simplest form [for example, $\frac{1}{4} \times \frac{1}{2} = \frac{1}{8}$].

ASSESSING MASTERY

Children can multiply, divide, add and subtract fractions when applied to various contexts, including perimeter and areas of 2D shapes. They can follow the correct order of operations when calculating with fractions. They can appreciate that there are multiple ways to solve a problem: for example, multiplying a fraction by 3 is the same as adding three of these fractions together.

COMMON MISCONCEPTIONS

Some children may not apply the order of operations when calculating with fractions. Ask:

- In $4 + 2 \times 3$, which operation do you do first? In $\frac{2}{3} + \frac{4}{5} \times \frac{1}{2}$, which operation do you do first? In $\frac{2}{3} + \frac{4}{5} \div 2$, which operation do you do first?

STRENGTHENING UNDERSTANDING

Use a bar model to help children to recognise the link between multiplication and repeated addition. Emphasise that there may be more than one effective method for solving a problem. If children use a different method from their partner, encourage them to discuss the methods to confirm that both are valid.

GOING DEEPER

Encourage children to always give their answer as a mixed number where possible and to simplify answers fully. When appropriate, challenge children to show two different ways of solving the problem and to explain why both methods work.

KEY LANGUAGE

In lesson: add, multiply, area, perimeter, square, triangle, isosceles triangle

Other language to be used by the teacher: order of operations, subtract, divide, mixed number, simplify

STRUCTURES AND REPRESENTATIONS

Bar model, number line

 In the eTextbook of this lesson, you will find interactive links to a selection of teaching tools.

Quick recap

Ask children to choose two fractions from this set:

$$\frac{1}{4} \qquad \frac{1}{6} \qquad \frac{2}{3} \qquad \frac{4}{5} \qquad \frac{10}{12}$$

Then ask them to find the sum, the difference and the product of their two fractions. For example, they could choose $\frac{1}{4}$ and $\frac{4}{5}$ and then work out $\frac{1}{4} + \frac{4}{5}$, $\frac{4}{5} - \frac{1}{4}$ and $\frac{1}{4} \times \frac{4}{5}$.

Discover

Unit 5: Fractions (2), Lesson 7

Mixed questions with fractions

Discover

WAYS OF WORKING   Pair work

ASK

- Question ① a): *How far did Luis walk each day? How can you work out how far Luis walked from Monday to Friday? What operation do you need to do to work out the total? Can you use a diagram to show how to work it out?*
- Question ① b): *How can you work out how far Luis walked in total in the week? How do you know from your total whether Luis achieved his target?*

IN FOCUS   Question ① a) can be solved using either repeated addition or multiplication: adding $\frac{2}{3}$ repeatedly 5 times or multiplying $\frac{2}{3}$ by 5. Encourage children to discuss how they could draw a diagram to show these two methods. Question ① b) builds on the learning taking place in part a) as it is also a problem that can be solved in more than way, using either addition or multiplication followed by addition. Question ① b) also requires a comparison of a mixed number with a whole number.

PRACTICAL TIPS   Use the context to explore further questions, such as: *What if his target was double?* Children could take part in a similar activity, to see how far they could walk in one week, which would link to other areas of the curriculum.

ANSWERS

Question ① a): Luis walked $3\frac{1}{3}$ km from Monday to Friday.

Question ① b): $4\frac{8}{9} < 5$, so Luis did not meet his target.

① a) How far did Luis walk from Monday to Friday?

b) Luis's target was to walk 5 km in total in the week.
Did he meet his target?

196

PUPIL TEXTBOOK 6A PAGE 196

Share

WAYS OF WORKING   Whole class teacher led

ASK

- Question ① a): *How does the bar model show how far Luis walked Monday to Friday? What would the bar model look like if you wanted to work out how far he walked Monday to Wednesday? Why can you work out the answer using addition or multiplication?*
- Question ① b): *What does the diagram show? What operation(s) do you need to do to find the total? Can you write a number sentence using just addition to show how to find the answer? Can you write a number sentence using multiplication and addition? Do you need to use brackets?*
- Question ① b): *Why does Flo use a common denominator to add the fractions?*
- Question ① b): *How do you know that Luis has not met his target?*
- Question ① b): *Can you use subtraction or division to show that Luis did not meet his target? Why not?*

IN FOCUS   Question ① a) shows two ways of finding the answer, repeated addition or multiplication. In contrast, question ① b) only shows one method: multiplication followed by addition. Discuss whether this could also be solved using just addition. Model the two methods, showing the calculations and matching bar models to emphasise why both methods are valid. Check that children understand that the final answer is 'yes' or 'no', not just the distance that Luis walked in the week.

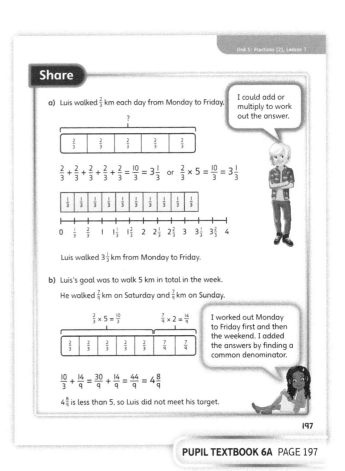

PUPIL TEXTBOOK 6A PAGE 197

Think together

WAYS OF WORKING Whole class teacher led (I do, We do, You do)

ASK

- Question ❶: *Is your answer in its simplest form?*
- Question ❷: *What is the same as question ❶? What is different? What do you do to multiply three fractions?*
- Question ❸ a): *What is the correct order of operations? Which part of the calculation do you need to do first?*
- Question ❸ b): *What do you think Alex did first?*

IN FOCUS Questions ❶ and ❷ both show how this skill is relevant to other areas of the curriculum, whilst still practising the calculations children have been doing in **Discover**. Question ❸ asks children to think about the misconception of completing the addition before the multiplication, which would result in an incorrect answer. Encourage children to explain their reasoning in full sentences.

STRENGTHEN In question ❶, suggest that children use a bar model to support their understanding. Discuss whether there are any alternative methods (repeated addition) and encourage them to reflect on which method they would have chosen. Model both methods using bar models to demonstrate that they give the same answer. In question ❸, encourage children to draw fraction grids to help them with the addition and multiplication.

DEEPEN Ask children to think about multiple ways of answering questions ❶ and ❷, and to explain which method they think is more efficient. In question ❷, encourage children to explore why multiplying by $\frac{1}{2}$ has the same result as dividing by 2.

ASSESSMENT CHECKPOINT Use questions ❶ and ❷ to assess whether children can solve problems using fractions. Check that they can decide what operation is needed, and recognise that the operation could be multiplication or addition when finding the perimeter. Check that they can convert from improper fractions to mixed numbers and give the answers in their simplest form. Use question ❸ to assess whether children can use the correct order of operations to calculate with fractions.

ANSWERS

Question ❶: $\frac{3}{4} \times 2 = \frac{6}{4}$, $\frac{1}{6} \times 2 = \frac{2}{6}$

$\frac{6}{4} + \frac{2}{6} = \frac{18}{12} + \frac{4}{12} = \frac{22}{12} = 1\frac{10}{12} = 1\frac{5}{6}$

The perimeter of the rectangle is $1\frac{5}{6}$ m.
Some children may simplify fractions earlier.

Question ❷ a): $\frac{1}{2} \times \frac{1}{4} \times \frac{1}{2} = \frac{1}{16}$

The area of the triangle is $\frac{1}{16}$ m².

Question ❷ b): The area of the triangle is $\frac{7}{36}$ m².

Question ❸ a): Jamilla's answer is correct.

Question ❸ b): Jamilla answered the question by completing the multiplication first and then adding $\frac{1}{5}$. Alex added $\frac{1}{5}$ and $\frac{3}{5}$ first and then multiplied. This is incorrect because the multiplication needs to be done first.

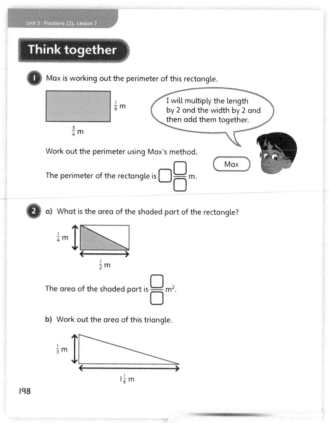

PUPIL TEXTBOOK 6A PAGE 198

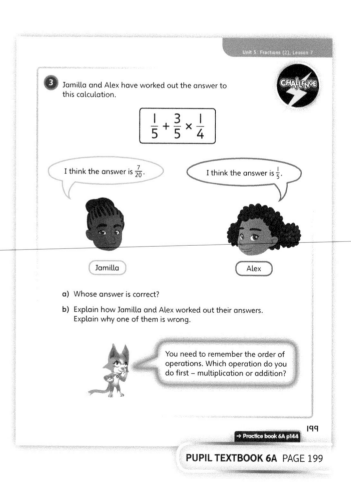

PUPIL TEXTBOOK 6A PAGE 199

Practice

WAYS OF WORKING Independent thinking

IN FOCUS Questions ① and ② aim to consolidate what children have covered in the **Textbook**, with question ① covering multiplication of a fraction by a whole number and question ② multiplication and addition of fractions. Questions ② onwards require children to decide which method to use. Question ⑥ incorporates both addition and subtraction, challenging children to think about how they can represent a whole 1 as a fraction.

STRENGTHEN In question ①, encourage children to model the problem with a bar model first. Then, they should solve the problem by repeated addition and also by multiplication, so they become familiar with both methods and understand why they give the same answer. Encourage children to use diagrams in all questions to help them to calculate with fractions, as well as helping them to see what operation is needed. Display a poster reminding them of the methods that they have used to calculate with fractions in this and the previous unit.

DEEPEN In question ④, ask: *What mistakes might happen with the order of operations when working out these calculations?* Ask children to create their own calculations for a partner to solve. Suggest that they include brackets in some of their calculations.

THINK DIFFERENTLY Question ⑤ introduces subtraction and division for the first time in the lesson, as children are given the perimeter and asked to find a missing side. They need to use their knowledge of isosceles triangles and squares to work out what numbers they need to multiply and divide by. Suggest that they write any information that they know on the diagram.

ASSESSMENT CHECKPOINT Use questions ② to ⑥ to assess whether children can decide what operation is needed and can complete a calculation involving fractions with more than one operation.

ANSWERS Answers for the **Practice** part of the lesson can be found in the *Power Maths* online subscription.

Reflect

WAYS OF WORKING Independent thinking

IN FOCUS This activity checks whether children can successfully follow the order of operations when calculating with fractions. They need to understand that when multiplication and addition are in the same calculation, the multiplication is completed first.

ASSESSMENT CHECKPOINT Assess whether children can explain, in full sentences, the mistake that has been made. Check that they can answer the question correctly.

ANSWERS Answers for the **Reflect** part of the lesson can be found in the *Power Maths* online subscription.

After the lesson ⏸

- Can children decide independently what operation is required?
- Do children recognise the correct order of operations?
- Can children complete a calculation that has more than one operation?
- Can children add, subtract and multiply fractions, and divide fractions by whole numbers?

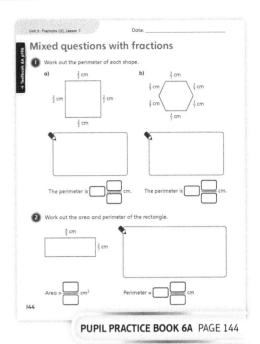

PUPIL PRACTICE BOOK 6A PAGE 144

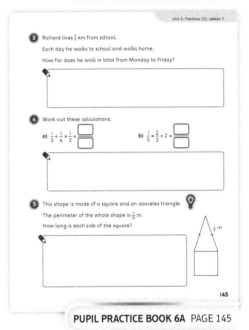

PUPIL PRACTICE BOOK 6A PAGE 145

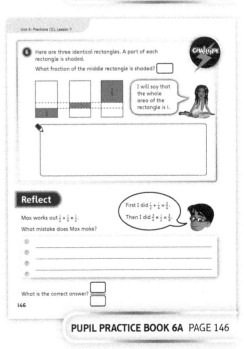

PUPIL PRACTICE BOOK 6A PAGE 146

Fraction of an amount

Learning focus

In this lesson, children will learn to find fractions of amounts in various contexts. They will use visual aids such as bar models to solve problems and support their understanding.

Before you teach ⏸

- Can children draw a diagram to represent a fraction?
- Can they divide an amount into equal parts?
- Can they multiply two integers?

NATIONAL CURRICULUM LINKS

Year 6 Number – fractions (including decimals and percentages)
Use written division methods in cases where the answer has up to two decimal places.

ASSESSING MASTERY

Children can find fractions of amounts involving unit and non-unit fractions by using a bar model. They can use the bar model to explain their understanding. Children can apply their skills of finding a fraction of an amount to unfamiliar problems and multi-step problems.

COMMON MISCONCEPTIONS

Some children may misinterpret what the question is asking. For example, if a question says, 'There are 30 students in a class. $\frac{2}{5}$ of the class are going swimming. How many children are not going swimming?', children may calculate $\frac{2}{5}$ instead of $\frac{3}{5}$. Ask:

- *What is the question asking for? Can you draw a bar model to show the number of children? How can you divide your bar into those going swimming and those not going swimming? Which part represents the children not going swimming?*

STRENGTHENING UNDERSTANDING

Give children strips of paper and counters to help them find fractions of an amount. Tell children that the strip of paper represents the whole and can be divided into whatever denominator the question asks for. Children can then share the counters, which represent the items in the question, equally between the parts. Encourage children to think about what one-quarter, one-third or one-half might be before asking for non-unit fractions.

GOING DEEPER

Add additional reasoning questions to the problems. For example, if the question was, 'There are 36 children in a swimming class. $\frac{1}{3}$ of the class has brown hair. How many of the class do not have brown hair?', ask additional questions such as: *How many more children who do not have brown hair are in the swimming class than children who do have brown hair?* to stretch children further. Where appropriate, ask children to show more than one method for finding the solution.

KEY LANGUAGE

In lesson: fraction of, equal parts, share

Other language to be used by the teacher: numerator, denominator, division, multiply

STRUCTURES AND REPRESENTATIONS

Bar model

RESOURCES

Optional: fraction strips, paper strips, counters, baskets, apples

 In the eTextbook of this lesson, you will find interactive links to a selection of teaching tools.

Quick recap 🔎

Ask children to solve these calculations. What is:

$\frac{1}{2}$ of 24?	$\frac{1}{3}$ of 24?	$\frac{1}{4}$ of 24?
$\frac{1}{6}$ of 24?	$\frac{1}{8}$ of 24?	$\frac{1}{12}$ of 24?

Discover

WAYS OF WORKING Pair work

ASK

- Question ① a): *How many baskets does each year have? What is different about Year 6? What diagram could you draw to help you?*
- Question ① b): *How is this question the same as part a)? How is it different? What diagram could you draw for this question? How many parts will it have? Did the children eat more apples in the morning or in the afternoon?*

IN FOCUS Questions ① a) and ① b) both illustrate the importance of reading the question carefully: part a) requires children to understand that the total needs to be divided into 5 equal parts, not 4; and part b) requires them to realise that they are being asked for $\frac{7}{10}$ of the total, not $\frac{3}{10}$. Question ① a) tests children's ability to divide by asking them to share the apples equally into 5 baskets. The question does not talk about fractions, just equal parts. A link could be made between 5 equal parts and fifths, and the fraction that Year 6 represents. Children may misinterpret the question and draw 4 boxes in their bar model, because there are 4 year groups. Emphasise the importance of labelling diagrams clearly to avoid misinterpreting the question.

PRACTICAL TIPS Carry out a similar question practically, by bringing in 5 baskets and a number of apples – a multiple of 5. Demonstrate the apples being shared equally among the baskets and then count the apples in 2 baskets. Ask children to draw a diagram to represent the baskets and the apples.

ANSWERS

Question ① a): The Year 6 children will get 80 apples.

Question ① b): The Year 6 children eat 56 apples in the afternoon.

Share

WAYS OF WORKING Whole class teacher led

ASK

- Question ① a): *Why does the bar model have 5 parts? How many parts does Year 6 have? Does there need to be an equal number of apples in each part? How many apples are there in 1 part? What fraction does each part represent? What fraction of the total apples does Year 6 have? How many apples does Year 3 get? What fraction is this?*
- Question ① b): *How many apples are in each part? What operations do you need to use for this question? Why does Flo say that she only needs to find $\frac{7}{10}$? Can you explain whether her method is more efficient?*

IN FOCUS Question ① b) draws attention to two possible approaches: find $\frac{7}{10}$ of 80, or find $\frac{3}{10}$ of 80 and subtract it from 80. Show how the bar model illustrates both these methods and discuss which is more efficient. Use Dexter's and Flo's statements to discuss what common mistakes may be made with these questions.

① a) The apples are shared into the baskets equally.

How many apples will the Year 6 children get?

b) The Year 6 children eat $\frac{3}{10}$ of their apples in the morning and the remaining apples in the afternoon.

How many apples do they eat in the afternoon?

200

PUPIL TEXTBOOK 6A PAGE 200

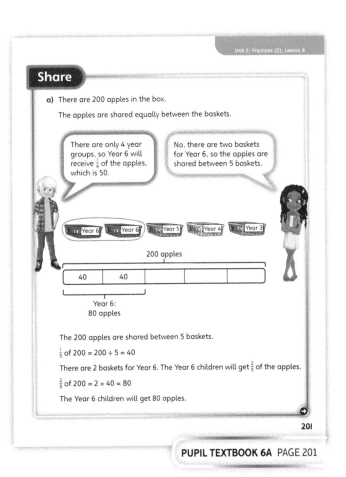

Share

a) There are 200 apples in the box.

The apples are shared equally between the baskets.

> There are only 4 year groups, so Year 6 will receive $\frac{1}{4}$ of the apples, which is 50.

> No, there are two baskets for Year 6, so the apples are shared between 5 baskets.

200 apples

| 40 | 40 | | | |

Year 6: 80 apples

The 200 apples are shared between 5 baskets.

$\frac{1}{5}$ of 200 = 200 ÷ 5 = 40

There are 2 baskets for Year 6. The Year 6 children will get $\frac{2}{5}$ of the apples.

$\frac{2}{5}$ of 200 = 2 × 40 = 80

The Year 6 children will get 80 apples.

201

PUPIL TEXTBOOK 6A PAGE 201

Think together

ASK

- Question ❶: *How many parts do you need to divide the bar into? How do you know?*
- Question ❷: *Can you explain what you need to find? What diagram could you draw to help you? What fraction of the children are not going on the trip?*
- Question ❸: *What has Richard found? How can you explain to Mo and Richard the mistakes they have made?*

IN FOCUS Question ❶ requires children to find a fraction of a quantity in its simplest form, where the fraction is a non-unit fraction. The scaffolding encourages children to find $\frac{1}{6}$ of the amount and then use that to find $\frac{5}{6}$. Question ❷ requires children to read the question carefully. The fraction in the question text is $\frac{5}{7}$, but the amount being asked for is the remaining amount, $\frac{2}{7}$. The question challenges children to think about how they could complete the questions without subtraction and therefore prompts them for two different methods. A bar model will help children to see that $\frac{2}{7}$ of the amount remains. Question ❸ is a two-step problem: children need to find a fraction of an amount and then a fraction of that number. A bar model will help them to understand what the question is asking and to identify the common mistakes illustrated by Mo and Richard.

STRENGTHEN To strengthen understanding, use the scaffolding provided in question ❶ as an example of how children could approach questions ❷ and ❸ through smaller steps. Encourage children to use bar models and to label their diagrams clearly and to shade what the question is asking for. This is particularly important for questions ❷ and ❸, which are more complex.

DEEPEN In question ❷, ask: *Could you have completed the question without subtracting? What are the similarities in the calculations for finding the number of children going on the trip and the number not going on the trip? What are the differences?* Ask children to create a question similar to question ❸ for a partner to solve.

ASSESSMENT CHECKPOINT Use questions ❶ and ❷ to assess whether children can use a diagram to find a fraction of an amount, labelling the whole, the parts and what the question is asking for. Use question ❷ to assess whether children can find a fraction of an amount when the fraction is not given directly.

ANSWERS

Question ❶: $\frac{1}{6}$ of 300 g is 300 ÷ 6 = 50 g

$\frac{5}{6}$ of 300 g is 5 × 50 = 250 g

250 g of flour is needed.

Question ❷: 8 children are not going on the trip.

Question ❸: Mo found $\frac{1}{2}$ of 36, without first finding out how many of the class are boys. Richard found $\frac{1}{3}$ of 36, which is the number of boys in the class, but has not continued the question to find out how many of the boys wear goggles. 6 boys wear goggles.

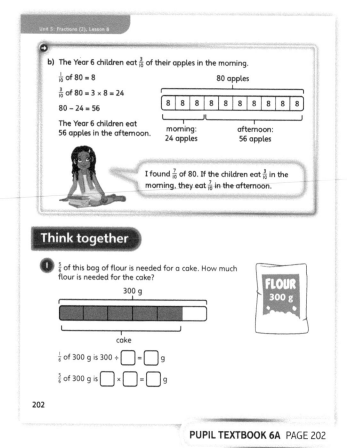

PUPIL TEXTBOOK 6A PAGE 202

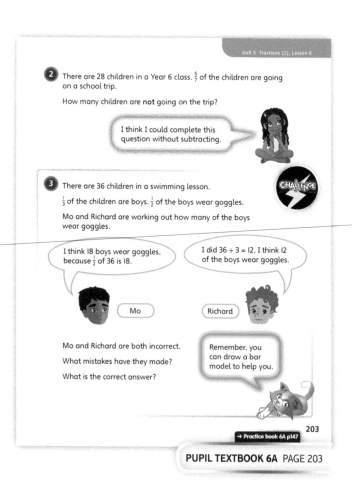

PUPIL TEXTBOOK 6A PAGE 203

Practice

WAYS OF WORKING Independent thinking

IN FOCUS Most of the questions are not scaffolded, reflecting that there are multiple ways of solving these problems. The emphasis is for children to find the most efficient method to solve each problem. In questions ① and ②, children can either first find the fraction given in the question and subtract it from the total or first work out the fraction represented by the answer. It might be useful to stop the class after a few problems and discuss the methods used. In question ⑥, the fractions in each comparison are of the same total, giving children the opportunity to realise that they can just look at the two fractions without finding fractions of an amount.

STRENGTHEN Read through a question together. Draw a simple bar model on the board, explaining each feature of the model. Emphasise that the bar is the total and they need to look at the denominator of the fraction to work out how many parts to divide the bar into. Agree that the numerator shows how many parts are represented. Read the question again, drawing attention to what is being asked for, and shade these parts on the diagram.

For questions that involve comparisons (for example, question ③), suggest to children that they draw one bar model above the other using bars of the same length.

DEEPEN When children have completed question ⑦, ask them to work out what fraction of the total Amelia gave to her mum. Ask them to explore the relationship between the fractions in the question and this fraction. They could extend this exploration to the problem in **Think together** question ③, which is a similar problem, and explain their findings using a bar model.

THINK DIFFERENTLY In question ⑥, children have to put < or > between two calculations. In part a) they may see straight away that $\frac{3}{7}$ of 70 is less than $\frac{5}{7}$ of 70 and will not need to draw a diagram to help them. Part b) will give them the opportunity to practise drawing fraction grids.

ASSESSMENT CHECKPOINT Use questions ① to ⑥ to assess whether children can solve problems involving finding fractions of amounts. Look for children using the most efficient methods.

ANSWERS Answers for the **Practice** part of the lesson can be found in the *Power Maths* online subscription.

Reflect

WAYS OF WORKING Whole class

IN FOCUS Children need to think about the questions they have answered and say which question they found the most challenging and why. Use this as a class discussion, so children can see that maths can be challenging and that everyone struggles sometimes. For those children who say it was all straightforward, discuss whether they think they used the most efficient method.

ASSESSMENT CHECKPOINT Assess whether children can describe how to find a fraction of an amount, looking for the most efficient method.

ANSWERS Answers for the **Reflect** part of the lesson can be found in the *Power Maths* online subscription.

After the lesson ⏸

- Can children find a fraction of an amount?
- Can they represent a problem using a bar model and use this to find the most efficient strategy?

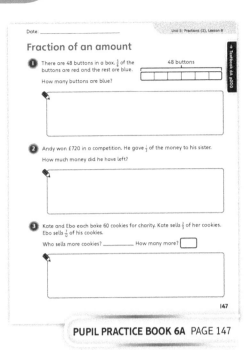

PUPIL PRACTICE BOOK 6A PAGE 147

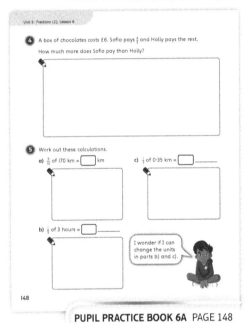

PUPIL PRACTICE BOOK 6A PAGE 148

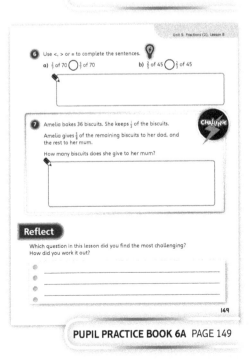

PUPIL PRACTICE BOOK 6A PAGE 149

Fraction of an amount – find the whole

Learning focus

In this lesson, children will solve problems involving finding fractions of amounts, including problems where children have to find the whole given information about a part.

Before you teach ⏸

- Can children find a simple fraction of an amount?
- Can they use bar models to represent fraction of amount questions?
- Do they know their times-tables off by heart to support them with the calculations?

NATIONAL CURRICULUM LINKS

Year 6 Number – fractions (including decimals and percentages)

Use written division methods in cases where the answer has up to two decimal places.

COMMON MISCONCEPTIONS

Some children may always just find a fraction of an amount regardless of the question. For example, for the question, '$\frac{2}{5}$ of a number is 10. What is the number?' they will work out $\frac{2}{5}$ of 10 = 4. Ask:

- *Is 10 the whole number or a fraction of it?*
- *Is $\frac{2}{5}$ of a number smaller or larger than the number? Will the answer be more or less than 10?*

STRENGTHENING UNDERSTANDING

In order to support children with understanding whether they are finding a fraction of an amount or the whole, ask children to underline the key information in the question. Draw a single bar and ask them if they know the whole bar or just part. Give them concrete objects such as counters to help. Divide the model into the number of parts. Write the information they know on the model (whole amount with a brace or the parts). The picture should then help them work out what steps they need to take.

GOING DEEPER

Give children more complex problems to solve, such as '$\frac{1}{3}$ of 60 = $\frac{2}{5}$ of ☐.' This requires them to both find a part and work out the whole from a part in the same question. Move on to questions like, 'Chris's height has increased by $\frac{1}{5}$. If his new height is 1·8 metres, what was his height before?'

KEY LANGUAGE

In lesson: fraction of

Other language to be used by the teacher: numerator, denominator, whole, part, multiply, divide

STRUCTURES AND REPRESENTATIONS

Bar model

RESOURCES

Optional: play coins, counters, multilink cubes

 In the eTextbook of this lesson, you will find interactive links to a selection of teaching tools.

Quick recap

Ask children to solve these calculations. *What is:*

$\frac{1}{2}$ of 24? $\frac{2}{3}$ of 24? $\frac{3}{4}$ of 24?

$\frac{5}{6}$ of 24? $\frac{3}{8}$ of 24? $\frac{11}{12}$ of 24?

Discover

Pair work

ASK

- Question ① a): *How does this question differ from the ones you were doing in the previous lesson? What do you know about £1·60? Can you represent the situation with a bar model? How can you work out how much pocket money Lee had?*
- Question ① b): *What information are you given? How many sweets are in $\frac{1}{5}$ of the jar? How did you work this out? How many sweets are in the whole jar? What other way could you get the same answer?*

IN FOCUS This lesson builds on the previous lesson, progressing to working out the whole amount given a fraction of the amount, using the same models. Question ① a) uses a unit fraction, so the whole can be found by multiplying by the denominator. Look for children who just find a quarter of the amount instead of multiplying up. Encourage children to draw a bar model to represent the situation. Question ① b) gives a non-unit fraction, so children can explore ways of working out the whole. For example, they may double and add on half or they could halve and then multiply by 5.

PRACTICAL TIPS Act out the scenario using toy coins and counters or cubes to represent the sweets.

ANSWERS

Question ① a): Lee had £6·40 to begin with.

Question ① b): There were 75 sweets in the jar when it was full.

Share

Whole class teacher led

ASK

- Question ① a): *What information were you given? How does the bar model show the situation? What mistake did Dexter make? What could you do to help you not make that mistake?*
- Question ① a): *Why do you multiply by 4? What method do you know of multiplying by 4?*
- Question ① b): *How is this question different from part a)? How is it the same? Why do you not just multiply by 5? What do you have to do first?*
- Question ① b): *Can you explain where Astrid has got her numbers from? What has she done?*

IN FOCUS In part a), focus on the difference between this type of question and the ones from the last lesson. By explaining what Dexter has done wrong, children can develop an understanding of how to use the bar model correctly. In part b), some children may want to multiply by 5, forgetting that $\frac{2}{5}$ is not a unit fraction. Discuss that this time they have to work out what $\frac{1}{5}$ of the jar is first, before they can multiply. The bar model helps explain why they first need to divide by 2. Use Astrid's method as a discussion point to explain other ways you could approach the question.

Fraction of an amount – find the whole

Discover

① a) Lee spends $\frac{1}{4}$ of his pocket money on sweets.
How much pocket money did Lee have to begin with?

b) The jar was full before Lee bought any sweets.
Lee bought $\frac{2}{5}$ of the jar.
How many sweets were in the jar when it was full?

204

PUPIL TEXTBOOK 6A PAGE 204

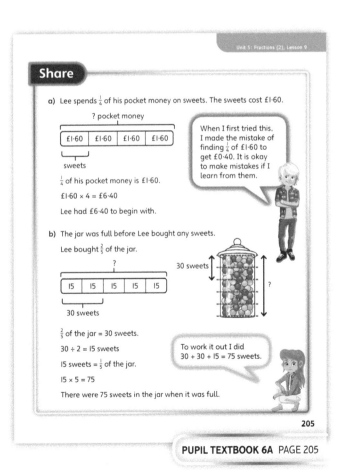

PUPIL TEXTBOOK 6A PAGE 205

Think together

Whole class teacher led (I do, We do, You do)

ASK

- Question **1**: *What information are you given? How can you use the bar models to find each unit fraction? Once you know the unit fraction, what do you need to do to find the whole?*
- Question **2**: *Do you know the whole rope length? How can you use a bar model to help you? How can you find the length of rope A? What about rope B?*
- Question **3**: *What information are you given? Are you given the whole? How could you work out the original number (the whole)? Once you have worked this out, what can you do then? How do you think Ash worked it out?*

IN FOCUS These problems are multi-step and require children to determine first whether they know the whole and have to find a fraction of the amount or whether they have to find the whole. Encourage children to see this as the first step.

Question **1** asks children to use a bar model to find the whole. They need to work out the unit fraction first before they can work out the whole. In question **2**, children need to realise that they have to work out both wholes in order to compare the lengths of rope. In this question they are working with decimals, so some children might need additional support.

Question **3** provides children with the opportunity to use their problem solving skills to find more than one way of working out the answer. For example, they may find the whole (45) and then work out $\frac{8}{9}$ of the whole. Ask what facts they know about the number; for example, can they find $\frac{1}{3}$, $\frac{1}{6}$, $\frac{1}{9}$ of the number? Some children may use their knowledge of equivalent fractions and realise that $\frac{2}{3} = \frac{6}{9}$ and so $\frac{8}{9}$ is $\frac{2}{9}$ more. Explore different ways of approaching the question.

STRENGTHEN Ask children to highlight the key information in the question. Ask if they know the whole or if they need to work out the whole. Provide them with concrete materials to help. Guide them through drawing a bar model to illustrate the question. This should help children work out what steps they need to take.

DEEPEN Question **3** asks children to make up their own problems for a partner to solve. Encourage children to create more complex problems than just '$\frac{3}{8}$ of my number is 15. What is my number?' Ask children to solve their partner's problems in more than one way, explaining why each method works.

ASSESSMENT CHECKPOINT Use questions **1**, **2** and **3** to assess whether children can solve a variety of multi-step problems involving fractions of an amount.

ANSWERS

Question **1** a): 32

Question **1** b): 200

Question **2**: Rope A is 0.9 m.
Rope B is 1.2 m.
Rope B is longer by 0.3 m.

Question **3**: $\frac{8}{9}$ of Amelia's number is 40.

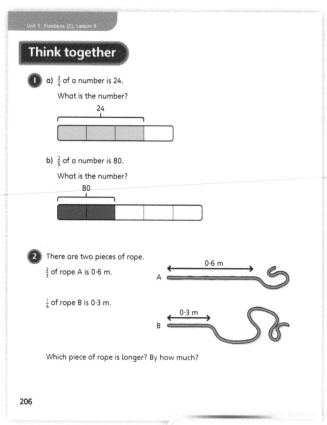

PUPIL TEXTBOOK 6A PAGE 206

PUPIL TEXTBOOK 6A PAGE 207

Practice

WAYS OF WORKING Independent thinking

IN FOCUS Questions ❶, ❷ and ❸ are simple problems where children have to find the whole, given a fraction of an amount. Question ❶ uses a unit fraction, so children simply have to multiply. In questions ❷ and ❸, children need to divide first. For all questions, encourage children to draw a bar model. Question ❺ provides more abstract practice. Each question steps up in difficulty, with part (d) requiring children to realise that there are several steps involved in finding the answer.

STRENGTHEN For questions that are more abstract (for example, questions ❷ and ❺), children might find it easier if they create their own context to fit the numbers. Provide them with counters or cubes to support their scenario.

DEEPEN Ask children to solve problems such as $\frac{1}{3}$ of $\frac{1}{4}$ of ◯ = 10. Ask them to represent this using a bar model.

THINK DIFFERENTLY Question ❻ requires children to find a mystery whole number when a given fraction of the total amount is a number with one decimal place.

ASSESSMENT CHECKPOINT Use questions ❷ to ❼ to assess whether children can solve multi-step questions on fractions of an amount, including ones where they have to find the whole.

ANSWERS Answers for the **Practice** part of the lesson can be found in the *Power Maths* online subscription.

Reflect

WAYS OF WORKING Independent thinking

IN FOCUS Children compare questions where they are given the whole and where they need to find the whole, requiring them to focus on the difference between the methods. Although the questions look similar, they have a different underlying structure.

ASSESSMENT CHECKPOINT Children should know when they are finding a fraction of an amount and when they are working out the whole. They should use a bar model or other pictorial representation to explain the key difference.

ANSWERS Answers for the **Reflect** part of the lesson can be found in the *Power Maths* online subscription.

After the lesson ⏸

- Can children find the whole, given a fraction of an amount?
- Can they interpret questions to work out whether they are finding a part or the whole?
- Can they solve more complicated multi-step problems involving fractions of an amount?

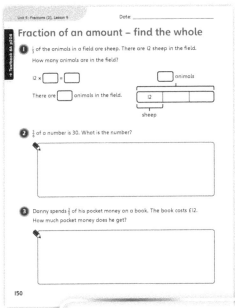

PUPIL PRACTICE BOOK 6A PAGE 150

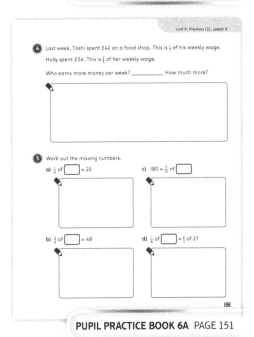

PUPIL PRACTICE BOOK 6A PAGE 151

PUPIL PRACTICE BOOK 6A PAGE 152

End of unit check

Don't forget the unit assessment grid in your *Power Maths* online subscription.

WAYS OF WORKING Group work adult led

IN FOCUS

- Questions ① to ⑦ are fluency-based questions to check children's understanding of methods for multiplying and dividing fractions. Many of the incorrect answers are likely to indicate that children have confused the two methods.
- Questions ⑧ and ⑨ are SATs-style questions. In question ⑧, children need to multiply the two fractions and then use their knowledge of simplifying fractions to ensure that the answer is given in its simplest form. Question ⑨ requires children to read the question carefully to identify what they need to find.
- Question ⑩ is a multi-step problem whereby children are given the whole from which they need to find the number of darts in each question. They then need to use their knowledge of the scores for each section to work out the total score.

ANSWERS AND COMMENTARY

Children should be able to multiply any fraction by a whole number and by any other fraction, and divide a fraction by a whole number. They should be able to solve simple and multi-step fraction problems, including problems on fractions of an amount where they are given the fraction and need to find the whole.

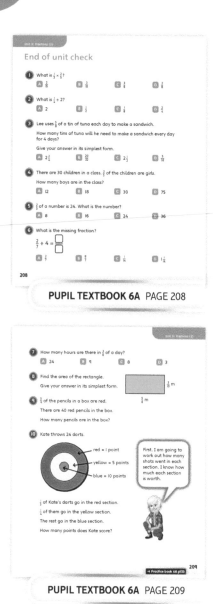

PUPIL TEXTBOOK 6A PAGE 208

PUPIL TEXTBOOK 6A PAGE 209

Q	A	WRONG ANSWERS AND MISCONCEPTIONS	STRENGTHENING UNDERSTANDING
1	A	B suggests numerators added instead of multiplied; C both numerators and denominators added; D denominators added.	To support children with multiplying fractions, divide a grid to show the fractions on adjacent sides. Shade the relevant squares and ask children what fraction this is. How do they know?
2	C	A suggests dividing 4 by 2; B denominator divided by 2; D numerator multiplied by 2.	
3	C	A is not fully simplified; B suggests both numerator and denominator multiplied by 4; D indicates denominator multiplied by 4.	To support children with dividing a fraction by a whole number, colour in the fraction on a fraction strip. Divide the shaded parts by the whole number. Discuss the size of of the parts and how many of them make up the answer.
4	B	A suggests calculating the number of girls. D suggests having found the whole if $\frac{2}{5}$ is 30.	
5	D	A suggests found $\frac{1}{3}$ of 24, and B $\frac{2}{3}$ of 24.	
6	C	A suggests that the child has not divided by 4; B suggests they have multiplied by 4 instead of divided; and D suggests having found the answer correctly, but put 1 in front of it.	For problems on fractions of an amount, clearly label a bar model with the known information, using a question mark to show the quantity that is being asked for.
7	B	C suggests $\frac{1}{3}$ found; D suggests $\frac{1}{8}$ found.	
8	$\frac{1}{4}$ m²	Some children may find the perimeter instead of the area.	
9	72	Some children may find $\frac{4}{9}$ of 40.	
10	138	Some children may not complete all steps of the problem.	

My journal

WAYS OF WORKING Independent thinking

ANSWERS AND COMMENTARY This journal activity brings together all the work that children have done on fractions in the last two units. Children often struggle to remember the different methods, and this journal attempts to try to recap all of them. Use the journal as a diagnostic tool to check:
- which children understand which questions
- the most common questions children are struggling with
- whether children are getting methods confused
- the methods that children use to get the answers
- whether children are giving their answers in the simplest form.

You might want to ask children to show their understanding of a method by drawing a diagram. You might also want to pair children up to teach each other and explain their method.

For the last two questions look for children following the correct order of operations.

$\frac{1}{5} \times 3 = \frac{3}{5}$ $\frac{1}{3} \div 4 = \frac{1}{12}$ $\frac{2}{3} \div 4 = \frac{1}{6}$ $\frac{7}{10} + \frac{2}{5} \times \frac{1}{2} = \frac{9}{10}$

$\frac{2}{3} \times \frac{3}{8} = \frac{1}{4}$ $\frac{4}{5} \div 2 = \frac{2}{5}$ $\frac{7}{10} + \frac{2}{5} = \frac{11}{10} = 1\frac{1}{10}$ $\frac{7}{10} \times \frac{2}{5} + \frac{1}{2} = \frac{39}{50}$

Power check

WAYS OF WORKING Independent thinking

ASK
- Are you able to multiply a fraction by a fraction?
- Can you divide a fraction by a whole number?
- Can you find a fraction of an amount?
- Given $\frac{2}{5}$ of a number is 8, can you find the number?

Power puzzle

WAYS OF WORKING Independent working

IN FOCUS This tests children's knowledge of all the work in this unit using problems in which each answer leads to the next. Emphasise the importance of double checking answers, as an error early on will lead to many wrong answers. Encourage children to start thinking about whether an answer is likely or not: complicated calculations or long decimal answers are likely to imply that they have made a mistake somewhere.

ANSWERS AND COMMENTARY

Question **1**:

A	B	C	D	E	F	G	H
36	18	27	15	4·5 or $4\frac{1}{2}$	8	2	$\frac{1}{20}$

Question **2**: Children first need to work out the distance between points A and B to identify the total that they are finding $\frac{2}{3}$ of. C = 150.

After the unit ⏸
- How useful did children find the diagrams in calculating with fractions?
- Which diagrams did they find most useful?
- Can they use diagrams to explain why a method works?

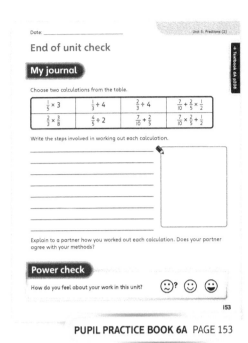

PUPIL PRACTICE BOOK 6A PAGE 153

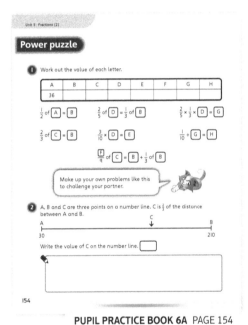

PUPIL PRACTICE BOOK 6A PAGE 154

Strengthen and **Deepen** activities for this unit can be found in the *Power Maths* online subscription.

Unit 6
Measure – imperial and metric measures

Don't forget to watch the Unit 6 video!

Mastery Expert tip! 'My class found it stimulating to investigate ways imperial units are used in real life. The more we investigated, the more we found – from 10-inch pizzas to quarter-pound burgers! This hook got children thinking about approximate metric equivalents of imperial units.'

WHY THIS UNIT IS IMPORTANT

The focus of this unit is to consolidate and apply understanding of units of measure, both metric and imperial. Children practise converting between units and apply conversions to problem-solving contexts. Children revise equivalences of metric units and work with numbers with up to two decimal places. They use reasoning to describe when a number should be multiplied or divided to convert and by how much. Children are introduced to the relationship between miles and kilometres, and how to apply the 5 : 8 ratio to convert between them. Previous work with imperial measurements is extended into problem-solving contexts.

WHERE THIS UNIT FITS

→ Unit 5: Fractions (2)
→ **Unit 6: Measure – imperial and metric measures**
→ Unit 7: Ratio and proportion

This unit builds on the concepts of imperial and metric measures from Year 5. Prior knowledge of prefixes of metric units is used as a reminder of the equivalences of different units before converting. Children will revise imperial measures and learn the relationship between miles and kilometres.

Before they start this unit, it is expected that children can:
- recognise the meanings of the different prefixes used in metric units (cent-, milli-, kilo-)
- express one metric unit in terms of another (for example, 1,000 g = 1 kg)
- convert simple measurements from one metric unit to another
- identify imperial and metric units of measure and convert them when given their equivalent values.

ASSESSING MASTERY

Children who have mastered this unit can choose appropriate metric units of measure, justify their choices, make sensible estimates of length, mass or capacity, and convert between units. Children can apply their knowledge of converting metric units to solve problems, choosing appropriate strategies. Children can state facts about different imperial and metric units and apply them when converting from one imperial unit to another, or between imperial and metric units.

COMMON MISCONCEPTIONS	STRENGTHENING UNDERSTANDING	GOING DEEPER
Children may confuse whether to multiply or divide to convert between units.	Cut lengths of string 1 m and 1 cm long. Ask children to make a 3 m length out of metres. Then ask whether they would expect to use more or fewer centimetre lengths to equal this.	Introduce multiple conversions to allow children to move between units several times. For example: convert 400,000 cm into metres, then kilometres, then miles.
Children may multiply decimals incorrectly by 10, 100 or 1,000 by adding zeros on to the end.	Use place value grids or digit cards to practise multiplying and dividing by 10, 100 and 1,000, by shifting the digits to the left and right.	State facts about a conversion and challenge children to derive what the mystery units might be. For example: 'There is a 2 in the first measurement but not in the second.'

Unit 6: Measure – imperial and metric measures

Use these pages to introduce the unit to children. You can use the characters to explore different ways of working too!

STRUCTURES AND REPRESENTATIONS

Bar model: This model provides a useful way of visualising the equivalence between two units of measure as they are converted (where each individual bar representing a value in one unit is shown as equal to a bar representing the equivalent value in another unit).

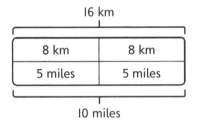

KEY LANGUAGE

There is some key language that children will need to know as part of the learning in this unit:

→ units (of measure/ment), metric, imperial, length, mass, volume, capacity, distance

→ measure, convert, equal, equivalent, approximate, smaller (unit), larger (unit), for every, ratio

→ millimetres (mm), centimetres (cm), metres (m), kilometres (km), grams (g), kilograms (kg), millilitres (ml), litres (l)

→ inches (in), feet (ft), yards, ounces (oz), pounds (lbs), pints, miles, gallons

→ digits, decimal

→ conversion table, conversion graph.

PUPIL TEXTBOOK 6A PAGE 210

PUPIL TEXTBOOK 6A PAGE 211

Metric measures

Learning focus

In this lesson, children will read, write and recognise all metric measures for length, mass and capacity. They will apply their understanding to make sensible estimations.

NATIONAL CURRICULUM LINKS

Year 6 Measurement

Use, read, write and convert between standard units, converting measurements of length, mass, volume and time from a smaller unit of measure to a larger unit, and vice versa, using decimal notation up to three decimal places.

ASSESSING MASTERY

Children can choose appropriate metric units of measurement for different scenarios where length, mass or capacity are measured, justifying their choices. Children can also make sensible estimates of length, mass or capacity using their prior knowledge about metric units to inform their estimates.

COMMON MISCONCEPTIONS

Children may think that 'the most appropriate unit' means 'the only possible unit' and so may think that large units can only be used to measure large distances and small units can only be used to measure small distances. Ask:
- *Are centimetres/grams/litres the only unit of measure you could possibly use in this situation?*

STRENGTHENING UNDERSTANDING

Provide practical opportunities for children to choose appropriate metric units to measure length, mass and capacity. Encourage them to discuss which unit of measure they think should be used in different scenarios and then to measure using those units. Set up objects and amounts of liquid for children to estimate their length, mass and capacity, and then provide opportunities for children to measure to check their estimates.

GOING DEEPER

Provide children with metric measurements and work inversely from these, estimating objects that may weigh or measure the same amount. Ask: *Can you find an object that you estimate weighs about half a kilogram? Can you find a container that holds over half a litre?* Children can then use subtraction to find the difference between their estimate and the actual measurement.

KEY LANGUAGE

In lesson: metric, units of measurement, estimate, kilograms, grams, litres, millilitres, centimetres, metres, kilometres, length, mass, capacity

Other language to be used by the teacher: larger, smaller, millimetres, volume

STRUCTURES AND REPRESENTATIONS

Diagrams of weighing scales, measuring jugs, rulers

RESOURCES

Optional: weighing scale(s), measuring jug(s)

 In the eTextbook of this lesson, you will find interactive links to a selection of teaching tools.

Quick recap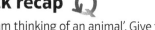

Play 'I am thinking of an animal'. Give the following clues and encourage children to make sensible guesses:

It is about 25 cm long. It weighs approximately 4 kg. [Cat]

Repeat with other animals, for example:

It is about 3 m tall. It can drink up to 100 l at once! [Camel]

It weighs less than 1 g. It is about 1 cm wide. [Bumblebee]

Discover

Metric measures

WAYS OF WORKING Pair work

ASK

- Question ① a): *Why do you need to use units of measurement when measuring things? What would happen if someone used the wrong unit of measurement?*
- Question ① b): *What information will help you decide which units to use?*

IN FOCUS When discussing question ① a), encourage children to suggest a variety of units of measurement for mass and capacity. Children may include imperial measures, so explain that during this lesson they will be focusing on metric units. Ask them which metric unit they would choose for the scales and jug in the picture and why. Check that they have some knowledge of what 1 g, 1 kg, 1 ml and 1 l are.

PRACTICAL TIPS Prepare a set of scales and a measuring jug where the metric units of measurement are covered over. Measure different objects or amounts of water and discuss the need to have more than just a number when expressing the mass or capacity.

ANSWERS

Question ① a): The unit of measurement on the scales is grams. The sugar weighs 150 grams.
The unit of measurement on the jug is litres. The jug contains 0·75 litres of milk.

Question ① b): It would be most suitable to measure the length of the tray in centimetres (cm).

Discover

① a) Look at the weighing scales and the measuring jug.
   What units of measure do you think they show?

   b) Alex wants to measure the length of the tray. What units would she use?

212

PUPIL TEXTBOOK 6A PAGE 212

Share

WAYS OF WORKING Whole class teacher led

ASK

- Question ① a): *What are the possible metric units? Can you think of anything measured in grams or kilograms that you could use to compare 150 g or 150 kg to? Can you think of anything measured in millilitres or litres that you could use to compare 0·75 ml or 0·75 l to?*
- Question ① b): *What metric units do you know that are used to measure length? Which is the most appropriate here? How do you know? How does knowing that the tray is almost the length of the ruler help you?*

IN FOCUS Spend time discussing the metric units of measure that children are familiar with for mass, capacity and length. Encourage them to reason on how they will apply the information given to help them decide which is the most appropriate unit to use for each item in the cookery scenario.

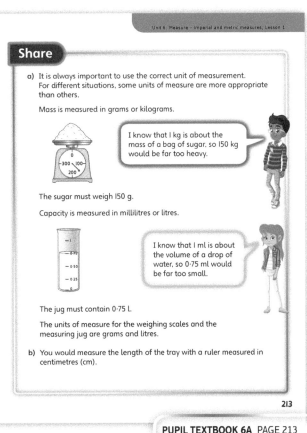

PUPIL TEXTBOOK 6A PAGE 213

Think together

WAYS OF WORKING Whole class teacher led (I do, We do, You do)

ASK

- Question ❶: *What facts do you know from the picture that will help you estimate the capacity of the glass? When estimating, why is it important to think about what you know and use it to help? Think of two ways to check your estimate and find the actual capacity of the glass.*
- Question ❷: *Which of the types of food are measured in units of mass and which in units of capacity? Can you think of any facts about mass and capacity that you can use to explain your choices?*

IN FOCUS Question ❶ provides a further opportunity for children to practise estimation using known facts. The two facts children should notice are that the bottle holds 330 ml of liquid and that, as there is some left over, the glass must hold slightly less than the bottle. Discuss with children what 'slightly less' might mean – is 150 ml a reasonable estimate? Why or why not?

STRENGTHEN To support understanding of question ❷, ask: *How can you sort the containers into two groups?* Discuss the fact that some contain liquids and some contain solids. Ask: *How does this affect the unit that is used to measure the object?* Invite children to sort the five possible units of measurement according to whether they measure mass or capacity. They can then use this information to help match the labels.

DEEPEN In question ❸, deepen children's understanding of 'appropriate units' by encouraging them to describe in words why one unit is any better than another. Ask: *If you want to measure to the nearest millimetre, but give the measurement in metres, what sort of number will your measurement be? How many decimal places will it have?*

ASSESSMENT CHECKPOINT Look for children who are confident suggesting the most appropriate metric units for measuring in different scenarios. They should understand the need to estimate based on what they know about metric units.

ANSWERS

Question ❶: The glass holds about 300 ml. (Accept any other reasonable estimate less than 330 ml and above about 250 ml.)

Question ❷: Water = 1·5 l, cereal bar = 30 g, flour = 1 kg, juice = 500 ml, pasta = 500 g

Question ❸ a): The length of your classroom: metres
The length of your shoe: cm
The mass of a pencil: grams
The capacity of a bath: litres

Question ❸ b): Centimetres are more appropriate because the spaghetti is much less than 1 m long.

PUPIL TEXTBOOK 6A PAGE 214

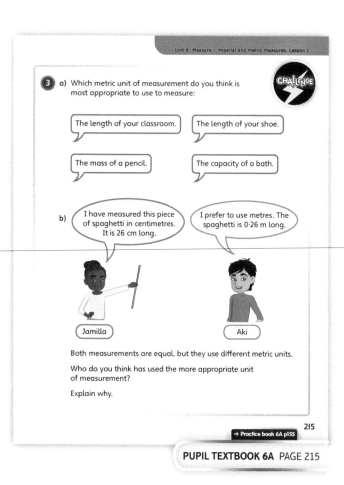

PUPIL TEXTBOOK 6A PAGE 215

Practice

WAYS OF WORKING Independent thinking

IN FOCUS In question ❶, the options children have to choose from do not allow for a variety of possible answers – only one unit is appropriate each time. The question applies children's knowledge of each metric unit of measurement and checks that they understand what each unit is worth and can choose the appropriate unit for different scenarios.

STRENGTHEN Strengthen children's estimation skills in questions ❷ and ❹ by providing them with concrete examples. In question ❷, provide children with opportunities to tip water out of containers like those listed in the table (without directly comparing or measuring them). In question ❹, provide some concrete examples for children to compare with each pair of estimates. Ask: *Given the measurement that you have been shown, which of the two is the better estimate?*

DEEPEN In question ❻, children consider a further metric unit of measure – the milligram (mg). Encourage children to use reasoning to justify their answer, looking at the prefix 'milli' and thinking about other units of measure that begin with this (for example, millilitre or millimetre). Extend the challenge by giving children time to investigate further, finding out how many milligrams are in 1 gram and the mass of different objects in milligrams.

THINK DIFFERENTLY The statements in question ❺ involve units of measure in various contexts, prompting children to think about the correct application of units of measure in real-life scenarios.

ASSESSMENT CHECKPOINT At this point in the lesson, children should be able to suggest appropriate metric units for measuring mass, length and capacity. They should be able to make estimates confidently based on their knowledge of these metric units.

ANSWERS Answers for the **Practice** part of the lesson can be found in the *Power Maths* online subscription.

Reflect

WAYS OF WORKING Independent thinking

IN FOCUS Children are asked to consider how appropriate the measures used on a given shopping list are. It may be beneficial to draw their individual reflections together into a wider discussion about what they have learnt about the use of appropriate metric units.

ASSESSMENT CHECKPOINT Look for children who write sentences demonstrating that they recognise metric measures of mass, length and capacity, can choose appropriate units for different situations, and can use prior knowledge of these to make reasonable estimates.

ANSWERS Answers for the **Reflect** part of the lesson can be found in the *Power Maths* online subscription.

After the lesson

- The next lesson will consolidate children's knowledge of converting between metric units. To what extent do you feel children's understanding of metric units is sufficient to prepare them for this next stage?

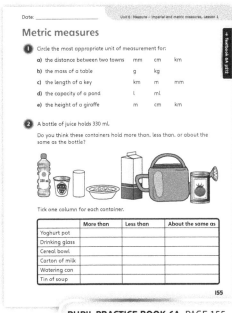

PUPIL PRACTICE BOOK 6A PAGE 155

PUPIL PRACTICE BOOK 6A PAGE 156

PUPIL PRACTICE BOOK 6A PAGE 157

Convert metric measures

Learning focus

In this lesson, children will convert between metric units of measurement, including measurements that involve decimals.

Before you teach

- Are children confident working with decimals?
- How can you improve the teaching of fluency through this lesson?

NATIONAL CURRICULUM LINKS

Year 6 Measurement

Use, read, write and convert between standard units, converting measurements of length, mass, volume and time from a smaller unit of measure to a larger unit, and vice versa, using decimal notation to up to three decimal places.

Solve problems involving the calculation and conversion of units of measure, using decimal notation up to three decimal places where appropriate.

ASSESSING MASTERY

Children identify the relationships between different metric units of length, mass and volume (for example, knowing that 1,000 g = 1 kg). Children apply this information to convert from one unit of measurement to another, choosing to multiply or divide according to the units involved.

COMMON MISCONCEPTIONS

Children may confuse whether to multiply or divide when converting. Ask:
- *If you convert from larger units to smaller ones, will you need fewer units to make up the amount, or more? Do you think you should multiply or divide?*

Children may multiply decimals incorrectly by 10, 100 or 1,000 by adding zeros to the end and not changing the place value of the digits. Ask:
- *What happens to the place value of each digit when a number is multiplied by 10/100/1,000? What happens if the number is divided by 10/100/1,000?*

STRENGTHENING UNDERSTANDING

Provide practical opportunities for children to measure length, mass and capacity of objects in given units. Split the group into two teams. Encourage one team to measure using one unit (metres, for example) and the other team to measure in another unit (centimetres, for example). Ask children to match their answers. Ask: *How do you know that 1·2 metres is the same as 120 centimetres? How could you use multiplication to show this?*

GOING DEEPER

Provide children with metre rulers and sticky labels. Ask: *What decimal parts of this metre do you know? How do you know what they are?* Ask children to label their rulers in tenths of a metre. Challenge children to link this understanding with knowledge of fractions. For example: *What is $4\frac{1}{4}$ metres in centimetres? What is $1\frac{36}{100}$ metres in millimetres?* Once they have found the answers practically, encourage children to consider how they might have found the answers by calculating.

KEY LANGUAGE

In lesson: metric, units (of measurement), kilograms, grams, metres, centimetres, millimetres, litres, millilitres, smaller, larger, convert, multiply, divide

Other language to be used by the teacher: digits, decimal

STRUCTURES AND REPRESENTATIONS

Place value grid

RESOURCES

Optional: digit cards, measuring jug and water, metre ruler, measuring scales with a marked dial, pictures of items from a building site

 In the eTextbook of this lesson, you will find interactive links to a selection of teaching tools.

Quick recap 🔁

Ask children to multiply each of these numbers by 1,000:

12   35   305   1·2

Then ask them to divide each of these numbers by 1,000:

5,000   2,500   12,345   6

Discover

Convert metric measures

WAYS OF WORKING Pair work

ASK

- Question ① a): *What other sorts of measurement might you see on a building site?*
- Question ① a): *Which of the two units of measurement labelled on the objects is probably not appropriate? Why is it better to use a different unit?*
- Question ① b): *What is the same and what is different about the numbers on the sign? How does this affect the way you convert between the units?*
- Question ① b): *If the sign was not there, would you have remembered the number of grams in a kilogram, millilitres in a litre and centimetres in a metre?*

IN FOCUS Use the picture to discuss the units that might be used to measure mass, length and capacity, and briefly assess children's current understanding. Begin by reinforcing the concept that measurements can be converted from one metric unit to another. Remind children of their work in the previous lesson on appropriateness of units and ensure they understand that a value of 40,500 g would be more appropriately expressed in kilograms.

PRACTICAL TIPS Show children images of objects that they might find on a building site (bags of sand, bottles of water, lengths of fencing, and so on) together with a numerical value for each one. Discuss what units of measurement children think they may have been measured in and what units they could be converted into. Children can then explore how they might convert from the existing unit into the new one.

ANSWERS

Question ① a): 45,000 g can be written as 45 kg.

Question ① b): 9·25 l can be written as 9,250 ml.

Discover

1,000 g = 1 kg
1,000 ml = 1 l
100 cm = 1 m

9·25 l

CEMENT 45,000 g

① a) Work out the mass of the cement in kilograms.

b) Work out the capacity of the bucket in millilitres.

216

PUPIL TEXTBOOK 6A PAGE 216

Share

WAYS OF WORKING Whole class teacher led

ASK

- Question ① a): *What rule do you use when converting from a smaller unit into a larger unit? Does this work for every unit of measurement? Can you think of another situation when you might use this rule?*
- Question ① a): *In this part of the lesson you are asked to divide by 1,000 and multiply by 1,000. What is the same about the strategy you will use each time? What is different?*
- Question ① b): *In this part of the lesson you are asked to multiply by 100. How is this similar to multiplying by 1,000? How is this different?*

IN FOCUS The nature of metric measurements is such that children will only ever need to multiply or divide by powers of 10 when converting between units. Take time to emphasise this useful point to ensure children can see the similarities in these operations.

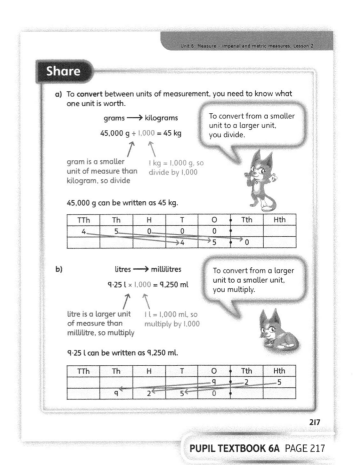

Share

a) To **convert** between units of measurement, you need to know what one unit is worth.

grams ⟶ kilograms

45,000 g ÷ 1,000 = 45 kg

To convert from a smaller unit to a larger unit, you divide.

gram is a smaller unit of measure than kilogram, so divide

1 kg = 1,000 g, so divide by 1,000

45,000 g can be written as 45 kg.

TTh	Th	H	T	O	Tth	Hth
4	5	0	0	0		
			4	5	0	

b)

litres ⟶ millilitres

9·25 l × 1,000 = 9,250 ml

To convert from a larger unit to a smaller unit, you multiply.

litre is a larger unit of measure than millilitre, so multiply

1 l = 1,000 ml, so multiply by 1,000

9·25 l can be written as 9,250 ml.

TTh	Th	H	T	O	Tth	Hth
				9	2	5
	9	2	5	0		

217

PUPIL TEXTBOOK 6A PAGE 217

Think together

Whole class teacher led (I do, We do, You do)

ASK

- Question **1**: *How could you ask the question in a different way? Can you explain why converting from millilitres to litres involves dividing by 1,000?*
- Question **2**: *Is this a multiplication or a division? How do you know what to multiply or divide by?*
- Question **2**: *Did you remember the number of centimetres in a metre, grams in a kilogram and millilitres in a litre, or did you need to look back at the sign in **Discover**? Do you need to multiply or divide in each case?*

IN FOCUS The two parts to question **1** give children real-life contexts to work with and, importantly, provide a degree of scaffolding to help them work through their choices step by step. Ask: *1) Should you divide or multiply – and why? 2) What should you divide or multiply by – and why?* Ensure that these sorts of questions are embedded and become second nature to children.

STRENGTHEN To support understanding of questions **1** and **2**, provide a measuring jug, metre ruler and measuring scales with a marked dial. Ask children to find evidence on these objects of what 1 litre, 1 metre and 1 kilogram are worth. Use this to strengthen understanding of what to multiply or divide by in each conversion.

DEEPEN When answering question **3**, begin by asking children to point to the important information in each statement. This is the operation (× or ÷) and the value being multiplied by or divided by. Children should use the given hints to help. Ask: *In each case, how does the operation give you a clue about the type of units the characters are converting? How does the number they are multiplying (or dividing) by help?* Challenge children to alter the statements so that the answers change and create similar problems for a partner to solve. Can they solve each other's problems?

ASSESSMENT CHECKPOINT At this point, children should be able to recognise the operation needed to convert from a smaller to a larger metric unit of measurement (÷) and vice versa (×). They should be growing in confidence when deciding on the correct value to multiply or divide by, based on what one unit is worth.

ANSWERS

Question **1** a): 1 l = 1,000 ml, so ÷ by 1,000
4,800 ÷ 1,000 = 4·8
The tin contains 4·8 litres of paint.

Question **1** b): 1 m = 100 cm, so × by 100   2·75 × 100 = 275
The hose is 275 cm long.

Question **2** a): 6,200 cm = 62 m

Question **2** b): 5,000 g = 5 kg

Question **2** c): 6.5 l = 6,500 ml

Question **3**: Olivia could have converted from kg → g, l → ml, km → m or from m → mm.
Ebo has converted from cm → m.
Children's answers should mention the direction of conversion and the value.

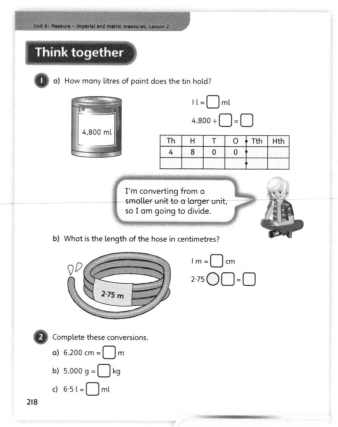

PUPIL TEXTBOOK 6A PAGE 218

PUPIL TEXTBOOK 6A PAGE 219

Practice

WAYS OF WORKING Independent thinking

IN FOCUS In question ❶, children convert from kilograms to grams and metres to centimetres, including some examples that involve decimals. At this stage they are provided with conversion prompts.

STRENGTHEN Question ❷ contains procedural variation for children to gain a more solid understanding of when to multiply and when to divide, and by how much each time. Encourage children to spot patterns in how the numbers change and strengthen children's understanding of these patterns by providing them with place value grids and digit cards. Ask them to make each number and then shift the digits according to whether they multiply or divide to convert. Children should notice that in each example in question ❷ a), the digits shift three places to the left and in each example in question ❷ b), the digits shift three places to the right. After modelling each calculation, encourage children to work separately to find each answer. Then, discuss the way that each answer changes (for example, when converting from l to ml, if the 1s digit changes in the question, the 1,000s digit changes in the answer).

DEEPEN In question ❺, children are not told the metric units that are being converted. By looking at the operations, children can derive whether the conversion is from a larger to a smaller unit or vice versa. By looking at the numbers, children can suggest which units are being converted. To deepen their understanding, they are prompted to consider whether two of the units might be the same. It is possible that they could both be centimetres because m → cm is a × 100 conversion and cm → mm is a × 10 conversion. Encourage children to explain their answer.

THINK DIFFERENTLY Question ❹ requires children to address two common misconceptions about converting between units of measurement: that when converting from a larger to a smaller unit, we divide (question ❹ b), and an incorrect multiplier being used (question ❹ a). In both cases, ask children how they can show that Lexi's working is incorrect.

ASSESSMENT CHECKPOINT At this point in the lesson, children should be able to convert confidently between metric units of measurement. They should be able to identify the operation needed and the correct value to multiply or divide by based on what one unit is worth.

ANSWERS Answers for the **Practice** part of the lesson can be found in the *Power Maths* online subscription.

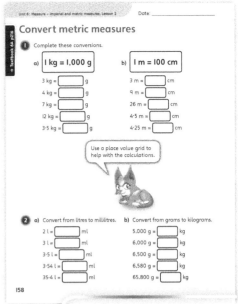

PUPIL PRACTICE BOOK 6A PAGE 158

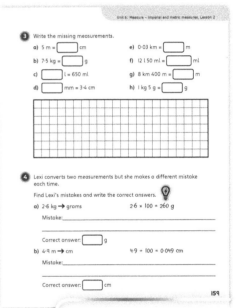

PUPIL PRACTICE BOOK 6A PAGE 159

Reflect

WAYS OF WORKING Independent thinking

IN FOCUS This question will reveal children's understanding of the nature of converting between metric units of measurement. Children should recognise that converting between such units always involves multiplying or dividing by 10, 100, 1,000, and so on, and a number's digits do not change when this happens.

ASSESSMENT CHECKPOINT Look for children who can explain that the digits simply shift when multiplying or dividing by 10, 100 or 1,000.

ANSWERS Answers for the **Reflect** part of the lesson can be found in the *Power Maths* online subscription.

After the lesson

- Are children confident converting between metric units of measurement?
- Can children apply these conversion skills in problem-solving contexts?

PUPIL PRACTICE BOOK 6A PAGE 160

Calculate with metric measures

Learning focus

In this lesson, children will solve a range of problems using all four operations in the context of metric measures.

Before you teach

- Are there ways you could make problem solving practical by linking it to other curriculum work?
- How might you scaffold questioning to help children reflect on their assumptions?

NATIONAL CURRICULUM LINKS

Year 6 Measurement

Solve problems involving the calculation and conversion of units of measure, using decimal notation up to three decimal places where appropriate.

ASSESSING MASTERY

Children can apply their knowledge of metric units and their conversion to solve problems. Children are able to choose appropriate strategies to calculate and convert using these units.

COMMON MISCONCEPTIONS

Children may solve a problem correctly but then express the answer with the wrong unit of measurement, neglecting to comprehend that they need to convert into another unit of measurement. Ask:
- *What do you notice about the units of measurement used in the question and those that you need to give in the answer?*

STRENGTHENING UNDERSTANDING

To help children identify the operation(s) needed to solve a problem, provide sticky labels or small pieces of paper and ask them to cover all the numbers in a question. This will focus their attention on the problem, helping them to comprehend it and to recognise whether they need to add, subtract, multiply and/or divide to find the answer. To further strengthen comprehension skills, encourage children to briefly act out the problem and discuss which operation it seems they should use.

GOING DEEPER

Children could be given very simple conversion problems with an unknown in them and asked to express the answer as an algebraic statement. For example: *The mass of one watermelon is x g. What is the mass of two watermelons in kilograms?* $(2x \div 1,000)$ Challenge children to give the unknown a value and to show how their statement could be used to find the answer.

KEY LANGUAGE

In lesson: how much more?, how much left?, units (of measurement), metres, centimetres, kilograms, grams, litres, millilitres

Other language to be used by the teacher: mass, length, capacity, metric, convert

STRUCTURES AND REPRESENTATIONS

Bar models

RESOURCES

Optional: measuring scales, measuring jugs, metre rulers that show two units of measurement (kg/g, l/ml, m/cm), a variety of containers with different volumes, sticky notes or small strips of paper

 In the eTextbook of this lesson, you will find interactive links to a selection of teaching tools.

Quick recap

Ask children to write each of these amounts in grams:

5 kg   12 kg   1·859 kg   2·75 kg   2·5 kg

Discover

Unit 6: Measure – imperial and metric measures, Lesson 3

Calculate with metric measures

Discover

WAYS OF WORKING Pair work

ASK

- Question ❶: *What metric units of measurement might you use to measure some of the objects in the picture?*
- Question ❶ a): *What do you notice about the way each character describes the capacity of their watering can?*
- Question ❶ a): *Which unit of measurement would you use to describe the capacity of a watering can?*

IN FOCUS The scenario in the image shows three characters describing capacity in three different ways (expressed in litres, millilitres and as a proportion of a measurement). Ensure that children recognise that they can still calculate with these measurements as long as they are converted into the same unit.

PRACTICAL TIPS Provide children with three containers. Discuss how to find the difference in capacity between two of the containers when you do not know how much water they hold. If the capacity is known, or if the measurements for different containers are in different units, does this change how children will approach the problem? Challenge them to act out different problems based on the scenario in the picture.

ANSWERS

Question ❶ a): Bella's watering can holds 150 ml more than Reena's.

Question ❶ b): There will be 650 ml left.

❶ a) How much more water does Bella's watering can hold than Reena's?

b) If the children use water from the container to fill their watering cans, how much water will be left in the container?

220

Share

WAYS OF WORKING Whole class teacher led

ASK

- Question ❶: *What do you need to convert to find the answers to the questions?*
- Question ❶: *What facts do you need to know to be able to answer both questions?*
- Question ❶ a): *Explain why converting from litres to millilitres involves multiplying by 1,000.*
- Question ❶ a): *Is it possible to work out the answer by converting Reena's capacity into litres? What would the calculation be? Do you think this is easier or more difficult than the method shown?*
- Question ❶ b): *Can you write the calculation you need to do as a number sentence?*

IN FOCUS It is important children recognise that they need to convert the units of measurement in order to solve both problems. However, neither problem specifies the units that the answer should be expressed in. Ask children to look carefully at the units in the question. Then ask: *Which units will you work out the answer in?* Challenge children to try working out the answer by converting to a different unit of measurement from the one shown. Ask: *Which method was more efficient?*

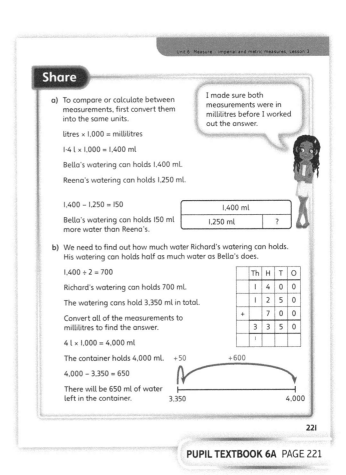

257

Think together

Unit 6: Measure – imperial and metric measures, Lesson 3

Think together

WAYS OF WORKING Whole class teacher led (I do, We do, You do)

ASK

- Question **1**: *Explain why you need to convert a unit of measurement in this problem. How would you write this problem as a number sentence?*
- Question **2** a): *What operation do you need to use to solve the problem? Can you think of a quick way to find out the number of 5s in 4,000?*
- Question **2** b): *What operation do you need to use to solve the problem? Can you sketch the problem? When do you need to convert to solve the problem?*

IN FOCUS In questions **1** and **2** a), children need to convert before calculating. However, in question **2** b), the information contains the same units of measurement. This means that children can decide when to convert. Allow children to make this decision themselves before addressing it when discussing question **3**.

STRENGTHEN Strengthen children's knowledge of unit equivalences by preparing sets of measuring scales, measuring jugs and metre rulers that show two units of measurement (kg/g, l/ml, and m/cm). Give children opportunities to discuss the scales and how to convert between the units. Ask: *Why do we have to convert?*

DEEPEN Question **3** encourages children to explore the timing of conversions. Ask: *What is the difference between the two methods?* Give children time to work out both answers. Ask: *Which method would you use and why?* Encourage children to identify other questions where they could convert before or after calculating the answer.

ASSESSMENT CHECKPOINT At this point, children should display confidence when applying their knowledge of metric measures to solve problems. They should be able to describe appropriate strategies for converting between units of measurement.

ANSWERS

Question **1**: 3 × 1,000 = 3,000 ml
3,000 − 600 = 2,400
There are 2,400 ml of water left.

Question **2** a): 4 × 1,000 = 4,000
4 kg of compost = 4,000 g of compost
4,000 ÷ 5 = 800
Each tree will get 800 grams of compost.

Question **2** b): 9·2 m − 8 m = 1·2 m
1·2 × 100 = 120
The hose needs to be 120 cm longer.

Question **3** a): Method 1: Convert before subtracting:
(6 × 1,000) − (1·4 × 1,000)
= 6,000 − 1,400 = 4,600 g
Method 2: Subtract before converting:
(6 − 1·4) × 1,000 = 4·6 × 1,000 = 4,600 g

Question **3** b): It does not matter when the units are converted.

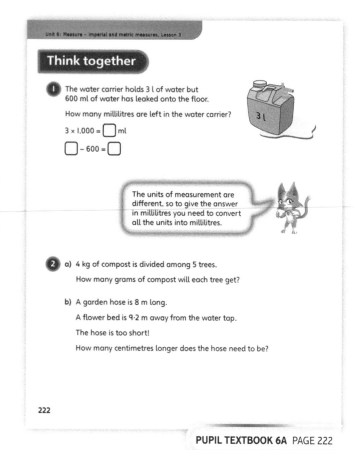

1 The water carrier holds 3 l of water but 600 ml of water has leaked onto the floor.

How many millilitres are left in the water carrier?

3 × 1,000 = ☐ ml

☐ − 600 = ☐

The units of measurement are different, so to give the answer in millilitres you need to convert all the units into millilitres.

2 a) 4 kg of compost is divided among 5 trees.
How many grams of compost will each tree get?

b) A garden hose is 8 m long.
A flower bed is 9·2 m away from the water tap.
The hose is too short!
How many centimetres longer does the hose need to be?

222

PUPIL TEXTBOOK 6A PAGE 222

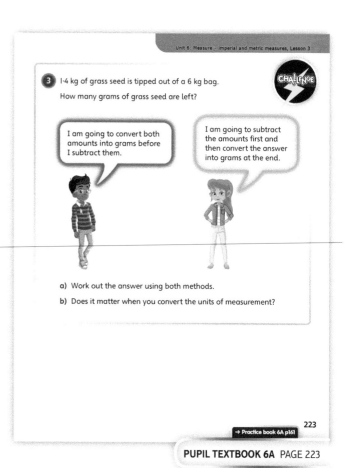

3 1·4 kg of grass seed is tipped out of a 6 kg bag.
How many grams of grass seed are left?

CHALLENGE

I am going to convert both amounts into grams before I subtract them.

I am going to subtract the amounts first and then convert the answer into grams at the end.

a) Work out the answer using both methods.

b) Does it matter when you convert the units of measurement?

223

→ Practice book 6A p161

PUPIL TEXTBOOK 6A PAGE 223

Practice

WAYS OF WORKING Independent thinking

IN FOCUS Question ❶ contains three problems with two units of measurement each. This means that children cannot leave the conversion until the end – they need to convert one of the measurements before they begin answering the question. Ask: *Which measurement do you need to convert? How will you convert it?* Encourage children to explain the operation they need to use to solve the problem.

STRENGTHEN To strengthen children's understanding when answering question ❹, encourage them to find visual ways of modelling the problem (perhaps by sketching it or modelling using strips of paper). It may not immediately be clear to children whether they are to add 12 cm or subtract it, so a labelled pictorial representation will aid their comprehension of the problem. A useful strategy is to encourage children to cross through any measurements they need to convert and write the new measurement.

DEEPEN Question ❺ provides a pre-algebra challenge to deepen problem-solving skills. If children require some prompting, begin by asking: *What is the difference in the number of bananas / apples between the two equations? What would you expect the mass of the fruit to be if the amounts in the first equation were doubled? What is the new difference between the two equations if the first equation is doubled?* Children should recognise that the second equation is double the first plus one apple. Ask: *How can you use the difference between these equations to calculate the mass of each piece of fruit?* They should deduce that they can use this fact to derive the mass of the extra apple and substitute this value into the first equation to calculate the mass of one banana.

ASSESSMENT CHECKPOINT At this point in the lesson, children should be able to apply their knowledge of metric measures in problem-solving contexts. They should be able to identify when a problem requires conversion between units and choose appropriate strategies to do this.

ANSWERS Answers for the **Practice** part of the lesson can be found in the *Power Maths* online subscription.

Reflect

WAYS OF WORKING Independent thinking

IN FOCUS This question reveals children's ability to apply metric conversions in real-life contexts. Ask: *Which units of measurement is your word problem going to feature? How will you show your partner that they need to convert to find the answer?* Encourage children to share their problems for peers to solve.

ASSESSMENT CHECKPOINT Check that children design word problems where units of measurement need to be converted (the question and the answer involve different metric units of measurement). Can children explain how they expect their problem to be solved?

ANSWERS Answers for the **Reflect** part of the lesson can be found in the *Power Maths* online subscription.

After the lesson

- Are children now able to apply their knowledge of metric conversions in problem-solving contexts?

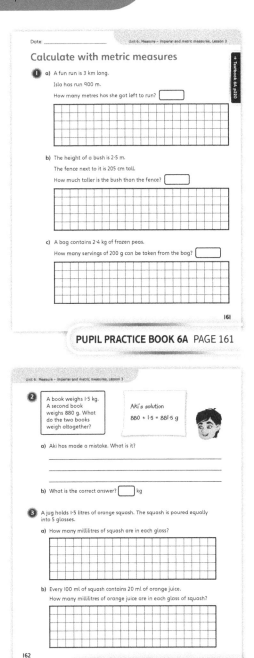

PUPIL PRACTICE BOOK 6A PAGE 161

PUPIL PRACTICE BOOK 6A PAGE 162

PUPIL PRACTICE BOOK 6A PAGE 163

Miles and kilometres

Learning focus

In this lesson, children will learn the 5 : 8 ratio between miles and kilometres. They will apply it to convert between these imperial and metric units of measurement.

Before you teach

- Are there any adaptations you can make to this lesson to link it with other curriculum work?
- How can you improve the teaching of reasoning through this lesson?

NATIONAL CURRICULUM LINKS

Year 6 Measurement

Convert between miles and kilometres.

ASSESSING MASTERY

Children know that a mile is more than a kilometre and that 5 miles are approximately the same as 8 kilometres. Children can apply this ratio to solve problems involving converting between miles and kilometres.

COMMON MISCONCEPTIONS

Children may think that 1 km is longer than 1 mile because they have confused the 5 : 8 ratio (thinking that the larger number implies a longer length), or because they know that 1 km = 1,000 m and believe this seems like a longer distance. Ask:

- *If 5 miles are the same as 8 kilometres, which is longer – a mile or a kilometre?*

STRENGTHENING UNDERSTANDING

Prepare strips of paper of equal length: some split into fifths and some split into eighths. Ensure children understand what each strip represents. Begin by cutting off one of the fifths and one of the eighths and comparing them. Ask: *Which is longer, 1 mile or 1 kilometre? How would you describe the relationship between them?* Children should be able to see that 1 km is over half a mile. Use the remaining strips to form bar models (for example, two strips of 5 = two strips of 8, so 10 miles are approximately 16 km). Ensure children recognise that these are approximate equivalences.

GOING DEEPER

Provide extension problems for children to explore. These may include multi-step problems or those that involve conversion between metric and imperial measures.

KEY LANGUAGE

In lesson: miles, metric, imperial, kilometres, convert, **conversion table**, bar model

Other language to be used by the teacher: distance, ratio, for every

STRUCTURES AND REPRESENTATIONS

Bar models, conversion tables

RESOURCES

Optional: plastic counters, graph paper, strips of paper

 In the eTextbook of this lesson, you will find interactive links to a selection of teaching tools.

Quick recap 🔍

Ask: *How far away is 1,000 m? How far away is 500 m? How far away is 20,000 m? How far away is 1 m?*

Prompt children to also think about these distances in kilometres.

Discover

WAYS OF WORKING Pair work

ASK

- Question ① : *What units of measurement are used on the signposts? Have you seen other units of measurement being used?*
- Question ① a): *What type of measure is a mile/kilometre? How long is a mile/kilometre?*
- Question ① b): *Where else might you find kilometres?*

IN FOCUS Children will have looked at a signpost and seen a distance or speed written in miles (or mph). Some children may have travelled in countries where the signposts are different because metric units (kilometres) are used. Discuss how road signs are a good example of real-life use of imperial or metric measures.

PRACTICAL TIPS Begin by getting children to express distances in terms of miles and kilometres. Share examples of maps, particularly those that show the local area and online maps that allow you to choose between a scale of miles or kilometres. Encourage children to measure short distances on maps, using the scale to give their answers in both miles and kilometres where possible. Use photos and online views of local streets to look at road signs in the area.

ANSWERS

Question ① a): Petite Ville is 10 miles away.
Grande Montagne is 25 miles away.

Question ① b): Various answers possible. See the conversion table in the **Share** section.

Share

WAYS OF WORKING Whole class teacher led

ASK

- Question ① b): *How has the conversion table been made? What patterns can you see? What will the next row in the table show? What will the tenth row in the table show?*

IN FOCUS The use of a conversion table may be new to children. Discuss how each column of the table has been populated and how this relates to the information shown on the bar model. Encourage children to spot the patterns between the rows and link this to multiplication facts and sequences. Discuss why this is a useful model for showing so many new facts when converting from kilometres to miles and vice versa.

Miles and kilometres

Discover

① a) What are the distances to Petite Ville and Grande Montagne in miles?

b) What other facts about miles and kilometres can you work out?

224

PUPIL TEXTBOOK 6A PAGE 224

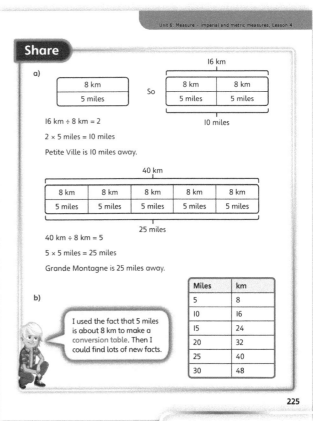

PUPIL TEXTBOOK 6A PAGE 225

261

Think together

Whole class teacher led (I do, We do, You do)

ASK

- Question **1** : *Do you expect the answer to be more or less than 56? The airport is 56 km away. How many miles are equivalent to 8 of those kilometres? How can you use this to help solve the problem? How would you complete the bar model? How does it help you to find the answer?*
- Question **2** : *Do you expect the answer to be more or less than 45? The railway station is 45 miles away. How many kilometres are equivalent to 5 of those miles? How can you use this to help solve the problem? What is similar about the strategies used to solve the two problems? What is different? How would you complete the bar model? How does it help you to find the answer?*

IN FOCUS Give children opportunities to describe how they would complete the bar models in questions **1** and **2** and why they think they are useful in helping to find the answer. In question **1** ensure that children understand that splitting the whole (56) into groups of 8 km enables them to calculate the number of groups of 5 miles that cover an equivalent distance.

STRENGTHEN To support understanding of the 5 : 8 ratio in question **1**, give children plastic counters and ask them to count in 8s until they reach 56, placing one counter down for each multiple of 8. Ask: *How many lots of 8 are there in 56?* Link this to the problem by asking how many lengths of 8 km are in a distance of 56 km. Ask: *How many miles is each one of these 8 km lengths?* Children should then count in 5s: one for each of the plastic counters they placed. Show how this can be modelled as a bar model by placing the counters on the bars.

DEEPEN Rather than using the 5 : 8 ratio, question **3** challenges children to consider a different way of converting between miles and kilometres. Encourage children to develop their proportional reasoning skills and make the connection that if 5 miles = 8 km then 1 mile = $8 \div 5 = \frac{8}{5}$ or 1.6 km and if 8 km = 5 miles then 1 km = $8 \div 5 = \frac{5}{8}$ or 0.625 km. Ask: *Which of the two facts tells you the value of miles in terms of kilometres? How will you convert 2 miles into kilometres using the fact to help? So, how will you convert 7 miles?* Challenge children to find the answers and discuss possible methods of multiplying 7 by 1·6 (for example, using short multiplication or multiplying 7 by 16 and dividing the answer by 10 to compensate).

ASSESSMENT CHECKPOINT At this point, children should be working with growing confidence when using the ratio of miles to kilometres to convert between the two units.

ANSWERS

Question **1** : $56 \div 8 = 7$ $7 \times 5 = 35$
   The airport is 35 miles away.
Question **2** : $45 \div 5 = 9$ $9 \times 8 = 72$
   The sign should show 72 km.
Question **3** a): Max travelled 100 km or 62.5 miles.
   Emma travelled 60 miles or 96 km.
   Max travelled the furthest.
Question **3** b): Zac is correct. The number of miles (7) needs to be multiplied by the number of kilometres in 1 mile (1·6) to find the equivalent number of kilometres.

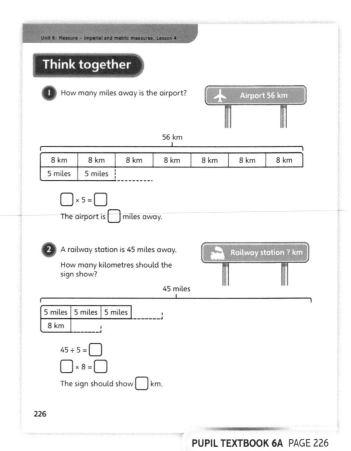

PUPIL TEXTBOOK 6A PAGE 226

PUPIL TEXTBOOK 6A PAGE 227

Practice

WAYS OF WORKING Independent thinking

IN FOCUS In questions ❶, ❷ and ❸ children are using the 5 : 8 ratio to convert between miles and kilometres, with either bar models or a conversion table, both only partly completed, to support them.

STRENGTHEN When answering question ❹, it is important that children are able to see how many groups of 5 or 8 are in each length of miles or kilometres. Strengthen their understanding of this strategy by providing them with small strips (3 or 4 cm long) of paper in two colours, but all the same length. Each group of strips should be labelled '5 miles' or '8 km'. Use these to help children make their own bar models. For example, ask: *How many '5 mile' strips will you need to model the length of the River Mersey? So, how many '8 km' strips will you need to match this? How many kilometres is this?*

DEEPEN In question ❺, children's understanding is deepened as they are asked to consider speed in relation to the two units of measurement they have been working with. Ask: *If the speeds had the same numerical value (for example, 50 mph and 50 km/h), which train would be faster? How can you compare two speeds when they use different units of measurement?* After converting one of the measurements to compare them, challenge children to try converting the other unit to check that their answer is correct.

ASSESSMENT CHECKPOINT At this point in the lesson, children should be able to convert confidently between miles and kilometres. They should understand how the relationship is different from converting between two metric units and they should be able to identify how to use related facts to help.

ANSWERS Answers for the **Practice** part of the lesson can be found in the *Power Maths* online subscription.

Reflect

WAYS OF WORKING Independent thinking

IN FOCUS This question assesses children's ability to use the ratio between miles and kilometres to find related facts. Ensure they understand that this relationship is approximate and so any facts are not exact.

ASSESSMENT CHECKPOINT Look for children who understand how they can use the fact that 5 miles is approximately 8 kilometres to derive other equivalent statements involving miles and kilometres. Children should have used the multiplicative relationship to find these (for example, 10 × 5 miles = 10 × 8 kilometres) rather than incorrectly attempting addition (for example, 10 + 5 miles and 10 + 8 kilometres).

ANSWERS Answers for the **Reflect** part of the lesson can be found in the *Power Maths* online subscription.

After the lesson ⏸

- Are children confident converting between miles and kilometres?
- Are children ready to apply their skills with other imperial units in Lesson 5?

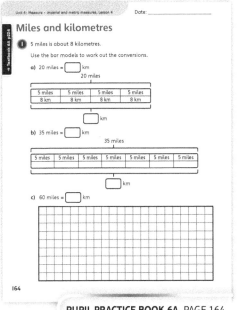

PUPIL PRACTICE BOOK 6A PAGE 164

PUPIL PRACTICE BOOK 6A PAGE 165

PUPIL PRACTICE BOOK 6A PAGE 166

Imperial measures

Learning focus

In this lesson, children will consolidate their knowledge of imperial measures, converting between two imperial units and between an imperial and metric unit of measurement.

Before you teach

- Based on previous lessons taught in the unit, are there any misconceptions to consider upfront?
- How can you develop and refine children's own representations for problems?

NATIONAL CURRICULUM LINKS

Year 6 Measurement

Use, read, write and convert between standard units, converting measurements of length, mass, volume and time from a smaller unit of measure to a larger unit, and vice versa, using decimal notation of up to three decimal places.

ASSESSING MASTERY

Children understand facts about different imperial and metric units of length, mass and volume (for example, knowing that 1 inch is approximately 2·5 cm). Children apply this information to convert from one imperial unit of measure to another (for example, converting from feet to inches) or between imperial and metric units (for example, converting from inches to centimetres).

COMMON MISCONCEPTIONS

Children may incorrectly multiply the digits on both sides of a decimal point separately by 10. For example, when calculating 2·5 × 10, they may end up with an answer of 20·50. Ask:
- *What is 2 × 10? How about 0·5 × 10? What do you need to be careful of when multiplying or adding decimals?*

STRENGTHENING UNDERSTANDING

Give children as much practical experience of measuring, comparing and converting imperial units to metric units as possible. Give them rulers that show inches and ask them to measure pieces of string that are a whole number of inches long. Ask: *How long is the string in inches? In centimetres? What do you notice?* Do the same with inches and feet. Where possible, provide measuring jugs and empty pint cartons and encourage children to physically explore the number of millilitres in 1 pint.

GOING DEEPER

Challenge children to make up conversion problems for peers to solve. These could be based on real-life measures, so children can check their answers by measuring. For example: *Sam is x inches tall. How many centimetres is this?* To extend this learning, encourage children to write problems that contain two steps.

KEY LANGUAGE

In lesson: imperial, metric, units of measurement, inches, **feet**, centimetres, metres, grams, kilograms, pounds (lbs), pints, litres, millilitres, ~~conversion graph~~, convert

Other language to be used by the teacher: length, capacity, mass, yards, gallons, ounces

STRUCTURES AND REPRESENTATIONS

Bar models, conversion graphs

RESOURCES

Optional: pint cartons, measuring jugs (millilitres), rulers (inches and centimetres), string, weighing scales (pounds and grams), small strips of paper in equal lengths

 In the eTextbook of this lesson, you will find interactive links to a selection of teaching tools.

Quick recap

As a class, count together up and down the 12 times-table.

Discover

Unit 6: Measure – imperial and metric measures, Lesson 5

Imperial measures

Discover

WAYS OF WORKING Pair work

ASK

- Question **1**: *What do you notice about the units of measurement used in the descriptions of the food on the specials board?*
- Question **1**: *Do you have an idea how much any of the measurements are worth? Can you show me with your hands, or point to something with a similar length / mass / capacity?*
- Question **1**: *Can you think of any other examples where imperial measurements are used in everyday life?*
- Question **1**: *When do you think it might be important to convert between imperial and metric measurements?*

IN FOCUS When discussing the **Discover** scene, ask children about their experiences with the units of measure that are mentioned on the specials board. Do they have a rough idea what each unit of measure refers to or do they just think of them as being the names of the foods?

PRACTICAL TIPS Provide opportunities for children to explore some of the everyday uses of imperial units. These could include empty pizza boxes, takeaway menus, pint containers, clothes labels, images of road signs, and so on. Talk about the differences between these units of measurement and metric units.

ANSWERS

Question **1** a): A 10-inch pizza is about 25 cm wide.

Question **1** b): Holly should weigh 1,125 grams of sugar.

1 a) 1 inch is about 2·5 cm.

How wide is the pizza in centimetres?

b) 1 lb is about 450 g.

How can Holly weigh $2\frac{1}{2}$ lbs of sugar with her scales?

228

PUPIL TEXTBOOK 6A PAGE 228

Share

WAYS OF WORKING Whole class teacher led

ASK

- Question **1** a): *Which is longer, an inch or a centimetre?*
- Question **1** a): *If you know that 1 inch equals 2·5 cm, how can you use this to convert from inches into centimetres?*
- Question **1** a): *What strategy would you use to work out the answer to 2·5 × 10?*
- Question **1** b): *How does the bar diagram represent the calculation you need to work out to find the answer? Both questions involve multiplying by $2\frac{1}{2}$. What is the quickest strategy you can think of to multiply a number by $2\frac{1}{2}$?*

IN FOCUS When considering questions **1** a) and **1** b), ensure children can explain how bar models can help them to represent both problems. Ask: *How can a bar model help? Which part of the bar model shows the equivalence between the units of measurement? How does the whole of the bar model show you the answer? What strategy will you use to calculate the answer?*

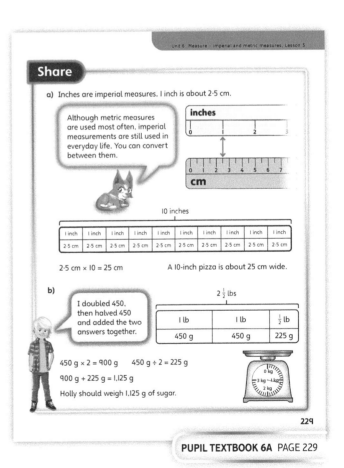

PUPIL TEXTBOOK 6A PAGE 229

Think together

WAYS OF WORKING **WAYS OF WORKING** Whole class teacher led (I do, We do, You do)

ASK

- Question **1**: *How do you think this graph has been plotted? Why is the graph a straight line? Would you expect it to carry on being straight? How can you use the graph to help convert between pounds (lb) and grams (g)?*
- Question **2**: *How could you draw a conversion graph to help answer this question? Estimate how many pints would be about the same as 1 litre. Could you calculate the answer to this problem by converting first? Is this an easier or more difficult method?*

IN FOCUS In question **1**, children are presented with a conversion graph. Talk about how the graph has been plotted and ask children to describe how they would go about drawing a similar graph. Ask children to read one fact from the graph and encourage them to describe how they found it. Challenge them to use the graph to find different equivalences (alternating between the number of pounds (lb) and the number of grams (g) as starting points).

STRENGTHEN To support understanding of question **2**, ensure children have opportunities to measure in both pints and millilitres. Can they prove that 1 pint is about the same as 560 ml? How many pints would they expect to be about the same as 1 litre? Can they prove this? Before starting the question, create a conversion table with children so they can refer to this when answering.

DEEPEN Question **3** deepens children's understanding by introducing a new unit of measurement – the foot. Ask: *How can you convert from feet into centimetres when the fact you are given only tells you how many inches are in one foot?* Children may suggest one of two methods: finding the number of centimetres in 1 foot and then multiplying by 6, or finding the number of inches in 6 feet and then multiplying by the number of centimetres in an inch. Challenge children to write these two methods as number sentences. They should be able to see that they both result in the same answer: $12 \times 2{\cdot}5 \times 6$ and $6 \times 12 \times 2{\cdot}5$.

ASSESSMENT CHECKPOINT At this point, children should be working with growing confidence when converting between two imperial units and between an imperial and a metric unit of measurement.

ANSWERS

Question **1** a): 4·5 lbs is about the same as 2 kilograms.

Question **1** b): Accept reasonable answers in the region of the following:
- 2 lbs is about the same as 900 g.
- 4,500 g is about the same as 10 lbs.
- 2·5 kg is about the same as $5\frac{1}{2}$ lbs.

Question **2**: $560 \times 4 = 2{,}240$ ml $2{,}240 \div 1{,}000 = 2{\cdot}24$ l

There are 2·24 l of milk in a four-pint carton.

Question **3**: Ebo's dad is 180 cm tall.

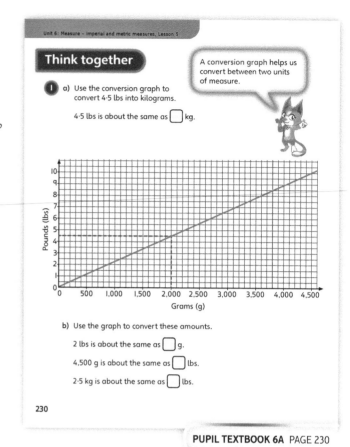

PUPIL TEXTBOOK 6A PAGE 230

PUPIL TEXTBOOK 6A PAGE 231

Practice

WAYS OF WORKING Independent thinking

IN FOCUS Inches and centimetres are helpful units to use with conversion graphs as the ratio between the two (1 : 2·5) lends itself to straightforward multiples (2·5, 5, 7·5, 10). As with previous conversion graphs, in question ① ask: *Why is the line straight? Do you think the line would be straight if the graph was extended and the line continued? Why? What do you notice about the scales on the axes? If you had been drawing the graph, which points would you have plotted to help you draw the line?*

STRENGTHEN When answering question ②, begin by asking children to write '1 kg' on one side of a strip of paper, and '2·2 lbs' on the other side. They should then take an identical strip of paper and again write '1 kg' on one side and '2·2 lbs' on the other side. Ask them to place two strips of paper side by side to model 2 kilograms and then to turn the strips over to show the equivalent number of lbs. Ask: *How would you add 2·2 and 2·2?* Encourage children to use vertical written methods to support their understanding as they will need to cross the 1s boundary later on. After trying several conversions, encourage children to attempt the conversion table. If further support is needed, explain that they should attempt the second and third columns, then the fifth column, before using what they have found to help with the others.

DEEPEN Question ⑤ deepens children's understanding of conversion as it tests their knowledge of converting between units of measure and also assesses their reasoning skills. They should understand that to compare these measurements they need to convert them all into the same unit. Ask: *Which unit does it make more sense to convert to? Why?*

ASSESSMENT CHECKPOINT At this point in the lesson, children should be able to use facts about imperial and metric units to convert confidently between imperial units or between imperial and metric units. They should apply these strategies to solve problems involving imperial and metric units.

ANSWERS Answers for the **Practice** part of the lesson can be found in the *Power Maths* online subscription.

Reflect

WAYS OF WORKING Independent thinking

IN FOCUS This question provides children with an opportunity to reflect on the different real-life examples of imperial measurements they have come across, allowing them to see the importance of being able to convert. Encourage children to consider the ease of converting by multiplying and dividing by 10, 100 or 1,000 and how the metric system has been designed to keep units with these equivalences. This is a useful way to bring the unit to a conclusion.

ASSESSMENT CHECKPOINT Look for children who can identify real-life applications of imperial units and who recognise the difference between imperial and metric conversions (namely that metric units are linked with powers of 10).

ANSWERS Answers for the **Reflect** part of the lesson can be found in the *Power Maths* online subscription.

After the lesson

- Are children confident converting between metric and imperial units of measurement?
- As this is the end of the unit, how successful do you feel the unit has been as a whole? Have children mastered the unit?

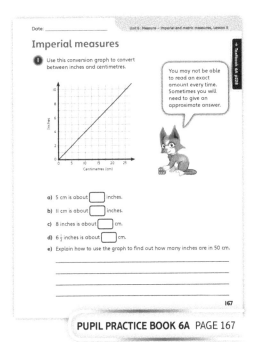

PUPIL PRACTICE BOOK 6A PAGE 167

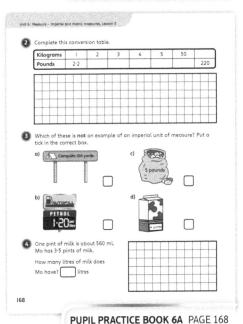

PUPIL PRACTICE BOOK 6A PAGE 168

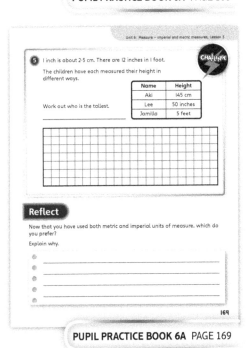

PUPIL PRACTICE BOOK 6A PAGE 169

End of unit check

> Don't forget the unit assessment grid in your *Power Maths* online subscription.

IN FOCUS

- Question **1** assesses children's ability to estimate using metric units of measure (mass).
- Question **2** assesses children's understanding of how to convert between metric units of measure (capacity).
- Question **3** assesses children's ability to solve problems involving conversion between metric units.
- Question **4** assesses children's ability to convert between miles and kilometres.
- Question **5** assesses children's understanding of approximate equivalences between some metric and imperial units.
- Question **6** is a SATs-style question. Knowledge of how to read scales is applied (as well as some knowledge of converting between metric units).

ANSWERS AND COMMENTARY

Children who have mastered this unit can choose appropriate metric units of measurement, justify their choices and make sensible estimates of length, mass or capacity. They can identify the relationship between different metric units of length, mass and volume and then apply this to convert from one unit to another. Children can apply their knowledge of converting between metric units to solve problems, choosing appropriate strategies. They recognise that a mile is longer than a kilometre and that 5 miles are approximately 8 kilometres, applying this ratio to convert between the two and to solve problems. Children can give facts about different imperial and metric units and apply them when converting from one imperial unit of measure to another, or between imperial and metric units of measure.

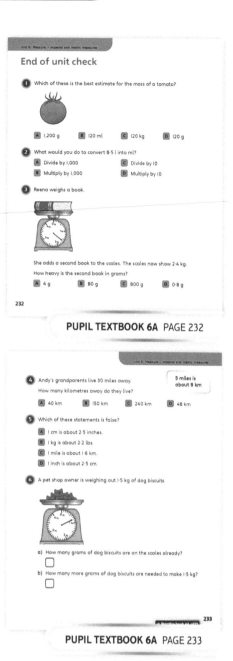

PUPIL TEXTBOOK 6A PAGE 232

PUPIL TEXTBOOK 6A PAGE 233

Q	A	WRONG ANSWERS AND MISCONCEPTIONS	STRENGTHENING UNDERSTANDING
1	D	A, B and C suggest a lack of understanding of metric units of measure.	Provide different representations, both pictorial and concrete, to allow children to build conceptual connections. • Weighing scales, rulers and measuring jugs are particularly useful for comparing imperial and metric units. • Provide everyday objects for children to estimate their mass, capacity or length. This will help children consolidate their understanding of the value of units. • Make or draw bar models where the bars show the equivalence of different units of measurement (for example, 1 kg across the top and 1,000 g across the bottom).
2	B	A suggests children have chosen the wrong operation. C and D suggest children think there are 10 ml in 1 l (or that the decimal point should be removed when converting).	
3	C	B and D suggest children realise the difference is 0·8, but do not know how to convert 0·8 kg into grams.	
4	D	A, B and C suggest children do not know how to convert between miles and kilometres using the 5 : 8 ratio.	
5	A	B, C and D suggest children do not recognise some approximate equivalences between imperial and metric units.	
6	800g 700g	Children may make a mistake when reading the scales or when counting on.	

My journal

WAYS OF WORKING Independent thinking

ANSWERS AND COMMENTARY Corrected statements may vary, depending on how children choose to correct each mistake.

Question **1** a): The mistake is that Lexi has divided by 100. She should have divided by 1,000 because 1 l = 1,000 ml. The correct answer is that 450 ml is the same as 0·45 l (also accept answers that state 4,500 ml is the same as 4·5 l).

Question **1** b): The mistake is that Max has found the difference between 250 and 1, ignoring the units of measure. The correct answer is that the difference between 250 g and 1 kg is 750 g (or 0·75 kg).

Question **1** c): The mistake is that Kate has incorrectly doubled 1·6 to get 2·12. Double 1 = 2, double 0·6 = 1·2, so double 1·6 should be 3·2. The correct answer is that 2 miles is equal to about 3·2 km.

If children are struggling to identify the mistake in each statement, ask:
• a): *What is 450 ml converted into litres?*
• b): *What is the difference between 250 g and 1 kg?*
• c): *If 1·6 km is about the same as 1 mile, how many kilometres are about the same as 2 miles?*

Power check

WAYS OF WORKING Independent thinking

ASK

• *What did you know about converting units of measure before you began this unit?*
• *What do you know now?*
• *If you saw a sign showing a distance in miles, would you be able to convert it into kilometres?*

Power puzzle

WAYS OF WORKING Pair work

IN FOCUS The purpose of this puzzle is for children to apply their understanding of conversion between metric units of measurement. Each answer corresponds to a letter of the alphabet and the letters form words. The answers have been devised to feature similar digits (for example, 56,000 and 5,600). This should encourage children to think through each conversion carefully.

ANSWERS AND COMMENTARY

a) Numbers: 56,000, 0·47, 470, 0·21, 39, 2·1, 470
   Letters: PASTIES

b) Numbers: 0·47, 56,000, 56,000, 5,600, 2·1, 56,000, 39, 2·1
   Letters: APPLE PIE

PUPIL PRACTICE BOOK 6A PAGE 170

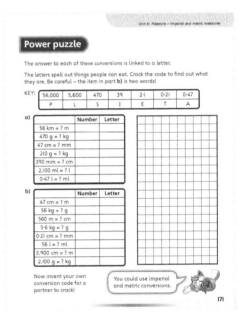

PUPIL PRACTICE BOOK 6A PAGE 171

After the unit ⏸

• When could you apply the concepts taught in this unit to other curriculum areas to keep children practising what they have learnt?

Strengthen and **Deepen** activities for this unit can be found in the *Power Maths* online subscription.

Published by Pearson Education Limited, 80 Strand, London, WC2R 0RL.

www.pearsonschools.co.uk

Text © Pearson Education Limited 2018, 2022
Edited by Pearson and Florence Production Ltd
First edition edited by Pearson, Little Grey Cells Publishing Services and Haremi Ltd
Designed and typeset by Pearson and Florence Production Ltd
First edition designed and typeset by Kamae Design
Original illustrations © Pearson Education Limited 2018, 2022
Illustrated by Diego Diaz and Nadene Naude at Beehive Illustration, Emily Skinner at
Graham-Cameron Illustration, Kamae Design and Florence Production Ltd
Images: The Royal Mint, 2017: 153
Cover design by Pearson Education Ltd
Back cover illustration Diego Diaz and Nadene Naude at Beehive Illustration
Series editor: Tony Staneff; Lead author: Josh Lury
Authors (first edition): Tony Staneff, Jian Liu, Josh Lury, Zhou Da, Zhang Dan, Zhu Deijang,
Tim Handley, Kate Henshall, Wei Huinv, Hou Huiying, Zhang Jing, Stephanie Kirk, Huang Lihua,
Yin Lili, Liu Qimeng, Timothy Weal and Zhu Yuhong
Consultants (first edition): Professor Jian Liu and Professor Zhang Dan

The rights of Tony Staneff and Josh Lury to be identified as authors of this work have been
asserted by them in accordance with the Copyright, Designs and Patents Act 1988.

First published 2018
This edition first published 2022

26 25 24 23 22
10 9 8 7 6 5 4 3 2 1

British Library Cataloguing in Publication Data
A catalogue record for this book is available from the British Library

ISBN 978 1 292 45062 9

Printed in the UK by Ashford Press Ltd

For Power Maths online resources, go to:
www.activelearnprimary.co.uk

Note from the publisher
Pearson has robust editorial processes, including answer and fact checks, to ensure the
accuracy of the content in this publication, and every effort is made to ensure this publication
is free of errors. We are, however, only human, and occasionally errors do occur. Pearson is
not liable for any misunderstandings that arise as a result of errors in this publication, but it is
our priority to ensure that the content is accurate. If you spot an error, please do contact us at
resourcescorrections@pearson.com so we can make sure it is corrected.